Sustainable Construction and Heritage Conservation in the Digital Age

Sujatavani Gunasagaran
Taylor's University, Malaysia

Kenn Jhun Kam
Taylor's University, Malaysia

IGI Global
Scientific Publishing
Publishing Tomorrow's Research Today

Vice President of Editorial Melissa Wagner
Director of Acquisitions Mikaela Felty
Director of Book Development Jocelynn Hessler
Production Manager Mike Brehm
Cover Design Jose Rosado

Published in the United States of America by
IGI Global Scientific Publishing
701 East Chocolate Avenue
Hershey, PA, 17033, USA
Tel: 717-533-8845
Fax: 717-533-7115
Website: https://www.igi-global.com E-mail: cust@igi-global.com

Library of Congress Cataloging-in-Publication Data

Names: Gunasagaran, Sujatavani A/P, 1976- editor | Kam, Kenn Jhun, 1986- editor
Title: Sustainable construction and heritage conservation in the digital age / edited by Sujatavani A/P Gunasagaran, Kenn Jhun Kam.
Description: Hershey, PA : IGI Global Scientific Publishing, 2026. | Includes bibliographical references and index. | Summary: "This book explores the integration of digital technology in construction and conservation practices. It examines the use of sustainable strategies for improved urban development, community preservation, and civil engineering"-- Provided by publisher.
Identifiers: LCCN 2025028144 (print) | LCCN 2025028145 (ebook) | ISBN 9798337352787 hardcover | ISBN 9798337352794 paperback | ISBN 9798337352800 ebook
Subjects: LCSH: Sustainable construction--Technological innovations--Case studies | Historic buildings--Conservation and restoration--Environmental aspects--Case studies
Classification: LCC TH880 .S8586 2026 (print) | LCC TH880 (ebook)
LC record available at https://lccn.loc.gov/2025028144
LC ebook record available at https://lccn.loc.gov/2025028145

British Cataloguing in Publication Data
A Cataloguing in Publication record for this book is available from the British Library.

Table of Contents

Detailed Table of Contents

The climate of Panzhihua, China, is characterized by dry, hot, and strong solar radiation. However, the existing regulations classify the region as a moderate climate zone at first, ignoring the local climate characteristics, resulting in a lack of effective guidance for housing construction. Due to the growing needs of residents, roofs in the area have adopted many immature climate response strategies. The research objective of this study is to conduct a literature review of existing roofing strategies, identifying and prioritizing the roles and importance of passive roofing in the region. The thermal performance of common passive roofs in this region, including GRs, CRs, VRs, and BRs, is analyzed, and the practical value of these passive roofs in the hot-dry region is demonstrated. This study will effectively guide the construction of residential roofs in the region, which is conducive to building energy conservation and resource utilization in the region.

The adaptive reuse of historically significant buildings represents a critical intersection between heritage conservation and contemporary development needs. This research explores the strategies for adaptive reuse. By examining the characteristics that define historical significance, this study contextualizes the adaptive reuse concept within the framework of heritage conservation. Economic considerations, such as cost savings and financial incentives, are analyzed alongside environmental benefits of sustainability and reduced resource consumption. The social and cultural implications are assessed, emphasizing the role of community engagement and the maintenance of cultural value. Architectural and structural challenges are identified, with a focus on integrating current building codes and standards while respecting historical integrity. This research aims to provide a comprehensive framework for adaptive reuse strategies, ensuring the sustainable preservation of historically significant buildings within the boundaries of established conservation guidelines.

Heritage buildings (HB) are crucial cultural and historical assets facing modern challenges. Retrofitting, combined with Historic Building Information Modeling (HBIM), offers innovative processes, tools, and datasets to enhance project performance and extend the lifespan of HBs. While global studies integrate HBIM-enabled retrofitting, its implementation in Malaysia remains limited due to a lack of procedural guidance and studies on its impact on enhancing performance in HB conservation projects. This research aims to propose an HBIM-retrofit framework to improve design performance and guide conservation practices. A systematic review based on content analysis of research publications and grey literature was conducted to identify HBIM-retrofit themes, leading to the development of a conceptual framework underpinned by task-technology fit theory. The framework would serves as preliminary guidance for practitioners to systematically adopt HBIM-retrofit procedures for innovative decision-making and sustainable outcome, thus advancing HB conservation practice in Malaysia.

Digital technologies are vital for boosting productivity and efficiency in Malaysia's construction industry. Despite a 74% awareness rate of Building Information Modeling (BIM) reported in the Malaysia BIM Report 2019, adoption remains limited. This study employs a quantitative methodology, using questionnaire surveys to assess the need for technology adoption among industry players. The findings aim to inform the development of a decision-making tool for selecting suitable technologies for construction projects. By identifying key challenges arising from limited digital adoption, the research proposes strategies to enhance technology integration. Expected outcomes include improved project outcomes, reduced cost overruns, minimized delays, and enhanced industry productivity. Ultimately, these advancements will contribute to Malaysia's GDP growth while addressing critical inefficiencies in the construction sector. Promoting digital adoption is essential for elevating the industry's performance and ensuring its long-term sustainability.

Internet of Things (IoT) provides real-time monitoring function for the construction players to assess accurate data for decision making. The technologies being used in Malaysia construction industry include sensors, drones, RFIDs and others. These technologies improve project's efficiency in several areas with its real-time report, yet adoption still low. Most studies focused on IoT implementation during the design phase, with few studies concentrating on the construction phase, particularly from the standpoint of contractors. Considering contractors make many decisions during the construction phase, IoT assist them for quick and precise decisions. This paper focus on the challenges in adopting IoT-based real-time monitoring technologies in the Malaysian construction industry. The research adopted qualitative method via semi-structured interviews with three G7 contractors in using those IoT and result in ten challenges. The findings are to increase industry participants' understanding and adoption of IoT in construction projects, hence enhancing project efficiency.

With the emergence of the Metaverse, the concept of the built environment has transformed, presenting architects and designers with both new opportunities and challenges. Unlike physical spaces, the Metaverse offers freedom from physical constraints and provides immersive user experiences. This research explores how

Kerala also faces many challenges like environmental degradation caused mainly through unsystematic waste management system. With the support of the Clean India Mission and in response to the Supreme Court verdicts, Kerala initiated the Suchitva Mission and Haritha Keralam Mission with the tagline of 'Clean Kerala' by effectively managing the waste. This mission functioned with the field work of Haritha Karma Sena (Green Task Force), mostly employed by uneducated women. They have provided awareness among the public and systematically managed waste with their committed work. However, it also has many drawbacks and failures with regard to its implementation. In line of above, this chapter tries to critically analyze the Haritha Karma Sena model of waste management in order to examine whether it is a suitable model of waste management.

Chapter 9
Water and Waste Management: Innovations for Sustainable Cities.................. 271
Treesa Sindhu P. Thomas, St. Teresa's College, Ernakulam, India
Leena Leon, Rajagiri College of Social Sciences, Kalamasseri, India

Urban water and waste management are critical in achieving sustainability, particularly in coastal cities that face unique challenges from rapid urbanization and climate change. These pressures demand innovative approaches to resource management that balance environmental stability, economic resilience, and social well-being. This article explores the intersection of cutting-edge technologies, adaptive urban planning, and participatory governance models, offering insights into how these approaches contribute to the Sustainable Development Goals (SDGs), specifically SDG 6 and SDG 11. The findings emphasize the importance of adopting holistic strategies that integrate zero-waste policies, enhanced recycling programs, extended producer responsibility (EPR), and community engagement. By leveraging innovation, circular economy principles, and active community participation, cities can build adaptable systems that ensure long-term sustainability, improve quality of life, and support global efforts toward environmental and social progress

Chapter 10
Innovations for Sustainable Coastal Cities: Strategies for Water and Waste Management.. 305
Treesa Sindhu P. Thomas, St. Teresa's College, Ernakulam, India
Leena Leon, Rajagiri College of Social Sciences, Kalamasseri, India

Coastal cities face unique challenges at the intersection of rapid urbanization and climate change, making sustainable water and waste management critical for their development. Innovative approaches to resource management that harmonize environmental stability, economic resilience, and social well-being by leveraging cutting-edge technologies, adaptive urban planning, and participatory governance models are essential for mitigating these issues. The adoption of holistic strategies,

including zero-waste policies, enhanced recycling programs, and extended producer responsibility (EPR) frameworks, as key drivers of environmental impact reduction and resource recovery. The transformative role of community engagement in fostering behavioural change and policy instruments such as pay-as-you-throw initiatives and public-private partnerships are identified as crucial enablers of scalable, sustainable solutions.

In recent years, the Kribi coast in Cameroon has experienced significant coastal erosion, resulting in the submersion of land parcels, including those with land titles. This study aims to create a spatial data infrastructure (IDoGek) to monitor and manage coastal changes from 2015 to 2023, enabling better planning and sustainable growth in the region. Using DSAS (ArcGIS plug-in), historical data and remote sensing analyses revealed a loss of over 7 hectares of land, with erosion rates averaging 1.1 meters per year. By 2034, up to 11 hectares could be lost without intervention, as some areas may retreat by 10.97 meters. This approach maps land use changes, tracks coastline positions, and assesses the kinetics of coastal erosion. The results emphasize the need for high-resolution imagery, offshore tide gauges, and integration

Preface

In an era defined by rapid urbanization, climate change, and growing societal expectations for sustainable living, the built environment stands at a critical crossroads. *Sustainable Construction and Heritage Conservation in the Digital Age* emerges from this context as a timely and essential reference, addressing the intricate challenges and transformative opportunities facing the construction and heritage conservation sectors today. As editors, we have witnessed the accelerating demand for cities that are not only functional and resilient but also deeply attuned to environmental stewardship and cultural preservation.

This volume brings together a diverse range of scholarly perspectives and professional insights to explore how sustainable urban development can enhance both resilience and quality of life. We believe that cities must do more than merely withstand environmental, social, and economic disruptions—they must evolve in ways that promote well-being, inclusivity, and long-term ecological balance. The chapters in this book examine forward-looking strategies, policies, and technologies that shape the urban fabric, focusing on practical approaches to building cities that are adaptable, livable, and sustainable for future generations.

The contributions span topics such as smart infrastructure, eco-urban design, digital innovations for community engagement, and green architecture—each emphasizing the pivotal role of interdisciplinary collaboration in crafting urban environments that honor both progress and heritage. Importantly, this book also underscores the critical importance of heritage conservation in the digital age, advocating for the thoughtful integration of tradition and innovation in the urban development process.

This volume begins with **Chapter 1**, which presents a critical review of passive roofing strategies in the hot-dry region of Panzhihua, China. The authors confront regulatory misclassifications that have long hindered climate-responsive housing design. By analyzing various passive roof types and their thermal performance, this chapter offers practical guidance for improving energy efficiency and resource utilization in rural construction, aligning with broader goals of sustainable development and environmental adaptation.

In **Chapter 2**, attention turns to Malaysia's architectural heritage through an exploration of adaptive reuse. The chapter offers a comprehensive framework for repurposing historically significant buildings, balancing cultural preservation with contemporary economic and environmental needs. The strategies discussed highlight the synergy between conservation practices and modern construction standards, promoting sustainable urban regeneration rooted in community and identity.

Chapter 3 builds upon this conservation theme by introducing a novel HBIM-retrofit framework tailored for heritage buildings in Malaysia. Leveraging Historic Building Information Modeling, the chapter emphasizes the importance of digital integration in enhancing design and project performance. It proposes a conceptual model that bridges the gap between technological potential and practical implementation, empowering stakeholders to make informed, sustainable decisions.

In **Chapter 4**, the focus shifts to digital technology adoption in Malaysia's construction industry. Using Analytic Hierarchy Process (AHP), this study identifies key barriers and strategic enablers to greater integration of digital tools like BIM. The findings support the development of a decision-making tool that can enhance efficiency, reduce delays, and promote sustainability—offering essential insights for modernizing industry practices.

Continuing the discussion on digital innovation, **Chapter 5** explores the adoption of Internet of Things (IoT) technologies for real-time monitoring during the construction phase. Drawing on interviews with G7 contractors, the chapter identifies ten core challenges that hinder IoT implementation, offering industry-specific insights for improving on-site decision-making and project outcomes.

Expanding into the virtual domain, **Chapter 6** investigates how design thinking must evolve to address the unique conditions of the Metaverse. Through interviews with designers and architects, this chapter proposes a new set of guidelines for creating immersive virtual built environments, ensuring the continued relevance of architectural practice in an increasingly digital world.

In **Chapter 7**, the potential of Virtual Design and Construction (VDC) in addressing construction waste management is explored through the Chinese context. With BIM identified as a key enabler, the study proposes a technology prioritization framework, providing policy and operational guidance for reducing waste and promoting sustainable construction through digital integration.

Chapter 8 turns to India, offering a critical analysis of the Haritha Karma Sena waste management model in Kerala. This grassroots initiative—led largely by underrepresented women—demonstrates both the promise and limitations of community-driven sustainability efforts. The chapter provides an important case study of how local governance and civic engagement can support cleaner urban environments.

The themes of innovation and sustainability continue in **Chapter 9**, which addresses water and waste management systems within the context of sustainable city planning. Drawing links to SDG 6 and SDG 11, the chapter emphasizes the role of circular economy models, zero-waste policies, and participatory governance in building adaptive, resilient urban systems.

A similar thread is woven through **Chapter 10**, which focuses specifically on the complex challenges facing coastal cities. This chapter presents forward-looking strategies for water and waste management, underscoring the importance of holistic policy frameworks, behavioral incentives, and public-private partnerships in crafting scalable, impactful solutions for vulnerable urban areas.

Finally, **Chapter 11** presents a geospatial analysis of coastal erosion in Kribi, Cameroon, demonstrating the value of spatial data infrastructures and digital shoreline analysis systems (DSAS) in environmental monitoring. By documenting significant land loss and projecting future trends, the chapter provides actionable recommendations for sustainable coastal planning and ecosystem preservation.

Together, these chapters represent a rich tapestry of perspectives, methodologies, and innovations. Each contributes to a deeper understanding of how sustainability, heritage conservation, and digital transformation intersect to shape the future of the built environment. As editors, we are proud to present this volume as a multidisciplinary contribution to ongoing global conversations about resilience, sustainability, and the digital age.

We hope this book serves as a valuable resource for academic researchers, architects, engineers, building technologists, and professionals across the construction industry. It is our intention that the insights presented here will not only inform but also inspire new pathways toward a more sustainable and resilient urban future.

Chapter 1
A Review on the Practical Value of Passive Roofs for Rural Housing in Panzhihua, China

Li Shun
https://orcid.org/0009-0008-9641-5943
Taylor's University, Malaysia & Panzhihua University, China

Siti Norzaini Zainal Abidin
https://orcid.org/0000-0002-2637-4794
Taylor's University, Malaysia

Myzatul Aishah Kamarazaly
Taylor's University, Malaysia

ABSTRACT

The climate of Panzhihua, China, is characterized by dry, hot, and strong solar radiation. However, the existing regulations classify the region as a moderate climate zone at first, ignoring the local climate characteristics, resulting in a lack of effective guidance for housing construction. Due to the growing needs of residents, roofs in the area have adopted many immature climate response strategies. The research objective of this study is to conduct a literature review of existing roofing strategies, identifying and prioritizing the roles and importance of passive roofing in the region. The thermal performance of common passive roofs in this region, including GRs, CRs, VRs, and BRs, is analyzed, and the practical value of these passive roofs in the hot-dry region is demonstrated. This study will effectively guide the construction of residential roofs in the region, which is conducive to building

DOI: 10.4018/979-8-3373-5278-7.ch001

energy conservation and resource utilization in the region.

1. INTRODUCTION

In hot-dry regions, the building envelope receives a lot of solar radiation, which causes thermal stress and indoor thermal discomfort. Today, however, residents rely on mechanical systems for cooling, resulting in huge energy consumption (Athmani, et al., 2022). According to statistics (Gan, et al., 2019; Gibberd, 2020; Harputlugil & de Wilde, 2021), buildings consume 40% of energy and produce more than a third of global greenhouse gases. These increased gases will trap more heat, leading to a greater "greenhouse effect" and an increase in global temperatures (Gibberd, 2020). In addition, in hot-dry regions, which account for one third of the world's land and 15% of the world's population, about one third of the energy used is used for cooling needs to mitigate indoor overheating risks (Athmani, et al., 2022; Baccega, et al., 2022). Refrigeration is a major basic need in this region for most of the year.

The roof is one of the most critical components of the building envelope (Rawat & Singh, 2022). According to some studies (Athmani, et al., 2022; Dristy, et al., 2020; Saurabh & Madhura, 2023), because roofs are exposed to sunlight throughout the day, especially in low-rise rural houses, they are a major source of thermal stress, and often cause uncomfortable indoor conditions. Panzhihua is located in the hot-dry region in southwest China, with strong solar radiation throughout the year, dry winter, and hot summer, and the roof faces great challenges (Panzhihua Climate, 2023). In Panzhihua area, most rural houses are low, with 1 or 2 floors, and the roof area accounts for a large proportion of the surface area of the enclosure structure, which bears strong solar radiation. Rural housing in China is mostly built by local builders based on their experience (Liu, 2021). Builder and owners pay more attention to economy and safety, and the lack of awareness of building energy efficiency leads to poor thermal performance of the building envelope, which requires more energy consumption to achieve indoor thermal comfort. Especially in hot-dry regions, buildings that lack energy-efficient design and use inefficient cooling systems consume a lot of energy (Rawat & Singh, 2022). As Panzhihua is located in Hengduan Mountain Area and is relatively remote, the builder's construction skills are not mature. In recent years, due to residents' lack of awareness of passive roofs, many of inefficient roof systems have been used in rural houses built in Panzhihua. Finally, rural energy consumption remains high. So, selecting low-energy technology considering the regional characteristics of rural housing in the region has become an urgent problem to be solved (Fu, 2020).

The purpose of this study is to analyze the importance of passive roofs in Panzhihua, demonstrate the practical value of passive roofs in the region, improve people's awareness of passive roofs, and promote thc sustainable development of roofs in the region. To achieve it, the following is the research question:

1. Why is passive roof important in Panzhihua?
2. What are the practical values of passive roofs in hot-dry climate and similar climates?

In summary, due to the significant environmental influences on roofs, such as solar radiation, air temperature, humidity, etc., the cooling energy consumption of buildings in this area is high, and roofs face numerous challenges. However, roof strategies to cope with hot and dry climates are not yet mature. Therefore, it is significant to analyze the practical value of passive roofs to promote the sustainable development of roofs in the region.

2. METHODOLOGY

This study employs qualitative research based on an extensive review of existing literature on rural building in Panzhihua and passive roofing technologies. In addition, literature on local climate and building codes were also analyzed. The focus will be on identifying strategies that are relevant to the hot-dry conditions of Panzhihua and assessing their potential for practical application.

The search was conducted through the following literature databases: ScienceDirect, Google Scholar, CNKI, and some related websites. For an overview of the search terms and search strings that were employed, see the Abbreviations. Filters in the search database using ScienceDirect and Google Scholar limit results to English scientific articles (Ingebretsen, et al., 2022). Due to building codes and research papers about the study site being in Chinese, the filters in CNKI's search database are used to limit results to Chinese articles. To further ensure that results were relevant to passive roofs, the selection of literature was limited to literature on green roofs, blue-green roofs, cool roofs, ventilated roofs, attic roofs, blue roofs, water-covered roofs, roof ponds, ventilated pitched roofs, and double roofs. Table 2 provides the research summary of the most important information of this study, and shows the main indicators considered when analyzing the existing literature.

3. LITERATURE REVIEW

3.1. The Climate of Panzhihua

Panzhihua is located in the Hengduan Mountains of southwestern China, at the border of southwestern Sichuan Province and northern Yunnan Province. It has a south subtropical dry hot valley climate. Furthermore, Szokolay (1980), in his "Environmental Science Handbook for Architects and Builders", argued that Panzhihua is a sub-dry-hot climate zone. From Figure 1, it can be seen that Panzhihua is hot from April to September every year, while the other months are either warm or comfortable, with a long dry season and a rainy season concentrated in the summer. It has a sunshine duration of about 2700 hours and a frost-free period of over 300 days a year. Drought and rainy seasons are distinct here, and evaporation is large. Its annual average temperature is 20.3 °C, hot in summer and warm in winter (People's Government of Panzhihua Municipal, 2019).

Figure 1. Annual climate of Panzhihua in 2023

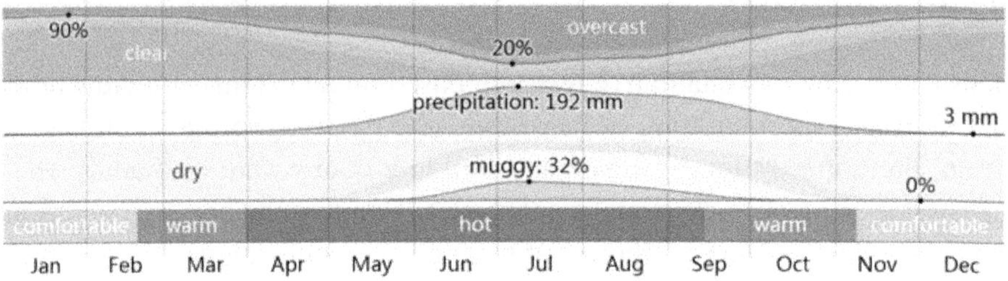

Source: Panzhihua Climate, 2019

Panzhihua is located in the mountainous region, the climate varies greatly at different altitudes, and the residents mainly live in the valley zone at lower altitudes. According to the official website of the Panzhihua government, the valley area below 1500 meters above sea level has a hot-dry valley climate (Panzhihua City Renhe District People's government office, 2022). As can be seen from Figure 2, the hot-dry valleys are in the red line, and these hot-dry regions are relatively dispersed and mainly distributed along the river valleys. The architecture in the region is mainly to solve the problem of shading and heat insulation.

Figure 2. Hot-dry regions in Panzhihua (Areas below 1500 meters)

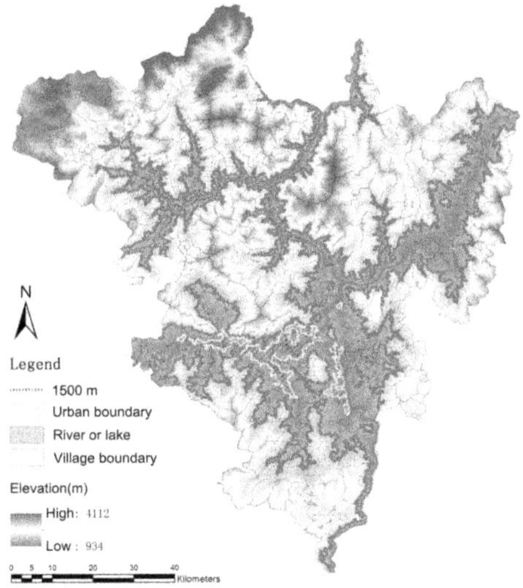

Source: Drawn up by the author.

3.2. Deficiencies in Standards and Guidelines

The main basis for the energy-efficient design of rural housing in China's rural areas is the "Design Standard for Energy Efficiency of Rural Residential Buildings (2012)". This standard proposes that the energy-efficient design of rural housing should be based on climate conditions, unique rural living patterns, economic conditions, appropriate building forms, energy-saving measures, and energy utilization methods to effectively improve the indoor living environment and reduce conventional energy consumption and greenhouse gas emissions. Although there are no specific energy-saving measures for hot-dry regions in this standard, reference can be made to its energy-saving measures for regions hot in summer and warm in winter. For rural housing in these regions, the following energy-saving measures should be adopted for passive roofs: 1. Light-colored finishes; 2. Insulated and ventilated roofs or passive evaporative roofs; 3. Roofs with plant shading.

Panzhihua is located in Sichuan Province. In Sichuan Province, the "Design Standard for Energy Efficiency of Residential Buildings in Sichuan Province (2019)" classifies Panzhihua as the temperate zone B. It only specifies the thermal performance of the roof of houses. When the average thermal inertia index of the roof is ≤ 2.5, the U-value should be ≤ 0.8. When the average thermal inertia index of the roof is > 2.5, the U-value should also be ≤ 0.8.

The "Technical Standard for Application of Energy-saving Technology for Civil Buildings in Panxi (2021)" developed by the Sichuan Provincial Department of Housing and Urban-Rural Development in 2022 provides construction guidelines for Panzhihua, the largest city in the hot-dry region of southwest China. The standard includes a small section on passive roofs, suggesting that for passive roofs of housing in Panzhihua with high solar radiation, low relative humidity, and not requiring high thermal insulation performance, inverted passive roofs with integrated insulation functionality such as overhead ventilated roofs, green vegetation roofs, and water storage roofs should be used.

In Li & Wan 's research (2014) on the building design of Panzhihua, which is located in the hot-dry region, a qualitative analysis method was used. The research proposed guidelines for the buildings here: increased shading, improved insulation performance, and the use of light-colored exteriors. However, no specific quantitative parameters or measures were provided.

In summary, the government has established various standards and guidelines to guide the construction of housing, but the content related to passive roofs is limited. The existing standards are brief, only proposing a few energy-saving strategies, and lacking practical guidelines that combine thermal comfort and cost. The thermal performance indicators of the roofs with different standards are too simple to effectively guide the design of passive roofs.

3.3. Challenges of Roof in the Region

As a building component, the roof is more susceptible than other building structures to climatic factors such as solar radiation, rain and snow (Ismail, 2013). A study (Li, et al., 2016) shows that roof is the main source of heat loss in residential building envelope, accounting for about 70% of the total heat loss. Because the roofs are more affected by direct solar radiation, and they are caused by better heat exchange positions between the interior and exterior Spaces of the building. According to Yu et al.,(2020), the heat transfer of the roof tends to be greater than that of any exterior wall. The roof of the building usually overheats in summer, resulting in a substantial increase in the air conditioning load, which has a negative impact on indoor thermal comfort. Therefore, the thermal performance of the roof is crucial for indoor thermal comfort. This mainly depends on the form of the roof and the materials used to minimize the amount of solar radiation absorbed by the roof (Omar, et al., 2017). A study (Sharifi & Yamagata, 2015) showed that the roof alone can carry up to 50% of the heat load of a single - or two-story building in the summer. Moreover, Elnabawi et al., (2022) found in their study that in hot, dry, warm, and humid climates, roofs increase the building cooling load by about 50-60%. Rawat & Singh (2022) see the roof as an important interface between indoor and outdoor environments, which must not only withstand various weather conditions but also be strong enough. During the hot summer months, roofs must be designed to minimize incompatibilities. Globally, heat gained through building shells has been a major concern in hot-dry climates. A study (Saurabh & Madhura, 2023) of hot-dry regions found that these regions usually have low-rise buildings, and the main heat gain is always through the roof. Rural houses in Panzhihua are basically low buildings with 1-2 floors, and the roofs also face the challenge of huge heat gain.

On the other hand, the most significant challenge affecting the long-term integrity of the building envelope is water intrusion (Andenæs, et al., 2021). In compact roofs, drying is often considered unfeasible because there is a steam protection layer on both the outside (roof membrane) and inside (steam barrier) of the roof. Furthermore, moisture promotes biological growth, which deteriorates materials and affects indoor air quality, may act as a solvent that affects material properties, may cause corrosion, and may impose mechanical loads due to frost expansion or weight. Therefore, it is particularly important to prevent moisture from entering the building structure.

Finally, in recent years, rural houses being built in Panzhihua have used immature roofing strategies. Residents' lack of awareness of reasonable roof selection strategies leads to the arbitrary adoption of various types of roofs, as shown in Figure 3. It can be seen that the roofs of rural housing in Panzhihua face many challenges.

Figure 3 . Various roofs of rural housing in Panzhihua

Source: Photographed by the author.

3.4. Passive Roof and Its Importance

3.4.1. Definition of Passive Roof

"Passive" is based on the concept of "adaptability", which belongs to the life sciences and refers to the phenomenon that an organism behaves in accordance with its environment (Song, et al., 2014). According to Asmat's doctoral thesis (2013), passive building strategy is a technology to improve the physical performance of buildings without using any mechanical equipment. In other words, passive cooling refers to the technology or design strategy used to cool buildings without consuming any energy. Passive cooling technology includes: night radiation cooling, roof evaporative cooling, roof pool, green roof, roof spray, roof insulation, night ventilation, sun shading, etc. (Maheshwari & Gehlot, 2017). Therefore, roofs using passive technologies are defined as passive roofs, including green roof (GF), cool roof(CF), ventilated roof (VF) and blue roof (BF).

3.4.2. The Role and Importance of Passive Roofs in the Region

Given the serious environmental problems caused by rising atmospheric temperatures associated with carbon emissions and the consequent rise in sales of air conditioning equipment, the demand for space cooling applications is increasing and the adoption of passive cooling technologies plays a vital role. In response to the global warming phenomenon facing the world today, Maheshwari & Gehlot (2017) proposes in their study that the implementation of passive cooling systems in building components may be a promising way to overcome these problems. Their research also found that among passive cooling technologies, roof cooling technology is very popular in hot areas for thermal comfort. Since the roof is the hottest part of the building, passive roofs are an appropriate technology to cool the building (Ismail, 2013). Although many buildings improve the thermal performance of the roof by increasing the thermal insulation capacity, it is not reasonable in many practical cases due to the limitation of roof thickness or economic reasons. Therefore, another technology is needed to reduce the heat flow of the roof, such as CF, GF, VF (Li, et al., 2016). In addition, a study by Baccega et al., (2022) found that among possible strategies, roofs utilizing natural ventilation techniques are particularly effective in hot climates. All these show the important role of passive roofs.

Moreover, because of the hot-dry climate in Panzhihua, the energy crisis is more obvious in summer because the cooling load of buildings is increasing, and the mechanical system is relied on to achieve indoor thermal comfort. As the main role of construction projects, architects should not rely solely on electricity and mechanical means generated by fossil fuels, but should invest time and energy in studying passive strategies to achieve the best combination of building low energy consumption and natural climate control (Dabaieh, et al., 2015). Therefore, people should pay more attention to passive cooling technology to prevent solar heat from entering the building, thus reducing its cooling demand.

Finally, the rural regions in Panzhihua are still backward mountainous areas, with relative poverty and inconvenient transportation. It is inevitable that large-scale and high-cost technologies will not work in these regions. Therefore, the development of sustainable passive roofs is of great significance for rural buildings in Panzhihua.

4. THE PRACTICAL VALUE OF PASSIVE ROOFS

There are four main types of passive roofs: green roofs, cool roofs, ventilated roofs, and blue roofs. In order to analyze the practical value of these passive roofs, 45 articles related to the research topic were screened, as shown in Table 1.

Table 1. Overview of studies

Passive Roofs	Key words	Description	Type of Study	Climate/Location	Ref.
GR	Green roof	Green roof refers to a kind of roof whose surface is covered with vegetation.	Simulation	Mediterr-anean climate	(Gagliano, et al., 2015)
GR	Green roof	Green roofs with thick and deep growing medium or substrate ranging from 15-20 cm with plantation of various plant species, shrubs and trees.	Simulation	/	(Gibberd, 2020)
GR	Green roof	There are mainly two types of green roofs, including intensive green roofs and extensive green roofs.	Literature review	Banglade-sh	(Dristy, et al., 2020)

continued on following page

Table 1. Continued

Passive Roofs	Key words	Description	Type of Study	Climate/Location	Ref.
GR	Green roof	Vegetation and its growth medium the roof on the top of the building is a green roof.	Literature review & Simulation	Indonesi-a	(Dewi & Paramitha, 2021)
GR	Green roof	Green roofs can not only be used for planting, but also absorb part of rainwater as a solution to control rainwater. In addition, they can reduce heat penetration into indoor space.	Simulation	The United Arab Emirates	(Jadaa, et al., 2019)
GR	Green roof	The green roof is usually composed of vegetation, substrate, filter layer, drainage material, insulation material, root barrier and waterproof membrane.	Literature review	/	(Alibaba, 2019)
GR	Green roof	Green roofs are broadly classified into intensive and extensive roofs.	Literature review	/	(Cascone, 2019)
GR	Green roof	Green roofs are a specific type of blue-green infrastructure and a well-known element of water-sensitive urban design.	Field study	Oceanic climate	(Richter & Dickhaut, 2023)

continued on following page

Table 1. Continued

Passive Roofs	Key words	Description	Type of Study	Climate/Location	Ref.
GR	Green roof	A green roof design consists of three layers: vegetation, substrate (growth media) and drainage.	Literature review	/	(Jusić, et al., 2019)
GR	Green roof	Green roof is an important system, which can improve the thermal performance of buildings, surrounding air quality and urban ecological environment without relying on additional mechanical equipment. In addition, it is an innovative scheme to solve the problem of rainwater.	Simulation	Türkiye	(Tuğba Ercan, et al., 2021)
GR	Green roof	A green roof, also known as a rooftop garden, is a nutrient layer that grows on the roof	Case study, interview and fieldwork	Malaysia	(Abdul Rahman, et al., 2022)
GR	Green roof	Green roofs, also known as rooftop gardens, are composed of growth media and vegetation on the roof.	Simulation	Kumasi,Ghana	(Nyame-Tawiah, et al., 2020)

continued on following page

Table 1. Continued

Passive Roofs	Key words	Description	Type of Study	Climate/Location	Ref.
GR	Green roof	The green roof is composed of several parts, including plant canopy, growth medium, waterproof membrane, protective membrane, drainage layer and filter layer.	Simulation	Panama	(Moreno, et al., 2021)
GR	Green roof	Green roof technology is an energy-saving building technology, which is used to help improve the sustainability of buildings. The main concept is to reduce the total amount of direct and indirect solar radiation on the roof.	Simulation	Dhahran, Saudi Arabia	(Mahmoud, et al., 2017)
GR	Green roof	/	Field study	Baghdad	(Hassan & Husain, 2020)
GR	Blue-green roof	Blue green roof is a kind of roof system, which is mainly used as part of the rainwater management strategy. Its vegetation and various basement layers can retain and store rainwater.	Literature review & Interview	Norway	(Andenæs, et al., 2021)

continued on following page

Table 1. Continued

Passive Roofs	Key words	Description	Type of Study	Climate/Location	Ref.
GR	Blue-green roof	The blue-green roof is a roof component. Because its vegetation and its substrate layer can store rainwater, the roof becomes part of the rainwater management strategy.	Literature review & Interview	Norway	(Andenæs, et al., 2020)
CR	Cool roof	The main refrigeration mechanism of cool roof includes high solar reflectivity and high thermal emissivity.	Simulation	/	(Gibberd, 2020)
CR	Cool roof	The cool roof is basically a roof with cool material (high reflective coating) on its outermost surface.	Simulation	Mediterr-anean climate	(Gagliano, et al., 2015)
CR	Cool roof	The cool roof is a kind of passive roof that can reflect most of the sunlight and keep the temperature of the roof surface lower than that of ordinary roof.	Literature review	Hot climate	(Rawat & Singh, 2022)
CR	Cool roof	The reflective coating of the cool roof can reflect 90% of the solar radiation back to the sky.	Literature review	Hot climate	(Rawat & Singh, 2022)

continued on following page

Table 1. Continued

Passive Roofs	Key words	Description	Type of Study	Climate/Location	Ref.
CR	Cool roof	Reflective materials for cool roofs are characterized by high infrared emission values. It can radiate into the sky at night and dissipate the accumulated heat without transferring it indoors.	Experiment	Mediterr-anean climate	(Romeo & Zinzi, 2013)
CR	Cool roof	The roof with high reflection and high thermal emissivity can reduce the heat transfer from outdoor to indoor. In addition, roofs with low thermal emissivity also have high solar reflectivity.	Literature review	Banglade-sh	(Dristy, et al., 2020)
CR	Cool roof	Cool roofs, also known as reflective roofs, basically reflect the heat of the sun and release it back to the sky, rather than transfer it to indoor space.	Case study	Hot and dry climate	(Athmani, et al., 2022)

continued on following page

Table 1. Continued

Passive Roofs	Key words	Description	Type of Study	Climate/Location	Ref.
CR	Cool roof	The cool roof reflects most of the incoming sunlight, reducing the heat transfer to the interior, making the roof surface cooler than the standard design roof.	Simulation	/	(Qin, et al., 2017)
CR	Cool roof	The cool roof mainly uses four materials: insulation material, albedo, air gap, and a water pond.	Simulation	Hot and dry climate	(Dabaieh, et al., 2015)
CR	Cool roof	Cool roofs use solar reflective materials to lower the temperature of the roof surface, just like white clothes.	Experimental Study	Hot climate	(Sedaghat, et al., 2023)
CR	Reflective roof	Reflective cooling with optimal roof pitch and ventilation helps to reduce and remove heat from the upper areas, thereby lowering the indoor temperature.	Literature review	/	(Roslan, et al., 2016)

continued on following page

Table 1. Continued

Passive Roofs	Key words	Description	Type of Study	Climate/Location	Ref.
CR	Reflective roof	The heat load of the roof can be reduced by applying a reflective coating on the roof. So reflective roofs basically use silver paint to reflect solar radiation.	Field study	Ujjain, India	(Gagliano, et al., 2015)
VR	Ventilated roof	Ventilated roof is a unique passive ventilation system, which can promote fresh air through the roof space and improve the air circulation of ceiling and building space.	/	Australia	(Lysagh, 2024)
VR	Ventilated roof	The ventilation layer of ventilation roof has a greater impact on the roof temperature delay, but the impact on the delay time is relatively weak. The thermal performance of ventilated roof is mainly affected by the thickness of air layer, roof slope and vent size.	Case study	Northeas-t cold area of China	(Yu, et al., 2020)

continued on following page

Table 1. Continued

Passive Roofs	Key words	Description	Type of Study	Climate/Location	Ref.
VR	Ventilated roof	The structure of a typical ventilated roof consists of an upper floor, a cavity and a lower floor. The cavity is used to circulate air.	Simulation	/	(Calati, et al., 2021)
VR	Ventilated pitched roof	In a ventilated pitched roof, the slats that support the tiles and the roof substrate form a hollow ventilation layer that allows air to circulate inside.	Case study	Continen-tal climate	(Baccega, et al., 2022)
VR	Ventilated pitched roof	In hot climates, tiled pitched roofs significantly reduce the transfer of outdoor heat to indoor Spaces.	Experiment & Simulation	Hot climate	(Bortoloni, et al., 2017)
VR	Attic roof	Ventilation roof has a small size thickness of the pipe, the air flow in the ventilation pipe is laminar flow of micro ventilation.	Simulation	/	(Ciampi, et al., 2017)
VR	Attic roof	/	/	America	(Lstiburek J, 2024)

continued on following page

Table 1. Continued

Passive Roofs	Key words	Description	Type of Study	Climate/Location	Ref.
VR	Attic roof	The heat gain to the indoor space is further reduced in the vented attic type buildings by allowing an outdoor air to pass through the soffit and exit at ridge in vented attics.	Literature review	/	(Iffa & Tariku, 2016)
VR	Attic roof	It is common practice for attic ventilated roof to design and build sloping roofs and to provide openings at the cornice and ridge to facilitate air circulation.	Simulation	Canada	(Tariku & Iffa, 2017)
VR	Double roof	Double roof with 6 inches ventilated air gap.	Field study	Penang, Malaysia	(Asmat, 2013)
VR	Double roof	Ventilated roof is a passive strategy that reduces the thermal gain of building envelope structures by creating air gaps on traditional roofs. Through the shading effect and free ventilation of air gaps, this type of roof reduces the heat flow entering the building.	Simulation	Djibouti	(Omar, et al., 2017)

continued on following page

Table 1. Continued

Passive Roofs	Key words	Description	Type of Study	Climate/Location	Ref.
BR	Blue roof	The blue roof consists of four main layers, including: aggregate layer, geotextile, drainage layer and waterproof membrane.	Experiment & Simulation	Philadelp-hia, America	(Rey-Mahía, et al., 2022)
BR	Blue roof	This water storage roof structure includes: 1. Open water surface; 2. Storage of water in or under porous media or modular surfaces; 3. Store water on deck surfaces or under coverings.	/	/	(Designingbuildings, 2024)
BR	Blue roof	The blue roof is a non vegetation rainwater collection system that can retain rainwater. Using a weir at the roof drain can generate temporary water accumulation and slowly release rainwater.	Field study	Seoul, Korea	(Shafique, et al., 2016)

continued on following page

Table 1. Continued

Passive Roofs	Key words	Description	Type of Study	Climate/Location	Ref.
BR	Water-covered roof	As pools or water-collecting roofs contain large amounts of water, they exhibit greater thermal inertia than inverted flat roofs.	Field study	Mediterr-anean climate	(Espinosa-Fernández, et al., 2020)
BR	Roof pond	Roof ponds can be divided into dry, wet and open types. The roof pond provides cooling benefits through indirect evaporative cooling and radiant cooling.	Literature review	Hot-arid climate	(Sharifi & Yamagata, 2015)
BR	Roof pond	A roof pond is built on the roof, which can be used for both passive heating in winter and passive cooling in summer. The roof pond is a unique passive system.	Field study	Ujjain, India	(Ingebretsen, et al., 2022)

4.1. Green Roof

Growing plants on the roofs of buildings is an old idea that people have used since ancient times. For centuries, people used to grow herbs and plants on their roofs for food, drinking, and even leisure. Germany is widely recognized as the birthplace of modern GFs. The development of GFs is to help prevent radiation on the roof and serve as a fire prevention measure. GFs are not only used for planting but also as a solution to control rainwater by absorbing part of the rainwater and preventing heat from penetrating into indoor Spaces (Jadaa, et al., 2019). GFs are generally divided into two types: extensive GF with soil thickness below 15cm, and intensive GF with soil thickness above 20cm (Dristy, et al., 2020; Cascone, 2019). A GF usually consists of several parts, including vegetation, substrate, filter layer, drainage material, insulation, root barrier, and waterproof membrane (Alibaba, 2019). The thermal performance of GFs is affected by changes in shading, transpiration, and wind (Yuan, et al., 2022). The GF reduces the roof temperature, alleviates temperature fluctuations, and stores energy. In hot weather, it can promote evaporation and transpiration, provide shade, and insulate the soil. By planting plants on the roof, GFs reduce heat transfer and increase thermal comfort. It reduces direct and diffuse radiation on the roof by limiting the thermal load and aging of roof components. In addition, the thermal performance of GFs is also related to the leaf area indexes. According to a survey (Nyame-Tawiah, et al., 2020) of GFs in Ghana, the soil depth ranges from 70 mm to 150 mm, LAl is 5, LAl is 5, soil depth is 200 mm, When LAl is 2 and the soil depth is 500 mm, the temperature is the lowest and the cooling effect is good, which can improve the thermal comfort of passengers.

The advantages of intensive GF and extensive GF are shown in Table 2. A study conducted by Gagliano et al., (2015) in the Mediterranean region shows that extensive and moderately insulated GF is the most effective technology, which can greatly reduce the cooling energy demand of buildings and help mitigate the heat island effect. According to Shafique et al.'s research (2016) on GFs and blue roofs in Seoul, it shows that GFs have many positive effects. These advantages include long-term retention and retention of large amounts of rainwater, absorption of noise pollution, reduction of building energy consumption through cooling effects, reduction of urban heat island phenomenon, creation of living environment for birds, and improvement of the quality of life in the area.

Table 2. GF types comparison with their pros (Gibberd, 2020)

Extensive Green Roof *(Thin growing medium; little or no irrigation; stressful condition for plants; low plant diversity.)*	**Intensive Green Roof** *(Deep soil; irrigation system; more favorable conditions for plants; high plant diversity; often accessible.)*
Light weight; roof generally does not require reinforcement.	Greater diversity of plants and habitats.
Suitable for large areas.	Good insulation properties.
Applicable for roofs with 0-30° slope.	Can simulate a wild garden on the ground.
Low maintenance and long life.	Can be made visually attractive.
Often no need for irrigation and specialized drainage system.	Often accessible, with more diverse utilization of the roof. i.e. for recreation,growing food, as open space.

continued on following page

Table 2. Continued

Extensive Green Roof *(Thin growing medium; little or no irrigation; stressful condition for plants; low plant diversity.)*	**Intensive Green Roof** *(Deep soil; irrigation system; more favorable conditions for plants; high plant diversity; often accessible.)*
Less technical expertise required.	More energy efficient and storm water retention capability.
Often suitable for retrofit projects.	Longer membrane life.
Can leave vegetation to grow spontaneously.	
Relatively inexpensive.	
Looks more natural.	
Easier for planning authorities to demand as a condition of planning approval.	

Mahmoud et al. (2017) conducted a comparative study on GFs and traditional roof systems in Saudi Arabian's hot climate. The research results indicate that GF systems have economic benefits, saving 15% -32% of energy annually, depending on soil thickness and thermal conductivity. The study also suggests that GFs not only provide energy-saving insulation, but also alter the temperature fluctuations experienced by the roof membrane. This reduces the thermal stress and thermal aging of the house facial mask, thus prolonging the life of the house facial mask. Planting a layer of vegetation on the roof of a building can also affect relative humidity. GF technology provides a wide range of benefits, such as improving air quality, creating green roof habitats, mitigating the heat island effect, and reducing the annual energy demand of buildings. In addition, it helps to improve the thermal comfort of building occupants, especially in climate conditions with high daytime temperatures and high solar irradiance.

It is worth noting that Hassan et al. (2020) conducted a comprehensive and in-depth analysis of the practical value of green buildings in a study conducted in Baghdad. They pointed out that in terms of environmental benefits, green roofs can help reduce surface high temperatures in summer and alleviate low temperatures in winter; It greatly helps to increase the green space area of cities, especially those lacking green spaces; It filters the air by absorbing carbon dioxide and releasing it into the atmosphere, which helps reduce environmental pollution; It can also collect dust from the air, so it can remove dust and dirt from the atmosphere; It can dispel fog, odor, and wind; It is a living area for small organisms, making up for the land used for construction; It is considered a refuge for birds; In addition, green roofs use recyclable materials, making them environmentally friendly. In terms of economic benefits, green roofs reduce the cooling and heating costs of buildings, achieving energy efficiency; It can grow vegetables and fruits cleanly and safely, becoming an economic resource and source of livelihood; It reduces the impact of environmental changes on cities, thereby reducing electricity consumption; It absorbs and stores rainwater and converts it into a favorable climate, regulates local humidity, maintains natural circulation of water vapor in the air, reduces the need for drainage, and increases. In terms of social benefits, the green garden on the roof serves as an outlet for the residents of these buildings, providing them with psychological comfort due to its aesthetic appeal; It reduces noise levels and their negative impact on people living in buildings; Provide vegetables to residents through green roofs.

Some studies (Dristy, et al., 2020; Dewi & Paramitha, 2021; Tuğba Ercan, et al., 2021) also have shown that the benefits of implementing GFs are manifold. In addition to reducing the urban heat island effect, it maintains air quality, absorbs pollutants from urban surface runoff stormwater, and acts as a water regeneration system. In addition, it can also act as a natural sound insulation material, absorbing heat from the sun rather than reflecting it; Therefore, it is actively involved in reducing the urban heat island effect and increasing the aesthetics of the roof surface. Its thermal insulation performance helps to reduce energy consumption costs, prevent the decay of waterproof roofing materials, and extend the service life of waterproof roofing materials. Moreover, according to Jusiki et al.(2019), GFs stabilize daily temperature fluctuations. Another study (Peñalvo-López, et al., 2020) in Spain showed that GFs improve the efficiency of photovoltaic systems installed on roofs by reducing the temperature fluctuations of the roof membrane throughout the year. The green vegetation on the roof uses 60% of the sun's radiation for photosynthesis. The air pollution generated by the production of GF materials is offset within 13 years of green roof operation. The study also found that GFs can increase fruit and vegetable production and also provide habitat for small organisms and insects. While the initial cost of a GF is high, the energy and costs saved over the life of the building mean that a GF is a more environmentally friendly option (Gibberd, 2020). A study (Rawat & Singh, 2022) also found that the economic initial cost of a GF is three to five times that of a conventional roof, but it saves energy due to increased shading in hot climates. The hotter and drier the environment, the more significant the role of green roofs in regulating building temperature.

Blue-green roof is another type of GF in which a layer of vegetation and its substrate layer are used to store precipitation, making the roof part of a stormwater management strategy (Andenæs, et al., 2021; Andenæs, et al., 2020). In addition to the retention (delayed runoff) capacity of a GF, the turquoise roof also has retention (rainwater evaporation/transpiration) capacity, or the additional water storage capacity required to maintain vegetation.

Finally, green roofs and blue-green roofs not only reduce human impact on the surrounding environment, but also have a positive impact on people's health and the appearance of their houes.

4.2. Cool Roof

CFs known as reflective roofs, this passive roof system is relatively simple and is particularly suitable for low-cost homes (Nyame-Tawiah, et al., 2020). In addition, CFs are also used as urban heat island mitigation strategies (Gagliano, et al., 2015). The main cooling mechanism of CFs is high solar reflectivity and high thermal emissivity. The range of values for these two parameters is from 0 to 1. The closer the value is to 1, the lower the temperature of the roof under sunlight (Sedaghat, et al., 2023).

Solar radiation control is the most important passive strategy in summer (Romeo & Zinzi, 2013). Qin et al. 's study (2017) found that CFs reflect most of the incoming sunlight, making the roof surface cooler than roof products with standard design. The surface temperature of a CF is usually lower than that of a conventional roof, and the daily heat gain of the roof is reduced. Research by Elnabawi et al., (2022) shows that when energy used for cooling is reduced, air pollution directly decreases, leading to fewer pollutants produced by power plants. In contrast, the indirect reduction in air pollution is due to a response to ozone formation, which accelerates smog formation at warmer temperatures. Therefore, as the temperature of the city decreases, the possibility of smog will decrease. In addition, CFs using white paint have a high water resistance in addition to their roof cooling ability, which can effectively form a physical barrier to water leakage and cracking (Dristy, et al., 2020). A survey (Rawat & Singh, 2022) suggests that CFs with adequate insulation are recommended in hot-dry climates. The study also found that reflective coatings keep the roof close to the internal air temperature compared to composite roofs, and the cost of painting the roof is not expensive and can reduce heat gain in the summer. CF applications do not require complex maintenance on the roof of a building and can be easily applied on an existing roof by traditional methods at low cost according to the user's requirements. Research by Athmani et al., (2022) suggests that the potential for CFs to reflect solar heat makes them most convenient when solar radiation intensity is high and diurnal variation is greatest. In addition, the benefits of CFs include reduced cooling energy consumption, carbon emissions, air pollutants, and urban heat island problems. The CFs are also appreciated for their low cost and energy-efficient retrofit, easy to install on the building envelope. Given these results, a CF not only has a positive impact on interior performance but also extends the life of the roof and cools the surrounding environment. Although CFs are considered to increase the demand for winter heating. However, if the location has no heating demand or low heating demand, a CF is the best choice for solving indoor overheating (Dabaieh, et al., 2015).

There are many studies on CFs in hot-dry climates or hot-arid climates. A study (Sedaghat, et al., 2023) shows that cool roof coatings can effectively reduce energy consumption by 1-3% during winter in Kuwait. After studying passive roofs in hot and dry climates in Egypt, Dabaieh et al. (2015) suggested that residential buildings choose a combination of CFs and natural ventilation to increase indoor thermal comfort. As a conclusion, this type of roof is suitable for low-cost residential buildings in a hot and dry climate. They suggest combining cold roof technology with energy-saving walls and combining the ratio of windows to walls with glass as a way to reduce heat load.

4.3. Ventilated Roof

Ventilated roofs include: attic roofs, double roofs, and ventilation pitched roofs. The roof method of ventilation enables the house to experience the effects of chimneys and wind pressure (Roslan, et al., 2016). A ventilated roof, if properly designed, can improve the thermal comfort of the attic and help reduce the summer heat load (Ciampi, et al., 2017). The special structure of the VF allows air to flow in the gaps between flat noodles. If air flow is driven by buoyancy, then a VF is considered a passive technology for reducing heat load (Calati, et al., 2021).

A website (Lysagh, 2024) describes the advantages of ventilated roof products, including 1. Improve roof ventilation through continuous airflow and reduce AC energy consumption; 2. Reduce condensation, humidity, and unhealthy mold and mildew; 3. Keep the space cool in summer and manage condensation in winter; 4. Provide NCC standard roof ventilation system; 5. Passive system without wear or maintenance of moving parts.

Iffa & Tariku (2016) studied attic roofs. The simulation results show that attic ventilation causes energy loss in the winter and is a benefit in the summer because it removes hot air from the attic space, reducing cooling loads and energy requirements. Ventilated lofts can also reduce ice dams that form near eaves and extend the life of roof tiles. The research results of Calati et al. (2021) indicate that due to a large amount of heat being carried away by the air flowing in the channels of VFs, the specific heat flux between VFs and non VFs is significantly reduced. Compared with traditional non breathable roofs, for insulated VFs, simulations show that the average heat load of indoor spaces is reduced by more than 70%; For uninsulated conditions, the heat load is reduced by about 60%. Rawat & Singh's research (2022) studied double roofing. The results show that the double roof reduces the conduction and convective heat transfer from roof to ceiling and is another way that the radiation barrier system blocks the radiation heat transfer between roof and ceiling. Omar et al. (2017) simulated the thermal performance of double roofs in tropical climates. The results indicate that in VFs, heat is carried away by the airflow through the double-layer skin, greatly reducing the heat flow into the building (saving almost 50% of energy efficiency), from 116 W/m^2 on standard roofs to 60 W/m^2 on ventilated but uninsulated roofs. This study suggests that VFs are cost-effective for developing countries like Djibouti.

According to Baccega et al. (2022), in hot climates, ventilated pitched roofs help dissipate excess heat, thereby reducing cooling energy requirements. The performance of ventilated pitched roofs covered with clay tiles can be further improved by a discontinuous covering of permeable air. In addition, it should be noted that the improvements achieved by optimizing the shape of the tiles to improve the breathability of the cladding do not increase the cost of a ventilated roof. In a traditional pitched roof, a system of "slats and anti-slats" covering the elements allows for the creation of ventilation ducts under the shingles and above the sheathed ventilation, as shown in Figure 4. In this volume, air flows from the eaves section (air intake) to the roof ridge, helping to dissipate excess heat in the summer. In tiled roofs, the permeability of overlapping tiles is an additional diffused intake/exhaust system (Bortoloni, et al., 2017).

Figure 4. Ventilated pitched roof. a) longitudinal section, b) cross section.

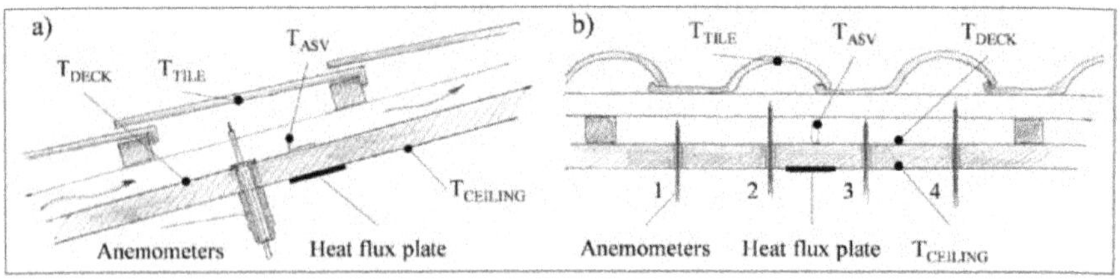

Source: Bortoloni, et al., 2017

Kurkinen's (2017) study of Northern Europe found that outside air creates condensation in cold areas of attics. Prolonged condensation can lead to structural damage and mold growth, which can have negative health effects (Liu, et al., 2024). However, according to Lstiburek's study (2024) of Phoenix, USA, condensation is not a big problem in hot-dry regions. Additionally, according to Iffa & Tariku's research (2016), in cold and coastal climates, ventilating the attic can cause snow and rain to enter the attic space, which can lead to mold and decay. Hurricane-prone areas and wildfire areas can suffer from overly pressurized building and roof fires, respectively. But Panzhihua, located in China's interior and with a hot climate, does not face these challenges.

4.4. Blue Roof

BFs are also known as water storage roofs or pond roofs. It is a roof structure for water storage that may include an open water surface, water storage in or under a porous media or modular surface, or water storage under a raised deck surface or covering (Andenæs, et al., 2021). BF consists of four main layers: gravel/aggregate layer, geotextile, drainage layer, and suitable waterproof roof membrane (Rey-Mahía, et al., 2022).

The BF is designed to bring water into contact with the outside air and solar radiation, producing a high evaporation rate, which greatly reduces the thermal load of solar radiation. Results of a study of BF by Espinosa-Fernandez et al. (2020) show that the external heat capacity of BF is 45% higher than that of stone or terrazzo roofs. BF utilizes water as the ideal thermal mass (due to its large volume heat capacity, as well as being inexpensive and non-toxic). Since high levels of air humidity reduce the potential for evaporation and radiative cooling, Sharifi & Yamagata (2015) recommend that BF should be adopted in dry climates. This technique is very useful for cooling buildings in arid areas with strong solar radiation and high wind speed (Gagliano, et al., 2015). Gupta & Tiwari's research (2016) shows that the use of roof ponds in arid areas can reduce indoor temperatures by 20 ° C. In addition, the BF serves as a form of sustainable urban drainage system in an attempt to alleviate urban flooding caused by stormwater runoff. It can collect rainwater for use, independently of or in addition to the tap water supply, such as toilet flushing or roof greening irrigation. It can also cool solar panels or reduce the load on mechanical refrigeration or cooling equipment. It can also be used for recreational purposes, such as rooftop swimming pools or water features (Designingbuildings, 2024). It is worth noting that compared to GFs, BFs have lower costs and can also reduce runoff in an area (Shafique, et al., 2016).

In summary, the BF is a unique passive system that can be used for both passive heating in winter and passive cooling in summer.

5. DISCUSSION

Through literature review, the importance of passive roofs in the Panzhihua region was discussed, and combined with research on passive roofs in areas related to dry and hot regions, the practical value of various types of passive roofs was analyzed. However, the application of these passive roofs in Panzhihua will still be constrained by local conditions.

According to a study on green buildings by Moreno et al. (2021), in hot climates in Phoenix, Los Angeles, Chicago, and Shanghai, changes in GF insulation and nighttime ventilation provided by fans can optimize the summer thermal performance of GFs. They also used simulation methods to determine the most important parameters of the vegetation layer: plant height, leaf area index, leaf reflectance, and stomatal resistance. Therefore, when using a GF, the thermal performance of the building can be optimized in combination with low-energy mechanical equipment. In addition, the benefits of these GFs also depend on the type of GF (extensive or intensive) and the number of layers of the GF, such as waterproof membrane, root barrier, drainage material, filter layer, insulation layer, vegetation and substrate (Abdul Rahman, et al., 2022). The thermal performance of a GF can be improved by adjusting the thickness of the soil; In this specific case study, it was found that the thicker the soil layer, the better the thermal performance. However, the thermal performance of a GF installed on an uninsulated roof is different from that of a well-insulated roof (Moreno, et al., 2021). Intensive roofs have higher water retention capacity due to thicker soil layers, but there are also structural limitations. GFs with soil thickness between 15 and 30 cm are called semi-intensive roofs, which can accommodate small plants and shrubs, but still require high maintenance costs and need to consider the carrying capacity of the building structure. The soil thickness of the extensive and semi-extensive GFs is 7-15 cm, and only small plants such as grass can be grown (Moreno, et al., 2021). This type of GF requires little maintenance, usually no irrigation, and is suitable for retrofitting older buildings due to its light weight. With low annual rainfall, hot, dry areas need to utilize native flora that can survive harsh environmental conditions. This type of vegetation can be found in these areas. Use plants with lots of mulch, which can be deciduous plants. In addition to maintaining normal growth in the hot summer sun, they can prevent the absorption of the sun's heat in the cold season. Another important aspect to consider is that in a wide GF system, the depth of the planting layer is shallower and the weight is lighter (Yuan, et al., 2022).

The performance of a CF is influenced by several parameters, such as the level of ceiling insulation, the type of reflective paint, climatic conditions, time of use, and the influence between buildings. In addition, due to the accumulation of dust / dirt on the roof, it can affect the ability of the cold roof surface to reflect sunlight, thus increasing solar heat storage. Therefore, the roof must be cleaned regularly. On the other hand, cool surfaces with high reflectivity cast intense radiation around them, which can be distracting and produce unpleasant brightness, especially when the weather is clear.

Although VFs may lead to phenomena such as ice dams, mold, and rot, these risks do not occur in Panzhihua area. In addition, attic ventilation can remove hot air from the attic space, reducing cooling loads and energy requirements. Some argue that while a VF is a benefit in the summer, it causes energy losses in the winter. Because the climate of Panzhihua is hot, the summer is long, and the winter is short and warm, so this defect of the VF has little impact here.

Although the BF has good thermal insulation performance, according to the study by Dabaieh et al. (2015), it is not ideal to adopt this idea in low-cost housing in Egypt. Because it requires a special roof structure to withstand the static load of water. In addition, it requires regular maintenance. Due to water scarcity in dry and hot areas, they recommend using grey water instead of fresh water if using cool rooftop ponds. Leakage can also be a major issue as it is difficult to locate cracks in the roof.

To sum up, although GFs, CFs, VFs and BFs can reduce the cooling energy consumption of buildings in hot and dry areas, and show their respective superior performance.However, the government, architects and scholars should pay attention to the combination with the characteristics of the region and formulate specific practical guidelines to promote and apply them to the rural housing in Panzhihua, so as to give full play to their potential.

6. CONCLUSIONS

Under the background of increasing building cooling energy consumption, this chapter defines the passive roof, demonstrates the importance of the passive roof in Panzhihua area, and its potential in significantly improving the thermal performance of rural residential buildings in Panzhihua. The practical values of GFs, CFs, VFs and BFs in hot and dry climates and similar climates are analyzed. By reducing reliance on mechanical cooling systems, these strategies can promote energy efficiency and environmental sustainability in the region. In addition, although these passive roofs have superior performance, they also have some defects. Therefore, the promotion of these roofs should be optimized in combination with local characteristics and roof characteristics. The literature selected in this chapter mainly comes from the hot and dry regions in the Middle East, and the number of studies on passive roofs in hot and dry regions in China is small. Therefore, in the analysis process, the results may be inaccurate. Further research will assess the risk of passive roofs to support the development of local roof guidelines in Panzhihua. Finally, through field study or simulation, it is also very meaningful to study the specific thermal performance of various passive roofs in this area.

REFERENCES

Abdul Rahman, A., Mohamed Zaid, S., & Mohammad Shuhaimi, N. D. A. (2022). Effects of Green Roof in Reducing Surface Temperature and Addressing Urban Heat Island in Tropical Climate of Malaysia. *Journal of Design and Built Environment, 22*(2), 1–20. DOI: 10.22452/jdbe.vol22no2.1

Alibaba, H. (2019). Green roof benefits, opportunities and challenges. *International Journal of Civil and Structural Engineering Research, 7*(2), 106–112.

Andenæs, E., Engebø, A., Time, B., Lohne, J., Torp, O., & Kvande, T. (2020). Perspectives on quality risk in the building process of blue-green roofs in Norway. *Buildings (Basel, Switzerland), 10*(10), 10. DOI: 10.3390/buildings10100189

Andenæs, E., Time, B., Muthanna, T., Asphaug, S., & Kvande, T. (2021). Risk reduction framework for blue-Green roofs. *Buildings (Basel, Switzerland), 11*(5), 5. DOI: 10.3390/buildings11050185

Asmat, B. I. (2013). *Cooling potential of passive roof treatments on the concrete flat roof in Penang,Malaysia*. Ph.D. Thesis. Universiti Sains Malaysia, Penang, Malaysia.

Athmani, W., Sriti, L., Dabaieh, M., & Younsi, Z. (2022). The potential of using passive cooling roof techniques to improve thermal performance and energy efficiency of residential buildings in hot arid regions. *Buildings (Basel, Switzerland), 13*(1), 21. DOI: 10.3390/buildings13010021

Baccega, E., Bortoloni, M., Bottarelli, M., Dino, G. E., & Zannoni, G. (2022). Improving the air permeability of ventilated roofs. *Future Cities and Environment, 8*(1), 6. DOI: 10.5334/fce.144

Bortoloni, M., Bottarelli, M., & Piva, S. (2017). Summer thermal performance of ventilated roofs with tiled coverings. *Journal of Physics: Conference Series, 796*, 012023. DOI: 10.1088/1742-6596/796/1/012023

Calati, M., Doretti, L., Gatto, C., & Mancin, S. (2021). 3D CFD Simulation of a New Ventilated Roof. *International High Performance Buildings Conference*, Purdue, Paper 358.

Cascone, S. (2019). Green roof design: State of the art on technology and materials. *Sustainability (Basel), 11*(11), 3020. DOI: 10.3390/su11113020

Ciampi, M., Leccese, F., & Tuoni, G. (2005). Energy analysis of ventilated and micro-ventilated roofs. *Solar Energy, 79*(2), 183–192. DOI: 10.1016/j.solener.2004.08.014

Dabaieh, M., Wanas, O., Hegazy, M. A., & Johansson, E. (2015). Reducing cooling demands in a hot dry climate: A simulation study for non-insulated passive cool roof thermal performance in residential buildings. *Energy and Building*, *89*, 142–152. DOI: 10.1016/j.enbuild.2014.12.034

Designingbuildings. (2024), Blue roof. Retrieved October 15, 2024, from https://www.designingbuildings.co.uk

Dewi, C. P., & Paramitha, P. A. (2021). An analysis of the efficiency of green roofs on cooling energy demand in residential building. *Proceedings of the 3rd International Conference of Green Civil and Environmental Engineering (GCEE 2021)*, Malang, Indonesia, n.d DOI: 10.1063/5.0072616

Dristy, N. F., Parvez, M., & Rahman, N. (2020). Potential measures of cool roof treatment for different building types in urban context of Bangladesh. *5th International Conference on Advances in Civil Engineering (ICACE-2020)*. CUET, Chattogram, Bangladesh, n.d

Elnabawi, M. H., Alhumaidi, A., Osman, B., & Alshehhi, R. (2022). Cool roofs in hot climates: A conceptual review of modelling methods and limitations. *Buildings (Basel, Switzerland)*, *12*(11), 1968. DOI: 10.3390/buildings12111968

Espinosa-Fernández, A., Echarri-Iribarren, V., & Sáez, C. A. (2020). Water-covered roof versus inverted flat roof on the mediterranean coast: A comparative study of thermal and energy behavior. *Applied Sciences (Basel, Switzerland)*, *10*(7), 2288. DOI: 10.3390/app10072288

Fu, Y. (2020). *Study on suitability evaluation system of low energy consumption technologies for rural housing envelope in southern Jiangsu province.* Doctoral dissertation, Southeast University, Nanjing, China.

Gagliano, A., Detommaso, M., Nocera, F., & Evola, G. (2015). A multi-criteria methodology for comparing the energy and environmental behavior of cool, green and traditional roofs. *Building and Environment*, *90*, 71–81. DOI: 10.1016/j.buildenv.2015.02.043

Gan, V. J. L., Wong, H. K., Tse, K. T., Cheng, J. C. P., Lo, I. M. C., & Chan, C. M. (2019). Simulation-based evolutionary optimization for energy-efficient layout plan design of high-rise residential buildings. *Journal of Cleaner Production*, *231*, 1375–1388. DOI: 10.1016/j.jclepro.2019.05.324

Gibberd, J. (2020). Green building technologies. In Yang, P. (Ed.), (pp. 482–510). Practice, Progress, and Proficiency in Sustainability.

Gupta, N., & Tiwari, G. (2016). Review of passive heating/cooling systems of buildings. *Energy Science & Engineering*, *4*(5), 4. DOI: 10.1002/ese3.129

Harputlugil, T., & de Wilde, P. (2021). The interaction between humans and buildings for energy efficiency: A critical review. *Energy Research & Social Science*, *71*, 101828. DOI: 10.1016/j.erss.2020.101828

Hassan, S. A., & Husain, S. I. (2020). The effect of green roof on improving the environmental efficiency for the residential buildings in Baghdad city. *Materials Science and Engineering*, •••, 928.

Iffa, E., & Tariku, F. (2016). A review of heat and moisture performance of attic roofs. *Applied Sciences (Basel, Switzerland)*, *2*(d).

Ingebretsen, S., Andenæs, E., & Kvande, T. (2022). Microclimate of air cavities in ventilated roof and façade systems in nordic climates. *Buildings (Basel, Switzerland)*, *12*(5), 683. DOI: 10.3390/buildings12050683

Ismail, A. (2013). *Cooling potential of passive roof treatments on the concrete flat roof in Penang, Malaysia*. Ph.D. Thesis. Universiti Sains Malaysia, Penang, Malaysia.

Jadaa, D. A., Raed, A. A., & Taleb, H. (2019). Assessing the thermal effectiveness of implementing green roofs in the urban Neighborhood. *Jordan Journal of Mechanical and Industrial Engineering*, *13*(3), 161–174.

Jusić, S., Hadžić, E., & Milišić, H. (2019). Stormwater management by green roof. *Acta Scientific Agriculture*, *3*(7), 57–62. DOI: 10.31080/ASAG.2019.03.0516

Kurkinen, K. (2017). Case study of a cold attic in a pitched roof with minimal ventilation. *Energy Procedia*, *132*, 466–471. DOI: 10.1016/j.egypro.2017.09.667

Li, D., Zheng, Y., Liu, C., Qi, H., & Liu, X. (2016). Numerical analysis on thermal performance of naturally ventilated roofs with different influencing parameters. *Sustainable Cities and Society*, *22*, 86–93. DOI: 10.1016/j.scs.2016.02.004

Li, H., & Wan, G. (2014). Analysis on city design strategy of dry hot climate region——Taking the city design of civic commerce office area in Panzhihua as an example. *Urbanism and Architecture*, *8*, 44–45.

Liu, Y. (2021). *Research on energy saving reconstruction design of rural residential roof based on the perspective of thermal comfort in winter——A case study of Yuncheng county*. Master dissertation, Qingdao University of Technology,Qingdao, China.

Liu, Y., Considine, B., & McNabola, A. (2024). Assessment of hygrothermal performance and mould growth risk in roofs of post-retrofitted non-dormer and dormer attic rooms with reduced ventilation. *Building and Environment, 250*, 111172. DOI: 10.1016/j.buildenv.2024.111172

Lstiburek, J. A. (2024). A crash course in roof venting: Understand when to vent your roof, when not to, and how to execute each approach successfully, 2024 (accessed on 15 Augest 2024). Available from http://beta.buildingscience.com

Lysagh. (2024). Vent-a-roof. Retrieved October 15, 2024, from https://lysaght.com/vent-a-roof

Maheshwari, D. G., & Gehlot, A. S. (2017). Evaluation of effectiveness of passive cooling (roof cool) tecnique in Indian climate. *n.d*, 3(4), 256-266.

Mahmoud, A., Asif, M., Hassanain, M., Babsail, M., & Sanni-Anibire, M. (2017). Energy and Economic Evaluation of Green Roofs for Residential Buildings in Hot-Humid Climates. *Buildings (Basel, Switzerland), 7*(2), 30. DOI: 10.3390/buildings7020030

Ministry of Housing and Urban-Rural Development of the People's Republic of China. (2012). *Design standard for energy efficiency of rural residential buildings*. China Architecture & Building Press.

Ministry of Housing and Urban-Rural Development of the People's Republic of China. (2019). *Standard for design of energy efficiency of residential buildings in moderate climate zone*. China Architecture & Building Press.

Ministry of Housing and Urban-Rural Development of the People's Republic of China. (2021). *General code for energy efficiency and renewable energy appliction in buildings*. China Architecture & Building Press.

Moreno, A., Austin, M. C., & Mora, D. (2021). A parametric study of implementing green roofs to improve building energy performance in tropical climate. *E3S Web of Conferences*, 312, 02004.

Nyame-Tawiah, D., Attuah, L., & Koranteng, C. (2020). Determination of the Effect of Green Roofs on Indoor Temperature by the Use of Simulation in a Tropical Landscape. *Journal of Scientific Research and Reports*, •••, 12–28. DOI: 10.9734/jsrr/2020/v26i430244

Omar, A. I., Virgone, J., Vergnault, E., David, D., & Idriss, A. I. (2017). Energy Saving Potential with a Double-Skin Roof Ventilated by Natural Convection in Djibouti. *Energy Procedia, 140*, 361–373. DOI: 10.1016/j.egypro.2017.11.149

Panzhihua City Renhe District People's government office. (2022), District situation overview. Retrieved October 15, 2024, from http:// http://www.screnhe.gov.cn

Panzhihua Climate. (2023), Weather by month, average temperature (China)—Weather spark. Retrieved December 25, 2023, from https://weatherspark.com

Peñalvo-López, E., Cárcel-Carrasco, J., Alfonso-Solar, D., Valencia, I., & Hurtado-Pérez, E. (2020). Study of the Improvement on Energy Efficiency for a Building in the Mediterranean Area by the Installation of a Green Roof System. *Energies, 13*(5), 1246. DOI: 10.3390/en13051246

People's Government of Panzhihua Municipal. (2019), An overview of Panzhihua city. Retrieved October 15, 2024, from http://www.panzhihua.gov.cn

Qin, Y., Zhang, M., & Hiller, J. E. (2017). Theoretical and experimental studies on the daily accumulative heat gain from cool roofs. *Energy, 129*, 138–147. DOI: 10.1016/j.energy.2017.04.077

Rawat, M., & Singh, R. N. (2022). A study on the comparative review of cool roof thermal performance in various regions. *Energy and Built Environment, 3*(3), 327–347. DOI: 10.1016/j.enbenv.2021.03.001

Rey-Mahía, C., Álvarez-Rabanal, F. P., Sañudo-Fontaneda, L. A., Hidalgo-Tostado, M., & Menéndez Suárez-Inclán, A. (2022). An experimental and numerical approach to multifunctional urban surfaces through blue roofs. *Sustainability (Basel), 14*(3), 1815. DOI: 10.3390/su14031815

Richter, M., & Dickhaut, W. (2023). Long-term performance of blue-green roof systems—Results of a building-scale monitoring study in Hamburg, Germany. *Water (Basel), 15*(15), 2806. DOI: 10.3390/w15152806

Romeo, C., & Zinzi, M. (2013). Impact of a cool roof application on the energy and comfort performance in an existing non-residential building. A Sicilian case study. *Energy and Building, 67*, 647–657. DOI: 10.1016/j.enbuild.2011.07.023

Roslan, Q., Ibrahim, S. H., Affandi, R., Mohd Nawi, M. N., & Baharun, A. (2016). A literature review on the improvement strategies of passive design for the roofing system of the modern house in a hot and humid climate region. *Frontiers of Architectural Research, 5*(1), 126–133. DOI: 10.1016/j.foar.2015.10.002

Saurabh, S., & Madhura, Y. (2023). A thermal study of roof shapes and materials in hot dry climates: Insights from a simulation from Rajasthan, India. *International Society for the Study of Vernacular Settlements*, 10(12), 597–609.

Sedaghat, A., Salem, H., Hussam, W. K., Mahdizadeh, A., Al-Khiami, M. I., & Malayer, M. A. (2023). Experimental Study on Cool Roof Methods for Two Identical Portable Cabins in the State of Kuwait. *ICMAME 2023 Conference Proceedings. International Conference on Mechanical, Automotive and Mechatronics Engineering,* Dubai, UAE., n.d

Shafique, M., Lee, D., & Kim, R. (2016). A Field Study to Evaluate Runoff Quantity from Blue Roof and Green Blue Roof in an Urban Area. *International Journal of Control and Automation*, *9*(8), 59–68. DOI: 10.14257/ijca.2016.9.8.07

Sharifi, A., & Yamagata, Y. (2015). Roof ponds as passive heating and cooling systems: A systematic review. *Applied Energy*, *160*, 336–357. DOI: 10.1016/j.apenergy.2015.09.061

Song, Q., Yang, L., & Yin, B. (2014). Building for the climate change:Passive building. *Huazhong Architecture*, *32*(04), 11–15.

Szokolay, S. V. (1980). *Environmental science handbook: For architects and builders* (2nd ed.). Architectural press.

Tariku, F., & Iffa, E., D. (2017). Temperature and air flow patterns in attic roofs. *Journal Of Architectural Engineering*, 23(3), 04017014-1-04017014-11.

Tuğba Ercan, M., Tuna Kayili, M., & Sultan Qurraie, B. (2021). The effects of green roof on heat loss and energy consumption in the buildings. *Computational Research Progress in Applied Science & Engineering*, *7*(4), 1–8. DOI: 10.52547/crpase.7.4.2422

Yu, J., Leng, K., Ye, H., Xu, X., Luo, Y., Wang, J., Yang, X., Yang, Q., & Gang, W. (2020). Study on thermal insulation characteristics and optimized design of pipe-embedded ventilation roof with outer-layer shape-stabilized PCM in different climate zones. *Renewable Energy*, *147*, 1609–1622. DOI: 10.1016/j.renene.2019.09.115

Yuan, J., Patra, I., Majdi, A., Acwin Dwijendra, N. K., Catalan Opulencia, M. J., & Chetthamrongchai, P. (2022). RETRACTED: Fundamental green roof performance of residential building in desert climate: In terms of sustainability and decrease in energy consumption. *Sustainable Energy Technologies and Assessments*, *53*, 102574. DOI: 10.1016/j.seta.2022.102574

Chapter 2
Repurposing Heritage:
Strategies for Active Use of Historical Buildings in Malaysia

TamilSalvi Mari
https://orcid.org/0000-0001-5442-3816
Taylor's University, Malaysia

Ariventhar Ayahvoo
https://orcid.org/0009-0006-5777-0868
Taylor's University, Malaysia

Sujatavani Gunasagaran
https://orcid.org/0000-0002-0377-2319
Taylor's University, Malaysia

ABSTRACT

The adaptive reuse of historically significant buildings represents a critical intersection between heritage conservation and contemporary development needs. This research explores the strategies for adaptive reuse. By examining the characteristics that define historical significance, this study contextualizes the adaptive reuse concept within the framework of heritage conservation. Economic considerations, such as cost savings and financial incentives, are analyzed alongside environmental benefits of sustainability and reduced resource consumption. The social and cultural implications are assessed, emphasizing the role of community engagement and the maintenance of cultural value. Architectural and structural challenges are identified, with a focus on integrating current building codes and standards while respecting historical integrity. This research aims to provide a comprehensive framework for adaptive reuse strategies, ensuring the sustainable preservation of historically significant buildings within the boundaries of established conservation guidelines.

DOI: 10.4018/979-8-3373-5278-7.ch002

1. INTRODUCTION

1.1 Adaptive Reuse in Architecture

Adaptive reuse involves repurposing old or abandoned buildings for new functions while preserving their historical value, bridging heritage conservation with modern development (Rodrigues & Freire, 2017; National Trust for Historic Preservation, 2023). This approach promotes sustainable urban growth by attracting investment, revitalizing underused areas, reducing environmental impacts, conserving resources, and lowering carbon emissions (Hosagrahar, 2017). It also fosters cultural continuity and social cohesion by maintaining a sense of place and encouraging community interaction (Yung & Chan, 2018; Hwang & Park, 2020). Despite challenges like regulatory hurdles and high costs, adaptive reuse remains a viable solution for cities seeking innovation while honoring the past (Smith, 2019).

While adaptive reuse dates back to ancient civilizations, its prominence grew in the post-World War II era, especially with preservation movements following widespread destruction in Europe. The 1964 Venice Charter emphasized conserving historic buildings, encouraging adaptive reuse (ICOMOS, 2014). The environmental movement in the 1970s positioned reuse as a strategy for sustainable development, conserving embodied energy (Shipley, Utz, & Parsons, 2006).

In recent years, adaptive reuse has gained attention due to environmental concerns, technological advances, and its potential for community and economic revitalization. It conserves embodied energy and reduces greenhouse gas emissions (Bullen & Love, 2019), with tools like Building Information Modeling (BIM) improving planning accuracy and efficiency (Ma & Li, 2020). These projects increasingly involve local communities, preserving historical identity while meeting modern needs (Langston & Shen, 2017), and stimulate local economies by transforming underutilized buildings into mixed-use developments (Yung & Chan, 2018).

1.2 Research Background

Due to their traditional design, materials, and historical significance, historical buildings are difficult to retro fit and have fewer options for modifications. Adaptive reuse interventions will be more difficult to implement if conservation aspects are taken into consideration (Kayan B. et al., 2018).

The impact of adaptive reuse on the sense of place and urban conservation in Melaka's historic town, where heritage shophouses have been repurposed as F&B establishments. Holistic approach that considers both heritage preservation and community needs is crucial in promoting sustainable and inclusive urban conservation practices (Man, A. W. K., 2023).

Because of erroneous conservation treatments, many have failed to achieve the intended goals for sustainable adaptive reuse practice. Businesses and building owners in George Town will have to deal with the various requirements and pressure in order to comply with the needs of maintaining the UNESCO status Zubir, Z. K. (2017).

1.3 Research Problem

1.3.1 Non-Sustainable Building Conservation Practice in Malaysia

Adaptive reuse can risk compromising the authenticity and integrity of historical buildings by potentially replacing original materials and altering historical context. Balancing modern functionality with heritage preservation is challenging, and inappropriate interventions can cause irreversible damage (Jokilehto, 2017).

The loss of original materials is a primary concern, as unique construction techniques and materials contribute to a building's historical value. Replacing these with modern alternatives can diminish the building's significance and structural integrity (Bullen & Love, 2019). Additionally, changes to a building's design or use can disrupt its historical context, impacting the surrounding environment and cultural landscape, especially in urban settings where historical buildings shape neighborhood identity (Yung & Chan, 2018).

The absence of comprehensive, practical guidelines for adaptive reuse further complicates these issues. While frameworks like the Venice Charter offer broad principles, they often lack specific guidance for diverse buildings and contexts, leading to inconsistent and potentially harmful practices (ICOMOS, 2014). Moreover, the involvement of multiple stakeholders—developers, architects, conservationists, and communities—can create conflicts that undermine preservation goals (Langston & Shen, 2017).

In conclusion, adaptive reuse offers sustainable benefits but poses significant risks to the authenticity of historical buildings. Addressing these challenges requires clear guidelines and effective collaboration among all stakeholders involved in the process.

1.3.2 Challenges from Conservation Guidelines

Conserving historic buildings in UNESCO Buffer Zones presents challenges for business owners, who struggle to meet diverse authority guidelines while ensuring investment success. Authorities often adopt a punitive approach, discouraging innovation and investment due to fear of penalties for deviations from guidelines (Zubir, 2017).

The lack of universal conservation guidelines exacerbates this issue. With each jurisdiction enforcing its own standards, business owners face confusion and inconsistency, leading to delays and increased costs. Standardized guidelines are essential to balance conservation and economic use, ensuring preservation without imposing unnecessary burdens (Plevoets & Van Cleempoel, 2019).

The adversarial nature of current regulatory frameworks creates distrust between business owners and conservation authorities, undermining collaboration. This often leads to suboptimal conservation outcomes, as property owners may bypass conservation efforts to avoid penalties (Shipley et al., 2020).

1.3.3 Lack of Systematic Maintenance Strategies

Maintenance practices for heritage buildings in Malaysia are inefficient, leading to deterioration despite their cultural and historical value. Current approaches are reactive, lacking systematic strategies and adequate resource allocation, resulting in higher costs and further degradation (Yung & Chan, 2018; Kayan, 2016).

Issues like limited funding, a shortage of skilled labor, and a focus on new developments hinder proper upkeep (Ahmad, 2019). Fragmented responsibilities among multiple agencies and a lack of specialized training in traditional materials and construction methods further complicate maintenance (Mohamed et al., 2017; Bullen & Love, 2019).

Malaysia needs structured maintenance strategies, with proper funding, training, and coordination to ensure the sustainable conservation of its heritage buildings (Yung & Chan, 2018).

1.4 Research Gap

Gap

In Malaysia, adaptive reuse is recognized as a sustainable strategy for heritage conservation, but research remains general, focusing on global environmental benefits without addressing local challenges, especially within Malaysia's multicultural context (Yunos et al., 2019). There is limited exploration of integrating modern building codes with heritage structures while preserving authenticity and structural integrity (Kamaruzzaman et al., 2020). While UNESCO sites like George Town and

Malacca are often discussed, studies rarely examine the impact of local policies on adaptive reuse (Salleh et al., 2021).

VOS viewer analysis highlights a scarcity of localized research, even in prominent cities like Malacca, and underexplores issues such as inconsistencies between modern and traditional management practices (Ahmad & Din, 2021) and the lack of financial incentives for private owners (Ibrahim & Rasli, 2020). Community engagement, crucial for cultural relevance and sustainability, is also overlooked (Nasir et al., 2020; Ibrahim & Rasli, 2020; Salleh et al., 2021). Technologies like Building Information Modeling (BIM) and energy-efficient retrofitting, which can enhance adaptive reuse, are underutilized (Salleh et al., 2021).

This research aims to integrate policy review, technological tools, modern management, and participatory strategies to advance adaptive reuse in Malaysia's heritage conservation efforts.

Figure 1. Network visualization generated on VOS viewer.

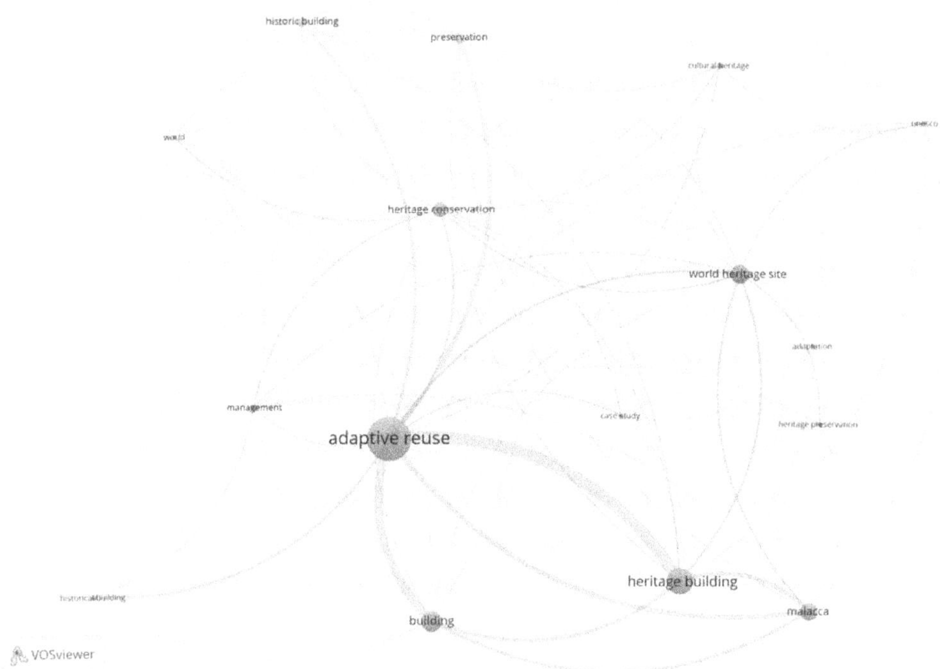

(Author, 2024)

1.5 Research Aim and Objectives

Aim

This research aims to formulate suggestions for adaptive reuse strategies that ensure the sustainable preservation of historically significant buildings in Malaysia.

By addressing gaps in existing conservation guidelines, the study seeks to develop well-founded recommendations that balance historical integrity with contemporary functionality. The ultimate goal is to enable the effective repurposing of these historic structures, thereby prolonging their lifespan and relevance in a rapidly modernizing urban landscape, ensuring their active use to prevent resource wastage, and preserving their heritage values.

Objectives

The following are the objectives:

1. To examine current practice of adaptive reuse as well as opportunities to improve it to suit the specific needs of historically significant buildings in Malaysia.
2. To analyze strategies and policies that currently exist for the repurposing of historical buildings within a UNESCO Heritage Site in Malaysia concerning its effectiveness and how local authorities have influenced these remaining regulations.
3. To evaluate and recommend adaptive reuse strategies that are compatible with the current building codes and standards while maintaining the true spirit of the historically important structures.

1.5 Research Questions

What are the strategies for repurposing historically significant buildings through adaptive reuse as a preservation strategy?

The sub-questions are also used to further support the main research question:

1. How can modern building codes and standards be integrated into historically significant buildings without compromising their integrity?
2. How do local policies and regulations influence adaptive reuse projects within the UNESCO Heritage sites in Malaysia?
3. What roles do heritage conservation guidelines play in the adaptive reuse process?

1.6 Limitation and Expected Outcomes

This study aims to highlight the importance of revitalizing Malaysia's historical buildings rather than leaving them underutilized in urban areas.

However, preservation regulations, zoning laws, and heritage designations often restrict the extent of modifications allowed. While these legal frameworks safeguard cultural and architectural values, they also limit adaptive reuse by making it difficult to incorporate modern amenities, ensure accessibility, and comply with safety or energy standards. Projects may face restrictions on adding features like elevators or reconfiguring interior layouts, and the approval process can be time-consuming, costly, and discouraging to investors.

Another key limitation is the shortage of skilled professionals in heritage conservation. Adaptive reuse demands expertise to balance preservation with modernization, but Malaysia lacks sufficient trained conservationists, architects, and engineers. This scarcity drives up costs, delays projects, and sometimes requires hiring international experts, further complicating logistics. The situation highlights the need for improved education and training programs to grow local capacity in heritage conservation.

1.7 Data Collection Method and Analysis

The primary sources for analysis compromised of the literature review on adaptive reuse by understanding of the strategies used successfully in adaptive reuse and how it is currently practiced in Malaysia. Strategies explained by authors, Ann Forsyth, Sheila Conejos, Craig Langston, Jim Smith, Esther H.K. Yung, Edwin H.W. Chan, John Pendlebury, Tim Guy Townshend, Rose Gilroy, Dr Craig Langston, Dr Hilde Remøy, and Dr Sara Wilkinson explain the latest strategies practiced around the world into implementing sustainable adaptive reuse. Studying these strategies allows for successful strategies adaptation into the Malaysian context.

Using relatable strategies, interview questions will be generated to further explore into this study by conducting interviews with industry professionals to further understanding whether the proposed strategies can be implemented into the Malaysian context.

1.8 Justification and Summary

This study is essential for developing effective adaptive reuse strategies that support the sustainable preservation of historically significant buildings in Malaysia. Adaptive reuse not only safeguards architectural and cultural heritage (Yung & Chan, 2018) but also aligns with sustainability goals by reducing material consumption and environmental impact. By giving new life to abandoned or underused structures, it contributes to urban renewal and community revitalization. However, without clear and actionable strategies, implementing sustainable and efficient adaptive reuse remains challenging in the Malaysian context.

This chapter established the foundation of the dissertation by outlining the research background, key issues, and justification. It introduced the research purpose, methodology, and structure of the study. The next chapter will explore definitions, strategies, benefits, challenges, and previous studies on adaptive reuse to identify existing research gaps.

2. LITERATURE REVIEW

Chapter two is the literature review study of adaptive reuse which firstly explains the understanding of the term in a Malaysian context in section 2.1 followed by the benefits in section 2.2, challenges in section 2.3 and case studies of buildings with significant historical values successfully converted into positive spaces through adaptive reuse in section 2.4. The seven strategies into implementing successful and sustainable adaptive reuse are further elaborated into section 2.5.

2.1 Adaptive Reuse practice in Malaysia

Adaptive reuse in Malaysia focuses on repurposing obsolete and historical buildings for modern use, preserving architectural heritage while meeting contemporary needs. This practice supports sustainable development and cultural conservation. Hamzah and Omar (2011) highlight its role in maintaining Malaysia's architectural identity amid urbanization.

2.2 Benefits of Adaptive Reuse in Malaysia

2.2.1 Preservation of Cultural Heritage

Adaptive reuse safeguards Malaysia's rich historical fabric. Projects in George Town and Melaka illustrate how restoring colonial-era buildings contributes to cultural continuity (Kamaruzzaman et al., 2019; Yunos et al., 2018).

2.2.2 Environmental Sustainability

By reusing structures, adaptive reuse reduces material demand and construction waste. Abidin and Hasan (2020) observed notable reductions in carbon emissions through conserved embodied energy, as seen in Kuala Lumpur's repurposed shophouses (Ali et al., 2020).

2.2.3 Economic Benefits

Often more cost-effective than new builds, adaptive reuse boosts tourism and local business. In Ipoh, heritage conversions have raised property values and revitalized the economy (Tharim et al., 2018; Yusoff et al., 2019).

2.2.4 Social and Community Development

Revitalized spaces encourage social interaction and community bonding. Projects like the Kuala Lumpur railway station transformation have fostered communal activities and cultural identity (Mohd Nasir & Salleh, 2020; Ibrahim et al., 2021).

2.2.5 Urban Regeneration

Repurposing derelict structures into productive spaces supports city revitalization. Penang's creative hubs exemplify how adaptive reuse attracts industries and rejuvenates urban areas (Rahman et al., 2019; Tan et al., 2020).

2.2.6 Educational and Cultural Awareness

These projects often incorporate educational elements. Adaptive reuse of the Sultan Abdul Samad Building, for example, serves as a heritage center that educates both locals and tourists (Jamaludin & Yahya, 2019; Zainal et al., 2018).

2.2.7 Conclusion

Adaptive reuse in Malaysia yields broad benefits cultural, environmental, economic, social, and educational while aligning heritage conservation with modern needs.

2.3 Challenges of Adaptive Reuse in Malaysia

2.3.1 Regulatory and Legal Challenges

Malaysia's heritage laws can restrict adaptive reuse. While the National Heritage Act 2005 mandates automatic protection for buildings over 100 years old, ambiguities remain for newer structures under Act 171 (Act 645, 2012). Current guidelines often lack clarity on sustainability integration (Kuala Lumpur City Hall, 2018).

2.3.2 Financial Constraints

High costs and limited funding hinder adaptive reuse. Specialized restoration work is expensive, and government support remains minimal (Hamzah et al., 2019; Ahmad & Din, 2021).

2.3.3 Technical and Structural Issues

Aging buildings require costly repairs, and integrating modern systems like HVAC or elevators without damaging heritage features presents difficulties (Rahman et al., 2020; Ibrahim & Rasli, 2019).

2.3.4 Social and Cultural Factors

Community resistance due to perceived cultural loss can impede projects. Ensuring inclusive stakeholder engagement is essential but challenging (Yunos et al., 2019; Mohamad et al., 2020).

2.3.5 Environmental and Sustainability Concerns

Balancing energy efficiency with preservation is complex. Modern sustainable practices can conflict with maintaining historical authenticity (Kamaruzzaman et al., 2019; Nasir et al., 2020).

2.3.6 Conclusion

Adaptive reuse faces regulatory, financial, technical, social, and environmental challenges. Addressing these requires collaboration, innovation, and supportive frameworks.

2.4 Strategies Used in Adaptive Reuse

2.4.1 Integrated Design Approach

A multidisciplinary planning strategy ensures modern interventions respect heritage value (Purnomo et al., 2019).

2.4.2 Stakeholder Engagement

Involving communities and professionals ensures cultural sensitivities are respected and project acceptance is enhanced (Mohamad et al., 2020).

2.4.3 Sustainable Practices

Green retrofitting like solar panels or rainwater systems reduces environmental impact while preserving building integrity (Kamaruzzaman et al., 2019).

2.4.4 Financial Incentives and Funding

Grants, tax reliefs, and subsidies can ease the financial burden. Such incentives are vital for project feasibility (Ahmad & Din, 2021).

2.4.5 Technological Integration

Technologies like BIM assist in precise planning and modernization without compromising heritage features (Ibrahim & Rasli, 2019).

2.4.6 Flexible Design Solutions

Adaptable interiors allow future reconfigurations, increasing long-term usability (Rahman et al., 2020).

2.4.7 Preservation of Authenticity

Using traditional craftsmanship and materials maintains a building's historical character (Nasir et al., 2020).

2.4.8 Conclusion

Adaptive reuse success depends on integrating design, community input, sustainability, financial strategies, technology, and authenticity preservation.

2.5 Theoretical Framework

Figure 2. Theories used in practicing adaptive reuse

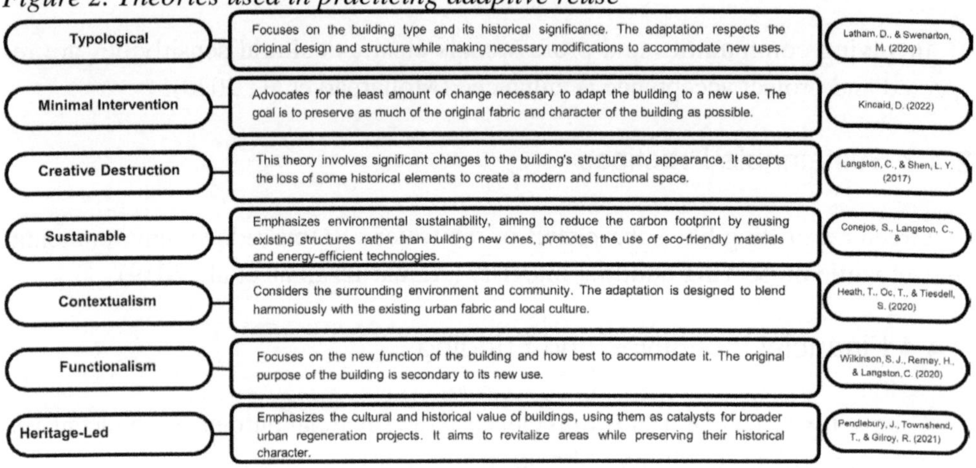

In order to determine whether the examined strategies might be applied to the practices inside Malaysia, the literature review primarily looks at the current conservation practices in Malaysia as well as the strategies for adaptive reuse that are currently in place. Malaysian historical or heritage buildings can use sustainable conservation techniques thanks to the methods derived from the research of these procedures. In order to ascertain whether conservators are using the methods taken from the literature study in their present conservation procedures, interview questions will be developed around these tactics.

RESEARCH METHODOLOGY

3.1 Introduction

This study uses a qualitative methodology to explore how building codes can be applied to historically significant buildings in Malaysia without compromising their integrity. This approach is ideal for understanding the complex interaction between adaptive reuse, heritage conservation, and modern regulatory demands. Semi-structured interviews are the primary data collection method, offering flexibility to explore participants' experiences in depth (Salleh et al., 2021). These interviews capture challenges faced by stakeholders, including architects, conservationists, and policymakers, particularly in UNESCO Heritage Sites.

Virtual interviews facilitate access to participants across Malaysia, including key heritage areas like Malacca and George Town (Ahmad & Din, 2021). Triangulation ensures validity by incorporating data from professionals, educators, and policy documents (Kamaruzzaman et al., 2020). Verbatim transcription ensures accuracy, and reliability is supported through thematic analysis and iterative coding to maintain consistency and depth (Ibrahim & Rasli, 2020).

3.2 Data Collection Method

3.2.1 Sampling Detail

This study employs purposive sampling, targeting participants with expertise in adaptive reuse and heritage conservation. These include conservation architects, heritage professionals, building owners, and practitioners from UNESCO-affiliated organizations in Malaysia. This ensures data relevance and richness, particularly through insights from those managing heritage sites in locations like Malacca and George Town, allowing the study to capture context-specific practices and challenges.

3.2.2 Data Collection

Data is collected via semi-structured interviews, allowing flexibility to explore participants' insights in depth. Interview questions are guided by the research objectives and literature, focusing on integrating modern building codes, the impact of policy, and conservation strategies. Digital tools such as video conferencing are used to enhance accessibility and convenience for participants across different regions.

3.2.3 Data Analysis

Thematic analysis is used to identify and interpret patterns within the qualitative data. This involves systematically coding interview transcripts, grouping codes into categories, and developing themes aligned with the research objectives. Emerging themes such as "regulatory compliance challenges" or "community-driven reuse" help highlight critical issues and opportunities in adaptive reuse practice.

3.2.4 List of Respondents

Table 2. List of respondents and field of expertise (Author, 2024)

No. Of Experts	Profession	Field of Expertise	Related Bodies
4	Conservation Architects	Conservation, Documentation	Architects
6	Heritage Professionals	Conservation, Heritage Preservation	Jabatan Warisan Negara, Badan Warisan Malaysia, Perzim Heritage Sdn Bhd
2	Heritage Building Owners	-	-

This study uses semi-structured interviews with experts in conservation and restoration to explore adaptive reuse strategies and the challenges faced in Malaysia. Unlike structured interviews, semi-structured formats allow flexibility in adjusting questions based on participant responses, enabling the emergence of new themes during the discussion (Alase, 2019). This approach is particularly useful for uncovering diverse perspectives and context-specific issues in heritage conservation. Open-ended questions encourage detailed responses, offering deeper insights into participants' experiences (Jamshed, 2020). Additionally, the ability to probe further into unexpected or insightful answers provides richer data essential for understanding the complexities of adaptive reuse within cultural and historical contexts (DeJonckheere & Vaughn, 2019).

3.2.5 Interview Questions

Interview questions were generated from the literature review, research objectives and research gap. The goals of the interview questions are to understand the current practices and strategies implemented into conducting adaptive reuse on historical buildings in Malaysia. This is to provide some insight before being able to formulate suggestions for adaptive reuse strategies that ensure the sustainable preservation of historically significant buildings in Malaysia.

3.2.6 Interviewee Selection

A purposive sampling method was used to select participants with relevant expertise in conservation and adaptive reuse, such as architects, policymakers, and conservation specialists in Malaysia (Robinson, 2020). These individuals were identified through online platforms, institutional directories, and professional networks, and invited to participate via email.

Semi-structured interviews explored themes like integrating updated building codes, the influence of local policies, and preserving historical integrity. This flexible approach allowed participants to offer deeper insights into the challenges and opportunities in Malaysia's adaptive reuse practices (Alase, 2019). The interviews provided valuable data for evaluating current practices, identifying gaps, and informing policy recommendations for enhancing adaptive reuse in Malaysia (Jamshed, 2020).

3.3 Thematic Analysis

Thematic analysis is a widely used qualitative method for identifying, analysing, and presenting patterns or themes within data. Its flexibility allows application across various research contexts and theoretical frameworks (Clarke & Braun, 2019).

The process begins with familiarization, immersing in the data to grasp its depth and noting initial ideas (Terry et al., 2019). Next is generating initial codes, where relevant features are systematically labelled to segment the dataset (Nowell et al., 2020). These codes are then organized into potential themes that align with the research objectives (Castleberry & Nolen, 2018).

Themes are refined and clearly defined to capture their essence and contribution to the overall research narrative (Terry et al., 2019). Finally, results are presented through a coherent, literature-supported analysis that conveys meaningful insights (Braun & Clarke, 2019).

Figure 3. Thematic Analysis Matrix Framework Labra, Oscar & Castro, Carol

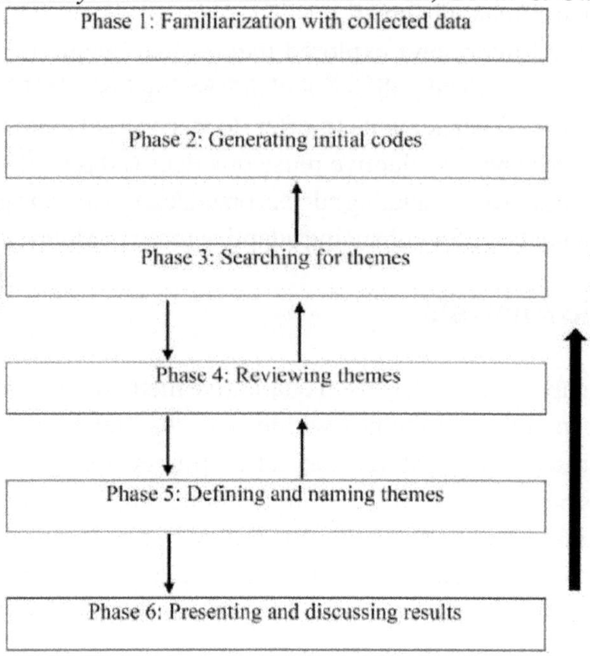

Wright, Robin & Chamblás, Isis. (2019

5. FINDINGS AND DISCUSSIONS

5.1 Current Sustainable Conservation Practices of Heritage Buildings in Malaysia

The current sustainable conservation practices of Heritage Buildings in Malaysia can be divided into four (4) main strategies as follows:

1. Community Intervention
2. Energy Efficiency Interventions
3. Digital Technology as Sustainable Practice
4. Low Carbon Restoration Works

5.1.1 Community Intervention

a) Emotional Connections and Long-term Sustainability (C1)

Interviews showed that involving local communities in adaptive reuse fosters emotional ties, pride, and ownership, supporting long-term project sustainability. These findings align with studies highlighting emotional bonds as key to encouraging conservation and reducing resistance (Ahmad et al., 2021; Salleh et al., 2021).

b) Understanding Community Interests and Needs (C2)

Engaging communities through workshops or discussions to identify suitable use, like cultural or recreational spaces is vital. Nasir et al. (2020) supports participatory methods in adaptive reuse to meet local needs, yet many frameworks still overlook marginalized communities and their diverse requirements.

c) Community Contributions for Heritage Value (C3)

Respondents highlighted the vital role of communities in preserving cultural and historical values during adaptive reuse, with local narratives guiding authentic, heritage-conscious design. This supports Ibrahim et al. (2020), who stress the influence of local stories in shaping reuse practices. However, regulatory and traditional integration challenges unique to Malaysia remain underexplored.

d) Catering Design to Community Needs (C4)

Interviews highlighted the importance of tailoring design changes to community needs, such as creating recreational spaces or ensuring inclusive access aligning with Terry & Braun (2020), who advocate user-centered design to enhance social impact. Respondents noted the challenge of balancing modern needs with strict conservation rules, requiring innovative, context-specific solutions.

e) Discussion

This study supports Ahmad et al. (2021) and Nasir et al. (2020), highlighting the value of community involvement in adaptive reuse. While emotional connections and community input align with global practices, Malaysia faces challenges due to cultural diversity and regulatory constraints. A mention of traditional practices in Melaka reveals a gap in integrating local customs. The findings emphasize adaptive reuse's potential for social equity and cultural preservation, while also pointing out misalignments between policy and community needs. Future research should focus on inclusive policies and digital tools to boost community engagement.

5.1.2 Energy Efficiency Interventions

a) Building Function and Structural Compatibility (C5)

Interviewees highlighted the challenges of integrating modern energy systems into heritage buildings without compromising their integrity, such as retrofitted HVAC systems and solar harvesting on fragile roofs. These findings align with recent studies on the technical complexities of energy upgrades in historic structures (Wang et al., 2020). Unique challenges also included the reliance on skilled craftsmen to make modifications while preserving historical authenticity.

b) Energy-Saving Systems for Carbon Reduction (C6)

Participants frequently discussed green materials and energy-efficient systems, like LED lighting, solar panels, and energy-efficient glazing, which reduce carbon emissions and align with global sustainability goals. Ibrahim et al. (2021) support this, noting that energy-efficient retrofits in heritage buildings reduce operational costs and environmental impacts, a need heightened in the post-COVID-19 context.

c) Passive Design Implementation (C7)

Passive cooling and lighting strategies, such as natural ventilation, improved insulation, and reflective materials, were highlighted as cost-effective, minimally invasive solutions. Smith et al. (2022) also identified passive design as the primary approach for maintaining heritage buildings' aesthetic integrity while improving energy performance. Interviews confirmed the relevance of these techniques, especially for heritage sites with strict conservation regulations.

d) Enhancing Building Function and Lifespan (C8)

Interviewees noted that energy efficiency interventions, such as upgrading public halls with cooling systems and thermal insulation, not only improved sustainability but also extended the functional life of heritage buildings. Ahmad and Din (2021) support this, emphasizing the adaptability of heritage buildings to modern needs while maintaining their relevance.

e) Discussion

This study supports Wang et al. (2020) and Ibrahim et al. (2021) in emphasizing that energy efficiency in heritage buildings reduces operational costs and enhances sustainability. Interviews highlight global trends in green technologies and passive design, showing how modernization can align with heritage preservation. However, integrating these systems in Malaysia faces challenges, such as maintaining authenticity and a shortage of skilled artisans, a gap in current literature. The post-pandemic rise in eco-awareness, noted by Smith et al. (2022), remains underexplored locally. These findings call for localized strategies, with tailored solutions and digital simulations needed to balance environmental and cultural priorities.

5.1.3 Digital Technology as Sustainable Practice

a) Recognizing Digital Technologies as a Means of Conservation (C9)

Interviewees emphasized the use of digital technologies, like 3D laser scanning and photogrammetry, to document heritage buildings. This approach provides accurate digital reproductions for planning, reducing invasive interventions. This aligns with research showing that digital documentation effectively preserves architectural details while minimizing physical threats during conservation (Wang et al., 2021; Terry et al., 2020).

b) Recognizing Digital Technologies for Documentation and Design (C10)

Participants highlighted the value of tools like Building Information Modeling (BIM) and augmented reality in improving documentation and design. These tools allow for visualizations of restoration proposals without altering the building. Ahmad and Rasli (2021) confirm that these technologies help ensure historical authenticity while maximizing functionality.

c) **Recognizing Digital Technology as a Means to Save Time and Effort (C11)**

Multiple interviewees remarked that digital means great streamlining in restoration efforts-mediating manual intervention and speeding decision-making processes. For example, cloud-sharing platforms allow stakeholders to cooperate in real-time, circumventing delays in communication. This supports the findings in Smith et al. (2022), emphasizing efficiency gains enabled by digital collaboration tools in large-scale conservation projects.

d) Digital Technology to Simulate and Study Historical Buildings (C12)

Simulations technologies, like energy modeling software, are essential tools for analyzing the structural and environmental performance of historic buildings and predicting the effects of proposed interventions on the building in terms of sustainability and safety. It is in line with Ibrahim et al. (2021), who prescribe digital simulations for balance-scales between preservation and modernization in such kinds of adaptive reuse activities.

e) Discussion

This study, aligned with Ahmad and Rasli (2021) and Wang et al. (2021), highlights the potential of digital technologies like laser scanning and BIM in enhancing historical building conservation. These tools improve accuracy, efficiency, and sustainability, but Malaysia faces challenges in accessibility and affordability. One interviewee noted the reliance on traditional craftsmanship due to limited digital integration, revealing a gap in research. This underscores the need for local capacity building. While digital documentation aligns with international standards, tools like augmented reality for stakeholder engagement remain underused. Future research should explore these tools' potential to boost community involvement and assess their cost-effectiveness in adaptive reuse.

5.1.4 Low Carbon Restoration Works

a) Building Materials Replacement for Conserving the Original Building Outlook (C13)

A few interviewees insisted that low-carbon materials should provide the original aesthetics of the heritage buildings. Reclaimed wood and carbon-neutral cement were indications of preserving historical attributes while achieving tremendous reductions in the carbon footprint of a project. This coincides with those of Wang et al. (2021), which point out the role of sustainable materials in conserving heritage. However, participants also mentioned that, in most cases, sourcing these materials can be a challenge, which requires innovative combinations of conventional and modern techniques.

b) Original Materials for Restoration Works (C14)

Most of the participants opted for salvaged as much original stuff-brick. wood as possible for preservation of historical integrity of the building. This reduces the waste of materials as well as emissions from production and transportation, corroborating Ibrahim et al (2020), who argued the need for the use of original materials in heritage conservation. However, few studies have focused on the challenges concerning sourcing and application of these materials within the Malaysian context.

c) Investments in Conservation are Long-term (C15)

Interviews highlighted the long-term benefits of low-carbon restoration, such as reduced operational costs and extended building lifespans, with examples like energy-efficient windows and super-insulating walls. This aligns with Ahmad and Salleh (2021), who argue that upfront investments in low-carbon technologies yield significant long-term savings. However, stakeholder acceptance of these costs depends on their perception of the balance between cost and cultural preservation.

d) Discussion

These findings align with Wang et al. (2021) and Ibrahim et al. (2020), emphasizing the dual benefits of low-carbon restoration, environmental sustainability and cultural conservation. Sustainable materials and preserving original components reflect global best practices. However, Malaysia faces challenges in sourcing conservation-compliant materials, with interviewees highlighting difficulties in obtaining locally reclaimed materials, as noted by Ahmad and Salleh (2021). Limited stakeholder awareness of the economic benefits of low-carbon technologies calls for education and policy incentives to encourage sustainable practices. Future research should focus on locally sourced materials and government support for low-carbon restoration.

5.2 Theories to Practice for Sustainable Conservation Practices

5.2.1 Community-Based Adaptive Reuse.

Engaging the community in decision-making fosters emotional connections, ensuring the long-term sustainability and relevance of heritage buildings (C1, C2, C3, C4). This aligns with participatory conservation theories emphasizing inclusivity and collective ownership (Ahmad et al., 2021; Nasir et al., 2020). Platforms like workshops and consultations can align adaptive reuse projects with local needs, guiding design alterations to ensure restored buildings serve as cultural and functional hubs. Involving the community in decisions fosters pride and ownership, ensuring the buildings' long-term relevance.

5.2.1 Energy Efficiency as a Dual-Goal Approach

Incorporating energy-efficient systems in heritage buildings reduces carbon emissions while enhancing functionality and lifespan (C5, C6, C7, C8). This aligns with sustainability frameworks that prioritize environmental impact reduction and adaptive functionality (Smith et al., 2022; Ibrahim et al., 2021). Passive design strategies like natural ventilation and insulation preserve aesthetic integrity, while modern technologies such as LED lighting and renewable energy sources lower operational costs. This approach ensures the sustainability and usability of heritage buildings, balancing modern efficiency with historical value.

5.2.2 Digital Technology as a Conservation Catalyst

Digital tools improve documentation, design efficiency, and decision-making in heritage conservation (C9, C10, C11, C12), aligning with theories promoting technology-driven accuracy (Wang et al., 2021; Ahmad & Rasli, 2021). Tools like BIM, laser scanning, and energy simulation software document and analyze adaptive reuse, while augmented reality enhances stakeholder engagement, ensuring proposals align with conservation goals. These innovations boost accuracy, efficiency, and sustainability in adaptive reuse projects.

5.2.3 Low-Carbon Restoration as a Long-Term Investment

Using sustainable materials and energy-efficient systems in restoration balances heritage preservation with environmental and economic sustainability (C13, C14, C15), aligning with global low-carbon frameworks (Wang et al., 2021; Ahmad & Salleh, 2021). Prioritize reclaimed, carbon-neutral materials and integrate renewable energy systems like solar panels and geothermal heating. Educate stakeholders on the long-term cost benefits of low-carbon technologies, including reduced operational costs, extended building lifespan, and minimal environmental impact.

5.2.4 Heritage Preservation through Integrated Approaches

Sustainable conservation balances cultural, environmental, and economic aspects through integrated planning, aligning with frameworks that view heritage buildings as cultural assets for sustainable development (Ibrahim et al., 2020; Smith et al., 2022). Form multi-disciplinary teams, including conservation architects, sustainability experts, and community members, to address preservation needs. Align efforts with global standards like the UN SDGs while adapting to local contexts, ensuring heritage buildings stay active, functional, and relevant.

5.2.5 Capacity Building and Policy Integration

Empowering stakeholders through education and policy support enhances the scalability of sustainable conservation, aligning with capacity-building models that stress knowledge and regulation for lasting impact (Ahmad et al., 2021). Offer training on digital tools and energy-efficient methods, advocate for supportive government policies, and partner with academic institutions to drive local research. This approach equips professionals with modern skills, encourages policy alignment, and fosters sustainable heritage management.

5.1 Recommendation of Strategies for Active Use of Historical Buildings in Malaysia

Figure 4. Community engagement and participation

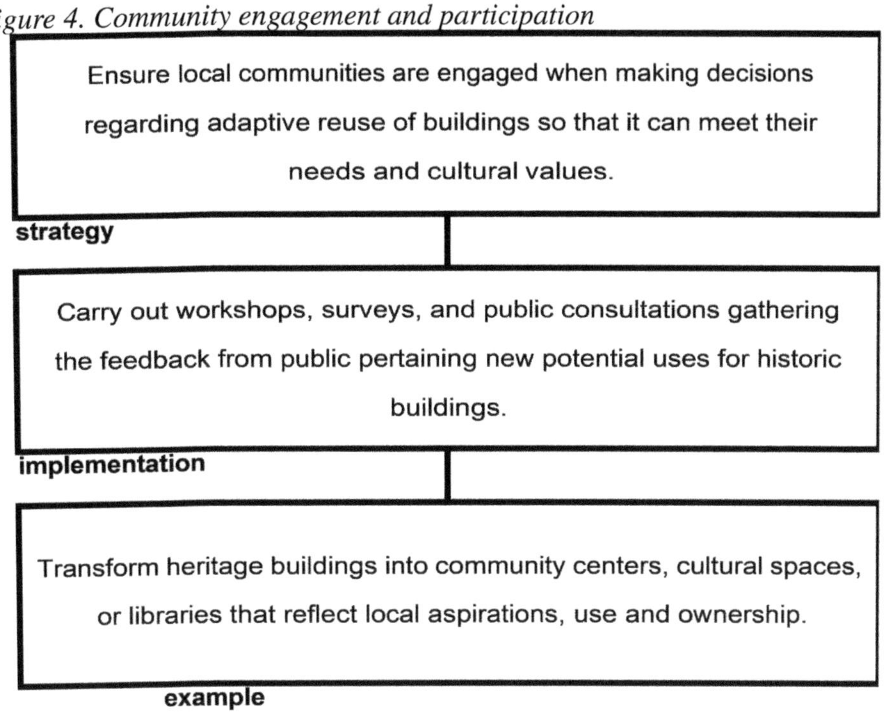

Community involvement ensures adaptive reuse of historical buildings retains cultural values and meets local needs. Through workshops, surveys, and public consultations, stakeholders capture community desires, fostering pride and ownership. This participatory approach promotes support and long-term sustainability, adding both cultural relevance and enduring value to heritage sites.

Figure 5. Integration of energy-efficient systems

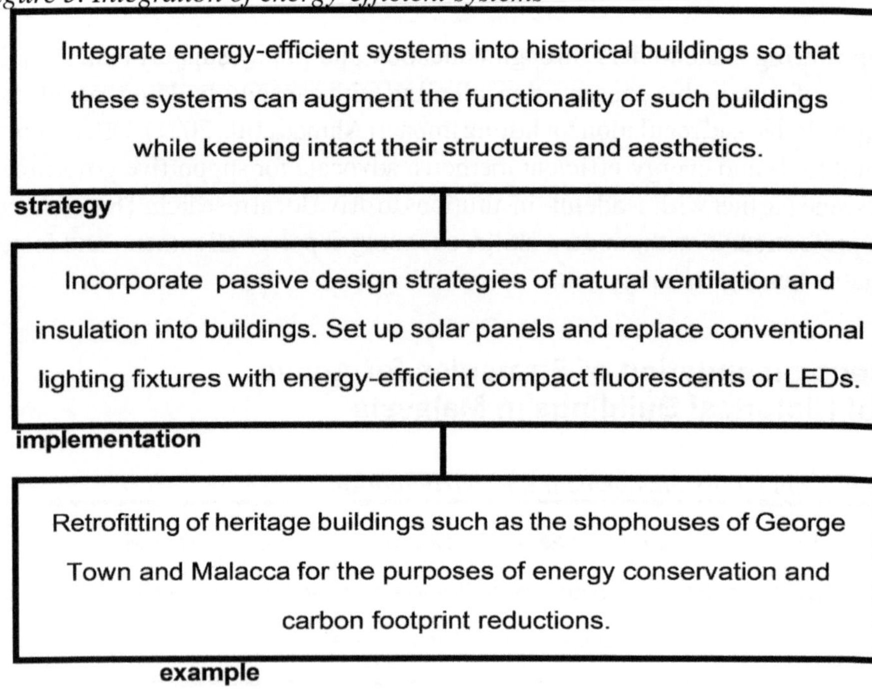

Incorporating energy-efficient systems into heritage buildings modernizes them without compromising historical significance. Passive design technologies, including improved insulation, natural ventilation, and daylight use, reduce energy consumption while preserving aesthetics. Advanced tools like LED lighting and solar panels cut operational costs and carbon emissions, ensuring heritage buildings remain functional, sustainable, and environmentally responsible in line with global energy efficiency trends.

Figure 6. Digital documentation and preservation

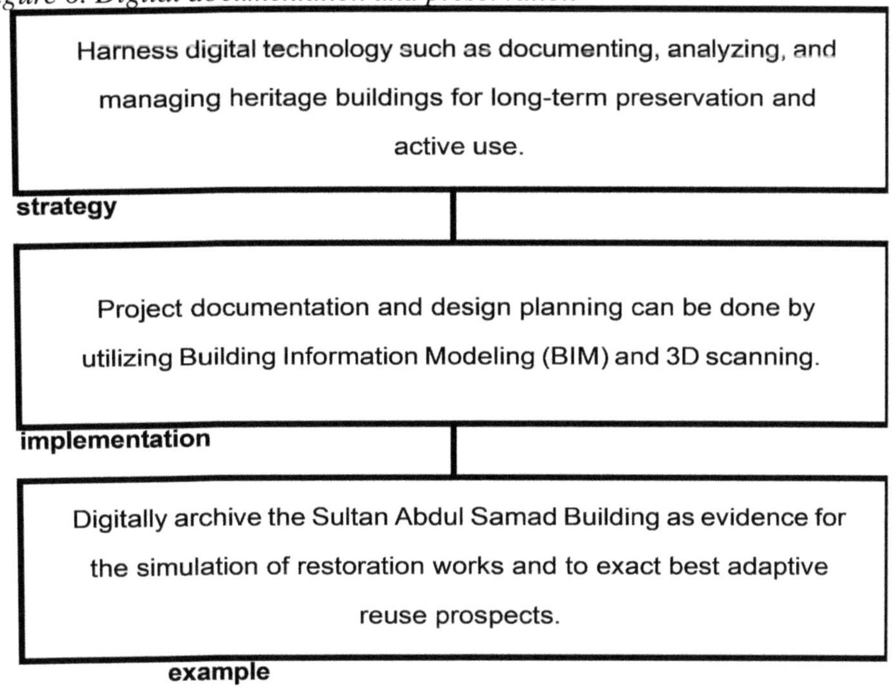

Digital technology has colorful powers in documenting, analyzing, and designing the conservation of historic buildings. Using tools like 3D scanners and Building Information Modeling (BIM) as powerful ways to hold structural and architectural detail recordings, restoration or adaptive reuse interventions can be very precise. Such methods allow collaboration among multiple stakeholders to visualize and evaluate the proposed change. For many valuable data on heritage buildings, it ensures future reference minimizing all possible physical risks in restoration operations and improving efficiency on such projects.

5.1.1 Figure 7. Low-Carbon Restoration Practices

Low-carbon renovation makes use of sustainable restoring materials that tend to preserve the authenticity of a historic building while trying to render it environment friendly. This includes reclaimed woods, salvaged bricks, and carbon-neutral materials such as lime plasters. Additional enhancement with the incorporation of renewable energy systems, for example, solar panels and geothermal heating, makes a building highly sustainable. As far as initial costs are concerned, these practices may cost more but in the in the long run benefits pay off with decreased expenses of operation, lower carbon footprints, and increased longevity of buildings.

5.1.2 Multi-functional Adaptive Reuse

Figure 7. Adaptive reuse plan.

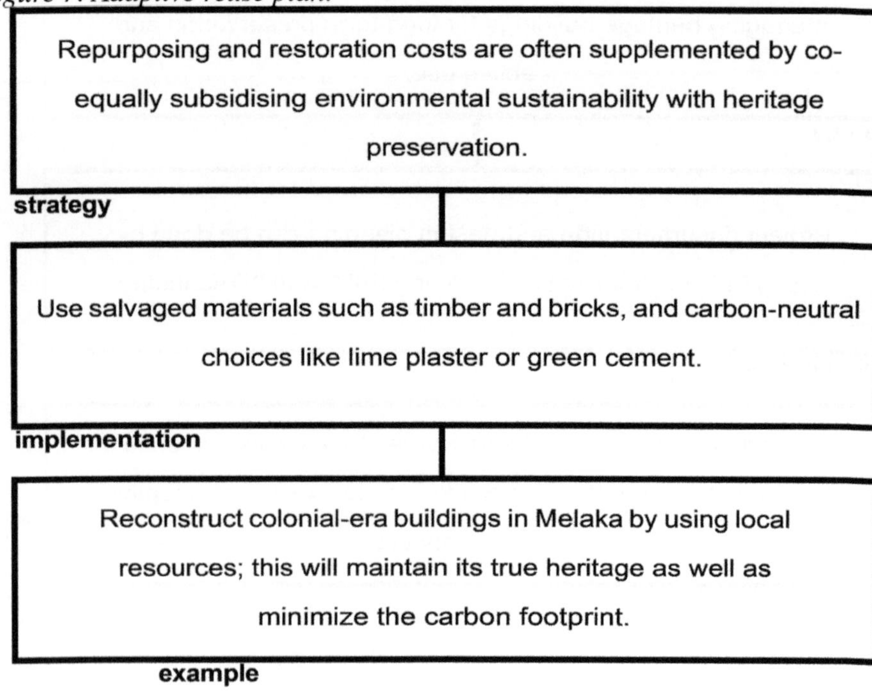

In fact, the multiple conversion of heritage buildings for different uses increases their utility besides answering the diverse demands of people and even of the economy today. One building, as an example, could host a cultural event, house co-working space, and include rest areas. Multi-functional designs cater to the new social needs while having access to the cultural integrity of these buildings. In other words, they not only yield revenue but reinvest heritage buildings with new life-glood in this modern age of vibrant community hubs.

5.1.3 Strengthened Policy and Financial Incentives

Figure 9. Strengthened policy and financial incentives.

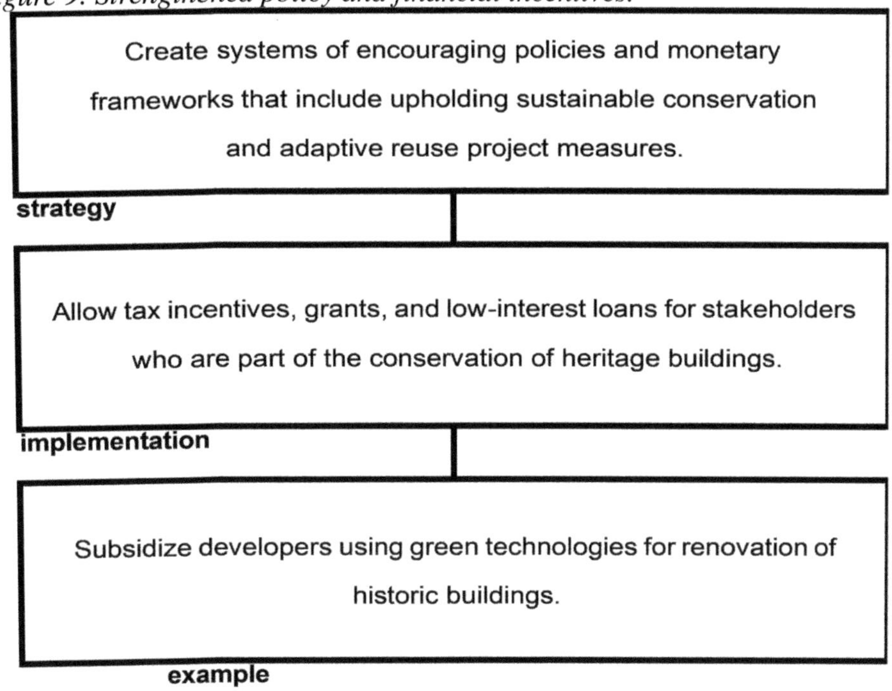

They include supportive policies and financial benefits for the perspective of sustaining heritage buildings. Federal grants, tax credits, or low-rate loans administered to builders or organizations that have adopted an environmentally- friendly sustainable restoration technique can encourage environmentally- friendly restoration work. An appropriate regulatory framework would facilitate the approval of adaptive reuse projects, thereby ensuring that conservation is feasible for an economically viable end. This strategy provides a bridge between the two opposing interests of the heritage preservation and modern development needs.

5.1.4 Capacity Building and Education

Figure 10. Capacity building and education.

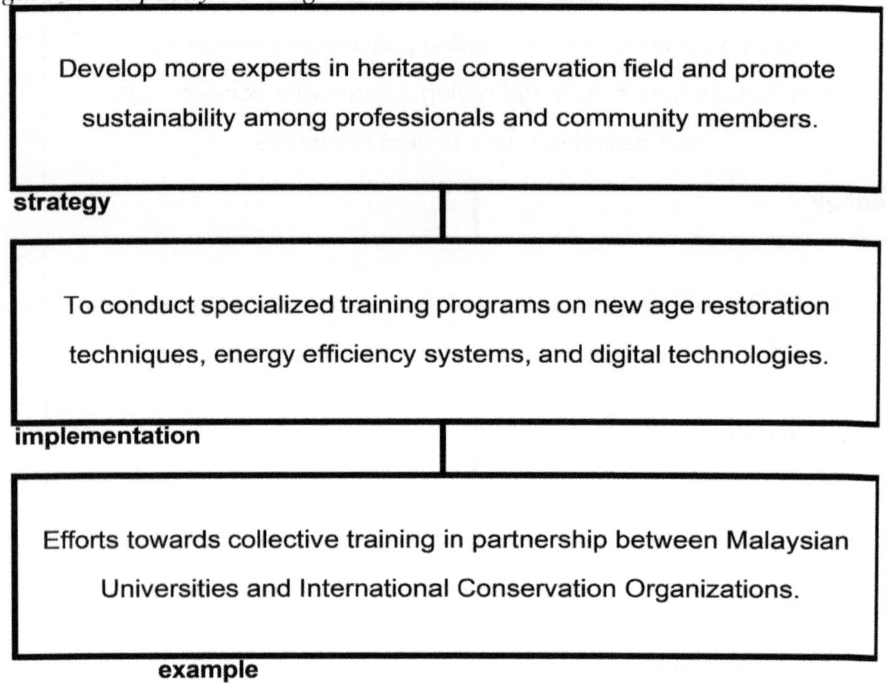

It is essential to equip professionals as well as communities with knowledge and skills necessary for heritage conservation in order to ensure the viability of adaptive reuse projects in the long term. Some of the training areas can be modern restoration techniques, energy-efficient technologies, as well as digital tools such as BIM. There is also the possibility of collaborating with universities, conservation bodies, and international organizations to develop their expertise. Such a strategy ensures that both professionals as well as local communities are empowered to participate meaningfully in the sustainable conservation effort.

5.1.5 Promote Heritage Tourism

Figure 11. Promoting heritage tourism.

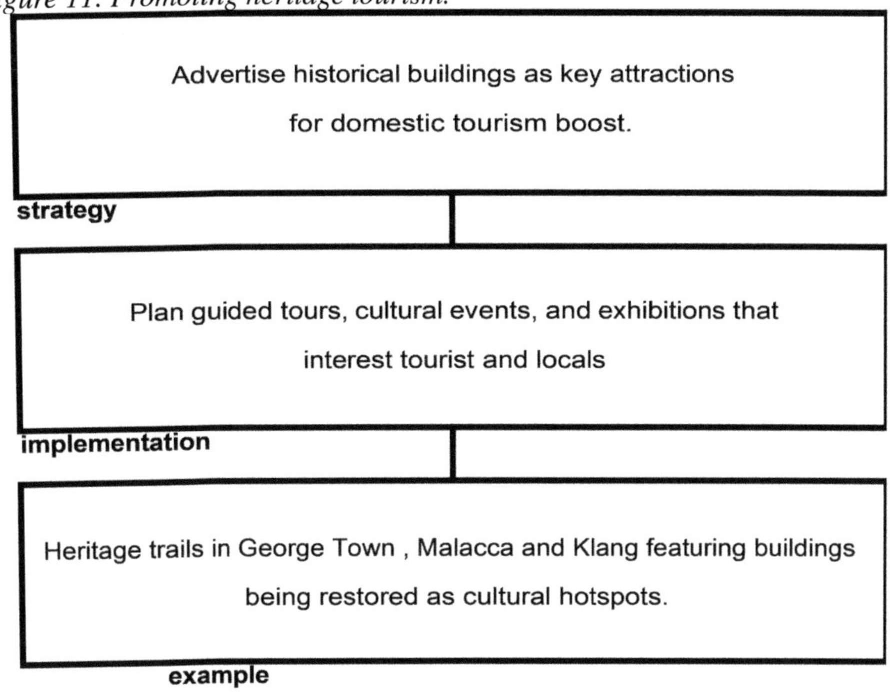

Cultural events, guided tours, and interactive exhibits are some of the possibilities that will make heritage buildings lively tourist attractions and thus conserve revenue in their active use. Heritage tourism has had many positive effects on the buildings themselves, providing income for their upkeep and restoration, and serving in the promotion and preservation of Malaysia's rich cultural history. These locations can be strategically marketed to enhance economic returns and attractiveness as part of heritage trails or eco-tourism packages.

5.1.6 Inclusive Design for Accessibility

Figure 12. Inclusive design for accessibility.

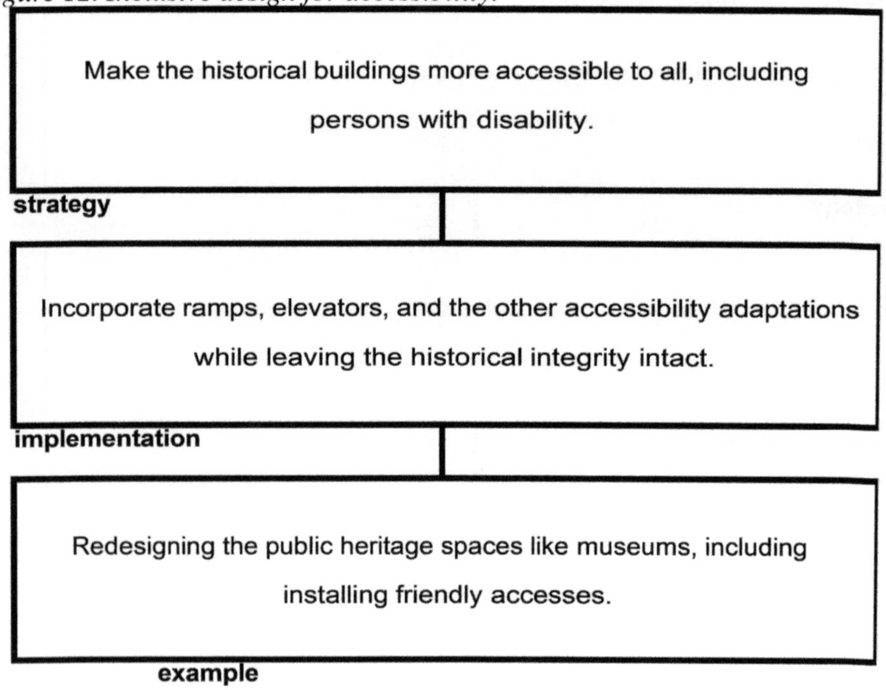

Accessibility to heritage buildings by all groups including persons with physical disabilities is extremely necessary for them to serve maximally in terms of use and people they can include. Such structural modifications may include introduction of ramps, elevators, and tactile signage, which could be incorporated without having any effect on historical integrity of the building facilities. Inclusive designing would indeed take those heritage buildings to be very functional as well as pretty appealing to a wider public user base as in a contemporary sense.

5.1.7 Collaboration with Private Sector

Figure 13. Collaboration with private sector.

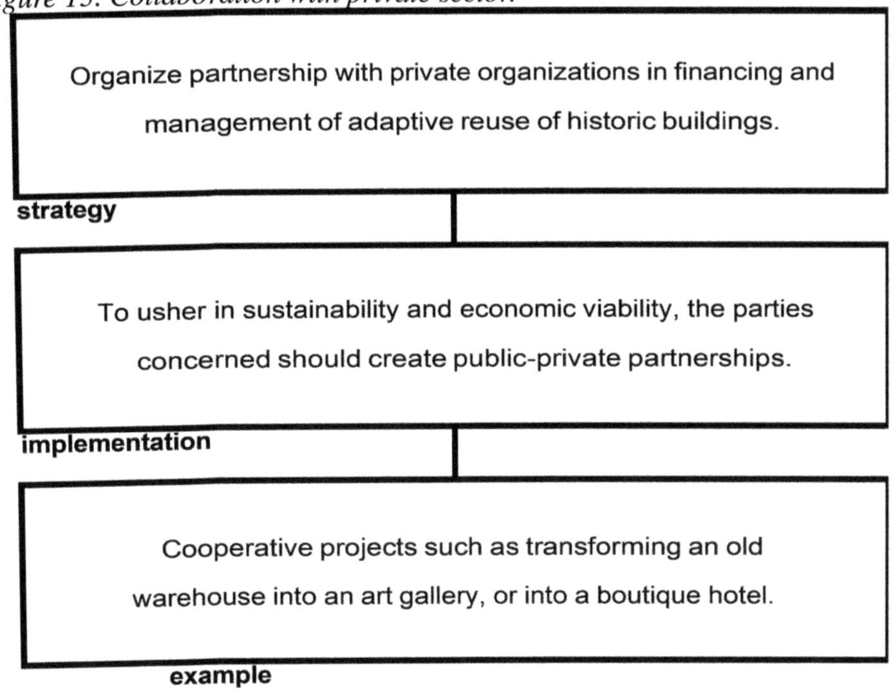

The public and private cooperation is an important approach in financing and using adaptive reuse for historical buildings. The private sector can help put together an investment package, expertise, and operational support to ensure that the conservation projects are successful. The combination of public heritage goals with private sector interests could thus create innovative, sustainable, and economically viable solutions for reuse. For example, to turn old warehouses into boutique hotels or art galleries enabling profit-making preservation.

6. CONCLUSION

This study explores sustainable adaptive reuse as a key strategy for preserving Malaysia's historic buildings, addressing whether heritage can align with modern needs. It examines community participation, energy efficiency, digital technologies,

REFERENCES

Abdullah, M. N., & Yusof, N. A. (2023). Challenges in retrofitting energy efficiency in heritage buildings: A Malaysian perspective. *Journal of Architectural Conservation, 29*(1), 20–35.

Abidin, N. Z., & Zainal, R. (2019). Heritage protection regulations in Malaysia: Challenges for adaptive reuse. *Journal of Cultural Heritage Management and Sustainable Development, 9*(3), 245–258. DOI: 10.1108/JCHMSD-09-2018-0065

Adams, W. C. (2015). Conducting semi-structured interviews. In Newcomer, K., Hatry, H., & Wholey, J. (Eds.), *Handbook of practical program evaluation* (pp. 492–505). Jossey-Bass. DOI: 10.1002/9781119171386.ch19

Ahamad, R., Rahman, A., & Tan, S. P. (2022). Sustainable heritage: The role of energy efficiency in adaptive reuse. *International Journal of Heritage Studies, 28*(4), 295–310.

Ahlqvist, T., & Rizzo, I. (2023). Community involvement in urban heritage: The case of adaptive reuse projects. *Journal of Urban Management, 12*(1), 45–58.

Ahmad, F., & Din, N. A. M. (2021). Financial challenges in heritage building conservation: The Malaysian context. *International Journal of Built Environment and Sustainability, 8*(1), 35–44. DOI: 10.11113/ijbes.v8.n1.494

Ahmad, M., & Rahman, A. (2022). Community engagement in heritage conservation: A pathway to sustainable practices. *International Journal of Heritage Studies, 28*(5), 399–412.

Ahmad, N., & Din, S. H. (2021). Economic challenges in heritage building conservation: A case for financial incentives. *Heritage Conservation Journal, 7*(2), 65–79.

Ahmad, N., & Rasli, N. (2021). Digital tools in heritage conservation: Bridging the gap between tradition and technology. *Journal of Sustainable Architecture, 15*(1), 34–50.

Ahmad, N., Salleh, M. F., & Ibrahim, R. (2021). Community participation in heritage conservation: A Malaysian perspective. *Journal of Heritage Conservation, 8*(2), 34–45.

Ahmad, S., & Hussain, N. (2021). The role of digital technology in heritage building conservation. *Journal of Heritage Management, 10*(2), 123–134.

Ahmad, S., & Hussain, N. (2021). The use of original materials in heritage building conservation: Challenges and approaches. *Journal of Heritage Management, 10*(2), 123–134.

Ahmad, Y. (2019). The challenges of preserving heritage buildings in Malaysia. *International Journal of Heritage Studies, 25*(3), 265–280.

Ali, Z., Nor, N. F. M., & Ismail, F. (2020). Heritage Building Reuse in Malaysia: A Case Study of Kuala Lumpur. *Journal of Design and Built Environment, 20*(2), 55–66.

Alshahrani, A., & Aoyama, Y. (2023). Augmented and virtual reality in architectural design: Engagement, visualization, and education. *Architectural Engineering and Design Management, 19*(2), 108–122. DOI: 10.1080/17452007.2022.2146554

Amato, A., Andreoli, M., & Rovai, M. (2021). Adaptive reuse of a historic building by introducing new functions: A scenario evaluation based on participatory MCA applied to a former Carthusian monastery in Tuscany, Italy. *Sustainability (Basel), 13*(4), 2335. DOI: 10.3390/su13042335

American Institute of Architects. (2022). Building reuse: A proven climate and economic strategy.

American Institute of Architects. (2023). The importance of building reuse in achieving carbon reduction targets.

Amin, N., Ahamad, R., & Lim, L. Y. (2022). Challenges in preserving Malaysia's built heritage amidst urban development. *Journal of Heritage Management, 7*(1), 45–60.

Bottani, E., D'Ambrogio, A., & Marini, M. (2021). Digital transformation in the construction industry: A review of innovations and their impact on energy efficiency. *Sustainable Cities and Society, 70*, 102928. DOI: 10.1016/j.scs.2021.102928

Braun, V., & Clarke, V. (2019). Reflecting on reflexive thematic analysis. *Qualitative Research in Psychology, 16*(2), 297–309.

Braun, V., & Clarke, V. (2019). Reflecting on reflexive thematic analysis. *Qualitative Research in Sport, Exercise and Health, 11*(4), 589–597. DOI: 10.1080/2159676X.2019.1628806

Braun, V., Clarke, V., Hayfield, N., & Terry, G. (2019). Thematic analysis. In Liamputtong, P. (Ed.), *Handbook of research methods in health social sciences* (pp. 843–860). Springer., DOI: 10.1007/978-981-10-5251-4_103

Brinkmann, S. (2018). The interview. In Denzin, N. K., & Lincoln, Y. S. (Eds.), *The SAGE handbook of qualitative research* (5th ed., pp. 576–599). SAGE.

Bullen, P. A., & Love, P. E. D. (2019). Adaptive reuse of heritage buildings: Sustainability, and social and economic considerations. *International Journal of Sustainable Built Environment*, *8*(1), 87–95.

Castleberry, A., & Nolen, A. (2018). Thematic analysis of qualitative research data: Is it as easy as it sounds? *Currents in Pharmacy Teaching & Learning*, *10*(6), 807–815. DOI: 10.1016/j.cptl.2018.03.019 PMID: 30025784

Cavallaro, C., Borsellino, V., & Gueli, A. (2023). Digital twin technology and its application in adaptive reuse of historic buildings. *Journal of Cultural Heritage*, *54*, 191–203. DOI: 10.1016/j.culher.2023.04.004

Cheng, L., & Tan, H. (2022). Integrating renewable energy into historic buildings: Case studies and best practices. *Heritage Management*, *10*(2), 112–127.

Clarke, V., & Braun, V. (2017). Thematic analysis. *The Journal of Positive Psychology*, *12*(3), 297–298. DOI: 10.1080/17439760.2016.1262613

Clarke, V., & Braun, V. (2019). Reflecting on reflexive thematic analysis. *Qualitative Research in Sport, Exercise and Health*, *11*(4), 589–597. DOI: 10.1080/2159676X.2019.1628806

Creswell, J. W., & Poth, C. N. (2018). *Qualitative Inquiry and Research Design: Choosing Among Five Approaches*. Sage Publications.

DeJonckheere, M., & Vaughn, L. M. (2019). Semistructured interviewing in primary care research: A balance of relationship and rigor. *Family Medicine and Community Health*, *7*(2), e000057. DOI: 10.1136/fmch-2018-000057 PMID: 32148704

Finlay, L. (2017). Engaging with the lived body: Reflections on the embodied nature of research. In *The Sage Handbook of Qualitative Research* (pp. 202–221). Sage Publications.

García, D., Melero, L., & Martorell, I. (2022). Data-driven sustainability: The role of big data analytics in the renovation of historical buildings. *Energy and Building*, *267*, 112145. DOI: 10.1016/j.enbuild.2022.112145

Guba, E. G., & Lincoln, Y. S. (2020). Fourth Generation Evaluation. Sage Publications. • Abidin, S. Z., & Hasan, N. A. (2020). Environmental Impact of Adaptive Reuse of Heritage Buildings in Malaysia. *Sustainability*, *12*(5), 1856.

Hamzah, N., Mohamed, S., & Rahim, A. H. A. (2019). The financial viability of adaptive reuse projects in Malaysia. *Journal of Building Performance*, *10*(2), 25–34. DOI: 10.21894/jbp.2019.10.2.3

Hasan, R., Yusof, N., & Idris, M. (2022). The role of local narratives in adaptive reuse: Enhancing community acceptance in heritage projects. *Journal of Urban Heritage*, *24*(3), 227–241.

Hashim, S. R., Mohamed, M., & Azhar, H. (2022). The role of 3D modeling in heritage conservation: A case study of Malaysian historical sites. *International Journal of Heritage Studies*, *28*(3), 213–227.

Historic England. (2024). New advice on adapting historic buildings for energy and carbon efficiency.

Hosagrahar, J. (2017). *Culture: Urban Future; Global Report on Culture for Sustainable Urban Development*. UNESCO.

Hwang, Y., & Park, K. S. (2020). Adaptive reuse as a sustainable practice for conserving cultural heritage: The case of the Bukchon Hanok Village in Seoul. *Sustainability*, *12*(3), 1103.

Ibrahim, M., Rasli, N., & Ahmad, S. (2020). Material reuse in heritage restoration: Challenges and opportunities. *Built Environment Journal*, *11*(3), 56–69.

Ibrahim, M., Salleh, M. F., & Kamaruzzaman, S. N. (2021). Energy efficiency interventions in heritage buildings: Challenges and opportunities. *Built Environment Journal*, *12*(1), 35–50.

Ibrahim, R., & Rasli, M. A. (2019). Technical challenges in the adaptive reuse of heritage buildings in Malaysia. *Engineering Journal (New York)*, *23*(4), 215–228. DOI: 10.4186/ej.2019.23.4.215

Ibrahim, R., Yunos, M. Y. M., & Hamid, A. F. A. (2021). Community Impact of Adaptive Reuse: The Case of the Kuala Lumpur Railway Station. *Journal of Cultural Heritage Management and Sustainable Development*, *11*(1), 47–63.

ICOMOS. (2014). The Venice Charter: International Charter for the Conservation and Restoration of Monuments and Sites.

Jamaludin, M., & Yahya, W. K. (2019). Educational Aspects of Adaptive Reuse Projects in Malaysia. *Procedia: Social and Behavioral Sciences*, *234*, 405–413.

Jamshed, S. (2020). Qualitative research method-interviewing and observation. *Journal of Basic and Clinical Pharmacy*, *9*(2), 87–88. PMID: 25316987

Jokilehto, J. (2017). *A History of Architectural Conservation*. Routledge. DOI: 10.4324/9781315636931

Kallio, H., Pietilä, A. M., Johnson, M., & Kangasniemi, M. (2016). Systematic methodological review: Developing a framework for a qualitative semi-structured interview guide. *Journal of Advanced Nursing*, *72*(12), 2954–2965. DOI: 10.1111/jan.13031 PMID: 27221824

Kamaruzzaman, S. N., Salleh, N. A., & Mahayuddin, S. A. (2019). The Impact of Adaptive Reuse on the Conservation of Heritage Buildings in Malaysia. *Journal of Building Appraisal*, *15*(1), 63–74.

Kamaruzzaman, S. N., Salleh, N. A., & Mahayuddin, S. A. (2019). Energy efficiency in the adaptive reuse of heritage buildings in Malaysia. *Energy and Building*, *183*, 284–295. DOI: 10.1016/j.enbuild.2018.11.039

Kayan, B. (2016). Maintenance management practices for heritage buildings in Malaysia: Addressing the gaps. *Journal of Building Appraisal*, *12*(4), 305–316.

Langston, C., & Shen, L. Y. (2017). Application of the adaptive reuse potential model in Hong Kong: A case study of Lui Seng Chun. *Facilities*, *35*(3/4), 230–245.

Lee, T., & Chua, A. (2023). Navigating challenges in heritage restoration: The impact of community knowledge on conservation practices. *Asian Journal of Architecture and Building Engineering*, *12*(2), 89–102.

Lee, T., Mohd Yusof, S., & Idris, M. (2023). The shift towards sustainability in adaptive reuse: Lessons learned from recent projects. *Journal of Urban Heritage*, *24*(2), 145–159.

Longhurst, R. (2016). Semi-structured interviews and focus groups. In Clifford, N., Cope, M., Gillespie, T., & French, S. (Eds.), *Key methods in geography* (3rd ed., pp. 143–156). SAGE.

Maguire, M., & Delahunt, B. (2017). Doing a thematic analysis: A practical, step-by-step guide for learning and teaching scholars. AISHE-J: The All Ireland Journal of Teaching and Learning in Higher Education, 9(3). Retrieved from https://ojs.aishe.org/index.php/aishe-j/article/view/335

MDPI. (2021). The role of community participation in the adaptive reuse of historic buildings. *Sustainability*, *13*(7), 3745.

Meleca, D. (2023). *Low Carbon and Future-Proofing Through Adaptive Reuse.* Architect Magazine.

Ministry of Energy. (2015). *Green Technology, and Water.* National Energy Efficiency Action Plan.

Ministry of Energy. Green Technology, and Water. (2015). National Energy Efficiency Action Plan. Ministry of Energy, Green Technology, and Water.

Mohamad, D., Yusoff, W. Z. W., & Majid, R. A. (2020). Stakeholder engagement in adaptive reuse projects in Malaysia. *Asian Journal of Environment-Behaviour Studies*, *5*(2), 43–54. DOI: 10.21834/ajebs.v5i2.155

Mohamed, R., Chan, E. H. W., & Yung, E. H. K. (2017). Integrated approach for heritage conservation in Malaysia: Towards sustainable maintenance. *Procedia Engineering*, *118*, 360–367.

Mohd Nasir, M. A., & Salleh, M. A. (2020). Social Sustainability through Adaptive Reuse: Malaysian Perspectives. *Social Indicators Research*, *147*(3), 875–894.

Mohd Yusof, S., & Idris, M. (2023). Craftsmanship in conservation: Navigating the challenges of energy efficiency in heritage buildings. *Asian Journal of Architecture and Building Engineering*, *12*(3), 91–105.

Nasir, M. A., Rahman, N. A. A., & Salleh, M. A. (2020). Sustainable practices in heritage building adaptive reuse: Malaysian perspectives. *Sustainability*, *12*(8), 3221. DOI: 10.3390/su12083221

Nasir, N. M., Rahman, R. A., & Salleh, A. F. (2020). Community perspectives on adaptive reuse of heritage buildings in Malaysia. *Heritage & Society*, *11*(3), 120–135.

National Trust for Historic Preservation. (2015). The Impact of Historic Preservation.

National Trust for Historic Preservation. (2023). Adaptive Reuse: A Sustainable Approach.

NewStudio Architecture. (2023). Adaptive Reuse: A Strategy for Holding History and Carbon.

Nor, R. M., Ibrahim, S., & Ng, K. H. (2021). Enhancing energy efficiency in heritage buildings using digital tools. *Energy Reports*, *7*, 33–42.

Nowell, L. S., Norris, J. M., White, D. E., & Moules, N. J. (2017). Thematic analysis: Striving to meet the trustworthiness criteria. *International Journal of Qualitative Methods*, *16*(1), 1609406917733847. Advance online publication. DOI: 10.1177/1609406917733847

Othman, A., & Mohd Yusof, S. (2020). Non-invasive techniques in the conservation of cultural heritage: A Malaysian perspective. *Asian Journal of Conservation Science*, *14*(1), 12–25.

Plevoets, B., & Van Cleempoel, K. (2019). Adaptive reuse as an emerging discipline: an historic survey. In *Adaptive reuse of the built heritage: Concepts and cases of an emerging discipline* (pp. 11–18). Routledge.

Purnomo, R. S., Susilawati, C., & Yusof, N. (2019). The role of integrated design approach for sustainable adaptive reuse. *International Journal of Building Pathology and Adaptation, 37*(1), 65–81. DOI: 10.1108/IJBPA-06-2018-0044

Rahim, A., Hassan, R., & Tan, S. P. (2021). Application of laser scanning technology in the preservation of historical buildings: A Malaysian case study. *Journal of Architectural Conservation, 27*(2), 147–161.

Rahman, A. S. A., Hamid, Z. A., & Hassan, M. H. (2019). Urban Regeneration through Adaptive Reuse in Malaysian Cities. *Journal of Urban Planning and Development, 145*(4), 04019020.

Rahman, A. S. A., Hamid, Z. A., & Hassan, M. H. (2020). Structural issues in the adaptive reuse of heritage buildings: Malaysian case studies. *Journal of Construction Engineering and Management, 146*(11), 04020129. DOI: 10.1061/(ASCE)CO.1943-7862.0001908

Rahman, F., & Wong, M. (2023). Energy-efficient interventions in heritage conservation: Bridging the past and future. *Heritage & Society, 16*(2), 76–89.

Rethinking The Future. (2023). Revitalizing aging infrastructure: Strategies for adaptive reuse of historic buildings. Rethinking The Future. Retrieved from https://www.rethinkingthefuture.com/revitalizing-aging-infrastructure/

Robinson, O. C. (2020). Sampling in interview-based qualitative research: A theoretical and practical guide. *Qualitative Research in Psychology, 17*(2), 1–22.

Rodrigues, C., & Freire, F. (2017). Adaptive reuse of buildings: an eco-efficient life cycle approach. In *Life cycle assessment: Theory and practice* (pp. 1–18). Springer.

Salleh, M. F., Kamaruzzaman, S. N., & Ibrahim, M. H. (2021). Policy and participatory challenges in heritage adaptive reuse in Malaysia. *Asian Journal of Heritage Studies, 15*(1), 10–25.

Santos, A. A., & Simon, A. (2022). The impact of community engagement on the success of heritage conservation projects: A case study of adaptive reuse. *Heritage, 5*(3), 540–559.

Shipley, R., Utz, S., & Parsons, M. (2020). Does adaptive reuse pay? A study of the business of building renovation in Ontario, Canada. *International Journal of Heritage Studies, 12*(6), 505–520. DOI: 10.1080/13527250600940181

Smith, J., & Chan, K. L. (2019). Energy analysis tools in adaptive reuse of heritage buildings: A comprehensive review. *Architectural Science Journal*, *22*(3), 245–261.

Smith, J., Tan, L. M., & Yusoff, M. (2023). Building Information Modeling (BIM) as a tool for adaptive reuse in heritage sites. *Architectural Review of Southeast Asia*, *15*(1), 45–60.

Smith, J. R., Wong, L., & Tan, K. (2022). Passive design strategies for energy efficiency in heritage architecture. *Energy and Building*, *247*, 111200.

Smith, R. (2019). *Adaptive reuse: Extending the life of old buildings*. Routledge.

Sulaiman, A., Mustaffa, M., & Rahman, F. (2023). Innovative engagement methods in heritage conservation: The role of virtual reality. *Heritage & Society*, *16*(1), 68–81.

Tan, H. W., & Lee, C. P. (2020). Sustainable approaches in historic building restoration: Techniques and implications. *Built Environment Review*, *15*(4), 180–195.

Tan, S. C., Lee, C. H., & Foo, T. S. (2020). Adaptive Reuse and Urban Regeneration: The Case of Penang, Malaysia. *Planning Practice and Research*, *35*(2), 172–188.

Terry, G., Braun, V., & Clarke, V. (2020). Digital preservation practices for heritage buildings: A qualitative perspective. *Heritage Studies Quarterly*, *8*(3), 110–130.

Terry, G., Hayfield, N., Clarke, V., & Braun, V. (2017). Thematic analysis. In Willig, C., & Stainton Rogers, W. (Eds.), *The SAGE handbook of qualitative research in psychology* (pp. 17–37). SAGE., DOI: 10.4135/9781526405555.n2

Terry, G., Hayfield, N., Clarke, V., & Braun, V. (2019). Thematic analysis. In Willig, C., & Rogers, W. (Eds.), *The SAGE Handbook of Qualitative Research in Psychology* (pp. 17–37). SAGE.

Tharim, A. H. A.. (2018). Economic Impact of Adaptive Reuse in Malaysian Cities. *International Journal of Built Environment and Sustainability*, *5*(1), 41–49.

Wang, X., Zhang, Y., & Chen, L. (2020). Balancing conservation and sustainability in heritage building retrofits. *Journal of Architectural Conservation*, *26*(2), 97–110.

Wang, X., Zhang, Y., & Chen, L. (2021). Leveraging digital technologies for sustainable heritage conservation. *International Journal of Heritage Studies*, *27*(5), 609–627.

Wong, L., & Cheng, S. (2023). Fostering cultural heritage preservation through community involvement: Case studies from Malaysia. *Heritage Management*, *10*(1), 15–30.

Yung, E. H. K., & Chan, E. H. W. (2018). Adaptive reuse: A sustainable solution for historic buildings in Hong Kong. *Procedia Engineering*, *118*, 352–357.

Yunos, M. Y., Nasir, N., & Othman, R. (2019). Evaluating the effectiveness of adaptive reuse projects in George Town, Penang. *Asian Heritage Review*, *6*(1), 78–95.

Yunos, M. Y. M., Nasir, N. F. M., & Othman, R. (2018). Preserving Heritage through Adaptive Reuse: A Case Study of George Town and Melaka. *Asian Journal of Environment-Behaviour Studies*, *3*(6), 75–89.

Yusoff, M. (2022). Digital innovations in architectural heritage: Malaysia's approach. *Heritage Conservation Journal*, *8*(3), 85–98.

Yusoff, M., & Ibrahim, R. (2022). Balancing heritage preservation and sustainability in Malaysian architecture. *Heritage Conservation Journal*, *8*(3), 85–98.

Yusoff, W. Z. W., Majid, R. A., & Ghazali, A. (2019). Economic Benefits of Heritage Building Conservation in Ipoh, Malaysia. *International Journal of Heritage Studies*, *25*(5), 455–470.

Zainal, A. (2021). Economic benefits of low-carbon restoration in heritage building projects. *Sustainable Architecture Reports*, *12*(1), 57–73.

Zainal, R., Ahmad, F., & Hamdan, H. (2018). Enhancing Cultural Awareness through Adaptive Reuse of the Sultan Abdul Samad Building. *Journal of Heritage Tourism*, *13*(4), 314–330.

Chapter 3
The Review of HBIM–Retrofit Framework for Enhancing the Performance of Heritage Building Conservation Projects in Malaysia

Khairool Aizat Ahmad Jamal
https://orcid.org/0000-0003-4674-581X
Taylor's University, Malaysia

Shahrul Yani Said
https://orcid.org/0000-0002-7727-5438
Universiti Teknologi MARA, Malaysia

Siti Norlizaiha Harun
Universiti Teknologi MARA, Malaysia

Ahmad Fahmi Zainazlan
https://orcid.org/0009-0002-8396-806X
Taylor's University, Malaysia

Noor Azeyah Khiyon
https://orcid.org/0000-0003-1477-3373
Universiti of Reading Malaysia, Malaysia

DOI: 10.4018/979-8-3373-5278-7.ch003

ABSTRACT

Heritage buildings (HB) are crucial cultural and historical assets facing modern challenges. Retrofitting, combined with Historic Building Information Modeling (HBIM), offers innovative processes, tools, and datasets to enhance project performance and extend the lifespan of HBs. While global studies integrate HBIM-enabled retrofitting, its implementation in Malaysia remains limited due to a lack of procedural guidance and studies on its impact on enhancing performance in HB conservation projects. This research aims to propose an HBIM-retrofit framework to improve design performance and guide conservation practices. A systematic review based on content analysis of research publications and grey literature was conducted to identify HBIM-retrofit themes, leading to the development of a conceptual framework underpinned by task-technology fit theory. The framework would serves as preliminary guidance for practitioners to systematically adopt HBIM-retrofit procedures for innovative decision-making and sustainable outcome, thus advancing HB conservation practice in Malaysia.

1. INTRODUCTION

Heritage buildings (HB) are iconic symbols of Malaysia's rich cultural heritage, reflecting the nation's unique background, character and preserving these buildings for future generations is essential. An initial study by the Department of Museums and Antiquities, identified over 39,000 HB constructed between 1800 and 1948 in 235 in urban and rural areas across Malaysia and significant concentrations of these buildings are found in urban areas like Kuala Lumpur, Georgetown, and Ipoh, which collectively boast over 7,334 HB within its core areas (Ipoh, 2020; Georgetown, 2021). Furthermore, recent studies highlight the distribution of HB in Malaysia shows more than 20,000 buildings are considered historical and worthy of conservation (Yusoff, 2021). Hence, immediate actions are necessary to protect these valuable assets from various threats and challenges.

HB holds substantial economic, historical, socio-cultural, educational, architectural, and utilitarian values. The inclusion of Malacca and Georgetown as UNESCO World Heritage Sites has increased demand for conservation projects which has led to heightened awareness and positive attitudes towards preserving HB (Halim, 2021; Azizi et al., 2016). Conserving HB involves enhancing the lifespan of existing structures, akin to refurbishment efforts but distinct from new construction (Harun, 2021). Subsequently, these buildings faced numerous challenges in the 21st century, including defects, neglect, deterioration, and declining performance. Furthermore, climate change has further impacted their performance necessitating more adaptable and strategic measures (Khalaf, 2021; Ali et al., 2018; Yau, 2017).

HB often exhibits poor environmental performance, with high energy consumption and carbon dioxide (CO2) emissions. Improving their environmental sustainability is crucial due to limited knowledge on energy conservation during their construction and deterioration (Blavier et al. 2023). Countries, including Malaysia, face pressure to meet the global sustainability objectives by the year 2030, requiring effective transformation across various sectors (UN, 2022; UNSDG, 2015). Malaysia has committed to producing greener buildings through policies and initiatives focusing on sustainable material optimization, resilience strategies, energy efficiency, renewable energy, and retrofitting approaches (IEA, 2017; LCCF, 2017). Retrofitting is a promising approach to improve their performance while aligning with global sustainability demands (Arayici, 2017). The building retrofit projects in Malaysia contributed 17 percent to the construction sector in 2005, rising to 22 percent in 2014 (CIDB, 2015).

In definition, retrofitting involves replacing and upgrading systems and technology in existing buildings to address technological and environmental obsolescence (Husin et al., 2019). These alterations aim to improve energy efficiency and environmental performance (ASHRAE, 2019). However, retrofitting should not affect the essential physical features and character of HB, as this would negatively impact their heritage values (Gigliarelli, 2022; McCaig, 2018), hence this necessitates adoption of a structured and careful interventions being applied in practice. Globally, case studies have shown substantial benefits from retrofitting. For example, in the UK, retrofitting can optimize building thermal value by up to 70% and significantly improve Energy Performance Certificate (EPC) ratings (Historic Scotland, 2021). Similar practices in Malaysia could enhance energy efficiency, building performance, occupant comfort, health satisfaction, and economic returns (Wise et al., 2021). However, there are challenges pertain to the retrofitting procedures in Malaysia. Due to the complexity and fragmented multidisciplinary culture, cost and resource constraints, and dependency on conventional approaches like manual measurement and CAD-based coordination, HB conservation projects often face underestimation and variations from the final account (Ramli et al., 2021; Biagini, 2016). Additionally, the lack of accurate and comprehensive information about existing HB complicates the assessment and planning of conservation interventions, leading to issues of uncertainty, inaccuracies, and unreliability (Baharuddin, 2022). This results in reduced performance, poor decision-making, disputes, and risks throughout conservation project lifecycle (Khalil et al., 2021; Rodrigues, 2019; Arayici, 2017).

Digitalization offers solutions to these challenges with the adoption of Information Communication Technologies (ICT) to improve the delivery of projects. However, this requires strategic approaches to support managerial issues, decision-making and business processes across project lifecycle (Lim et al., 2021; Khalil, 2021; Baik, 2017). Subsequently, industry practitioners and researchers advocate the application of centralized and integrated ICT systems, such as Building Information Modelling (BIM), as a decision support tool for prioritizing heritage-built conservation (Alshawabke, 2024). The Construction 4.0 Strategic Plan identifies BIM as one of the 12 key technologies that could significantly change the industry, alongside 3D laser scanning, photogrammetry, virtualization, and cloud collaboration (Jiancheng et al., 2023). Ever since year 2019, public projects valued at RM100 million above have been mandated to use BIM, including HB conservation projects (CIDB, 2021), moreover plannings are in place for SME projects above RM10 million to adopt BIM in projects (Allison, 2024).

BIM, centered on a universal and integrated platform, facilitates collaborative working practices, optimizing building performance and creating significant resource savings throughout the planning, design, construction, and operation phases of buildings. Recent developments have embraced the extensive capabilities of BIM in supporting works in HB conservation, termed Heritage-BIM (HBIM). HBIM offers new possibilities for capturing, mapping, and analyzing multiple dimensions of building data, thus enhancing the management of HB. It provides rapid data acquisition, efficient documentation, simulation, and visualization benefits, improving project performance in terms of time, cost, environmental impact, and communication (Diara, 2022). HBIM can also offers integrated solutions in HB retrofitting with much research suggested to explore potential of HBIM-retrofit as a sustainable approach to tackle contemporary environmental issues (Mari et al., 2024; Gagrealli, 2022; McCaig, 2018). However, application in Malaysia is still relatively new which requires further guidance to expedite the HBIM-retrofitting solutions (Idris et al., 2024).

2. RESEARCH PROBLEM AND GAP

Global concerns about climate change and resource depletion necessitate improving energy performance in HB. Although recent initiatives like the Green Building Index (GBI) standards for HB show potential progress, the rating tool requires comprehensive procedural guidance for project application. The absence of such guidance results in practitioner's lack of understanding of the complex retrofitting process, moreover, existing HB guidelines are emphasizing on preserving authenticity over adaptation (Jamal et al., 2024; Azis, 2022; Khalaf, 2021), hence making the retrofitting procedures unclear (Zaid, 2021). This gap leads to inconsistent practices, risking uncontrolled renovations and inefficient alterations (Penjor et al., 2024). Despite the benefits of retrofitting, many of the HB conservation projects face significant challenges, including delays, cost overruns, inadequate information, and technical difficulties (Baharuddin, 2022; Ramli et al., 2021; Nawi, 2020). The lack of standardized technical procedures, along with scarce accurate information, time constraints, unclear standards, and a fragmented work culture, complicates conservation efforts (Shan et al., 2022). Thus, many studies suggest better data management, comprehensive conservation knowledge, and adaptation to technological advancements to improve retrofit project performance (Mari et al., 2024; Martinelli et al., 2022; Ali, 2018).

Innovative technologies, including ICT and digital applications, can boost project's productivity. BIM, crucial for enhancing design and construction performance, has been actively promoted by the Malaysian government (CIDB, 2021). Despite these efforts, BIM adoption in Malaysian HB conservation remains slow due to a lack of guidance on HBIM for HB conservation, leading to limited understanding (Lovell et al., 2023; Yusoff, 2021). Although BIM implementation is well-documented in national guidance and protocols, it focuses on new buildings rather than HB, unlike other developed countries (Kermanshahi, 2020). Therefore, to expand the adoption of the HBIM-Retrofit model requires further studies, which would benefit the HB conservation sector (Idris et al., 2024; Azira, 2022). While there is much international research on HBIM-retrofit applications exists, none have studied their impact on enhancing HB conservation project performance especially within the Malaysian context, highlighting an opportunity for further study (Ismail et al., 2021).

Furthermore, many precedent theories postulate substantial improvements and benefits can be attained with the implementation of ICT/HBIM technology in design and construction project delivery, however very few and limited research that evaluates the fitment of HBIM attributes with the historic-retrofit characteristics and its relationship towards an enhanced project performance (Wong, 2020). Hence this gap signifies an area of opportunity to be studied as much research postulates the benefits of BIM technology could positively motivate utilization and creates high impact to the performance of HB practice (Ali, 2018; Ismail et al., 2021; Wong, 2020). Hence, through the issues and problem discussed, the study proposed two research questions to be explored as follows:

1. What would be an appropriate HBIM-retrofit concept and characteristics specifically for the historic building conservation projects in Malaysia?
2. What is the relationship between the HBIM-retrofit characteristics and the design performance of historic buildings conservation in Malaysia?

3. LITERATURE REVIEW

3.1 Definition of Historic Building Retrofit

The current national heritage conservation guideline lacks a specific definition of retrofitting. Instead, related terms like refurbishment, restoration, rehabilitation, and adaptation are used (Harun, 2021). These terms can be confusing, as they often overlap with retrofit (Sopery, 2023), creating a "state of pleasant confusion" (Wilkinson, 2011). Therefore, clear distinctions are needed to define retrofitting properly. Retrofit and refurbishment are often used interchangeably but differ. Douglas (2006) defines retrofit as "any work to a building above and beyond maintenance to change its capacity, function, or performance" or to "adjust, reuse, or upgrade" it for new conditions. Eames et al. (2014) describe retrofit as adding components not originally fitted during construction, often involving new systems or building fabric upgrades like insulation, walls, or windows. Historic England (2015) defines retrofit as improving a building's energy and environmental performance through technical interventions, often requiring lifestyle changes and a systematic upgrade, repair, and maintenance programme. The main goal of HB retrofit is to enhance energy performance while maintaining service quality and indoor comfort with lower energy consumption (Moran et al., 2013; Ma et al., 2012). This involves modifications to walls, doors, windows, floors, and roofs to reduce heating, cooling costs, and emissions (Tobias, 2004). Other interventions include improving natural daylighting, artificial lighting controls, air quality, ventilation, and adopting renewable energy. Wilkinson (2011) notes retrofitting can apply to an entire building or part of it, across within-use and across-use retrofits. A "retrofit event" refers to any permitted alteration, extension, upgrade, change of use, or renovation. Multi-tenant buildings often experience multiple events. According to Almeida (2014) and Aikiyuori (2006), HB should be retrofitted if it faces (1) deterioration, (2) change of use, (3) economic optimization through energy efficiency, (4) subjective decision-maker factors, or (5) changes in circumstances.

3.2 Historic-Building Information Modelling (HBIM)

BIM has been a wide subject matter within the realm of new construction, with BIM protocols being developed to make the construction process more productive and efficient. Unlike the new-build construction sector, where BIM has been applied widely for several years at an international level, with scores of relevant publications and online content, BIM for heritage assets (historic buildings and sites) is a relatively new field of academic research and appears as less popular in terms of adoption by the heritage conservation professionals (Antonopolou, 2017).

The significant difference between BIM and HBIM are in need to be defined in this research to provide a distinct context in heritage conservation (Edwards, 2017). Hence, the following discusses related literature review regarding HBIM to form the fundamental of the research context. Hence, by exploring, appraising and synthesis the definition, concept and use of HBIM in a defined heritage retrofit life cycle, the design benefits of HBIM and the evolution of HBIM in recent years.

3.3 Definition of Historic Building Information Modelling (HBIM)

The term HBIM comes with the adoption and widespread development of 3D laser scanning and photogrammetry technology, in which it has influenced innovative practice in the recording of cultural heritage sites and building. As many benefits can be identified through the adoption of BIM to improve and enhance a certain process within project lifecycle and combined with documentation technology has influenced HBIM to become a well-known platform that aids in the conversation of HB (Khoedir, 2016). According to Arayici (2017), the term Heritage-Building Information Modelling has only begun to be used in the latter part of the last decade, since Building Information Modelling (BIM) superseded 3D digital modelling and computer aid design (CAD).

It is also sometimes defined as Historic Building Information Modelling (HBIM) which according to Murphy (2009) is somewhat a narrower term. HBIM is defined as an ecosystem for modelling HB and their elements using data from laser scanning and photogrammetric technique. According to Arayici (2017), the significant difference of HBIM compared to BIM, is the incorporation of cultural heritage information which adds layers in the existing BIM process. This cultural information or elements includes the cultural significance of elements, historic significance analysis, data for authentic replication, particular process, methods for documenting authenticity, recording requirements (Khalil, 2021). Hence, the cultural significance information can be analyzed as a basis of decision-making for conservation interventions. Such data can be developed to build richness that could support many aspects of conservation works (Arplan, 2024).

In contrast with the use of BIM for new construction, the process of implementing HBIM involves a reverse engineering solution in which parametric objects representing architectural elements are mapped onto laser scan or photogrammetric survey data. To obtain the final product, several stages must be completed, beginning with the collection and processing of laser/image survey data, followed by the identification of historic details from architectural pattern books, the construction of parametric historic components/objects, and finally the correlation and mapping parametric objects onto scan data and the detailed development of measured drawings and documentation (Dore & Murphy, 2012). The product is the generation of comprehensive 3D model that includes inclusive details of the object's surface with proper construction techniques, materials, analytical construction and many more.

3.4 HBIM-Retrofit on Performance in Historic Building Conservation Project

The use of HBIM-retrofit brings many benefits and impacts to improve the project performance of historic conservation practice. This can be related through its high level of automation, ability to associate broad to specific range of information in digital environment, and this ability would impact in many aspects of design project performance. Furthermore, much of the past research have studied the influence and impact of BIM on performance of construction project (Van Tam, 2022; Cao & Wang, 2017). Many aspects could be such as the cost factors, Celoza (2021) evaluated the impact of BIM-related contract practices on project performance which concludes that the BIM association to the process relates to positive project cost outcomes. Hence, this drives the research to further determine the HBIM-retrofit would further influence the design performance of historic conservation projects in Malaysia.

For many years, many studies categorized the three success criteria corresponding to the Iron Triangle (namely cost, time and quality) and were considered as ideal elements to achieve success in a project. However, through time there is evidence of other research studies that include other aspects of key performance criteria such as safety, environment, and customer satisfaction, relationship and communication (Atkinson, 1999, Suk, 2012). In a review paper by McGinley (2022), summarized the key criteria to measure the design performance specifically for retrofits of existing domestic buildings, which classified three important criteria namely the economic (cost-based), social and environmental. These performance attributes have started to become a focus to be measured due to the increasing attention and awareness on the sustainability concerns and agendas (Orihuela, 2017).

3.4.1 Time

With the vast integrated knowledge areas in project management, delay and cost overrun are the most common issues in construction projects. Tahir (2018), in a review study, highlights design project coordination is among the leading factors to the cause of delays in design and construction projects, which often emphasized poor management of design changes, slow decision making, improper planning and design inefficiencies as the main causes of delays in projects (Samarghandi, 2016; Shehu, 2014). The use of BIM in historic retrofitting has been shown to have a significant impact on the time and efficiency of the process. Li (2014) and Selim (2018) both emphasize the importance of BIM in promoting information integration, collaboration, and error reduction in the reconstruction and restoration of historical buildings. In a study by Biagini (2016), significant time savings were gained in the measurement of as-built state of HB for restoration, with 3D laser scanning due to the speed and accuracy it provides to support the complex modelling BIM geometries, hence increasing design lead time of the project. Ghosh (2015) further emphasizes the need for process reengineering to accommodate the new value stream created by BIM, which can potentially improve labor productivity. Gigliarelli (2017) presents an integrated approach using BIM, numerical simulations, and decision support systems to address energy efficiency in HB, further demonstrating the potential time-saving benefits of BIM in this context.

3.4.2 Cost

Cost feasibility is a major factor influencing retrofit decisions (Liu, 2022). Minimizing investment cost is a key objective in designing retrofit solutions and evaluating their outcomes. However, estimating costs for historic retrofits is challenging due to wide variations based on location, building type, rehabilitation schemes, performance targets, and triggered building code requirements (Egbelakin, 2008). This complexity is worsened by limited data on existing buildings and reliance on conventional, inefficient 2D and paper-based coordination methods (Biagini, 2016; Gholami, 2013; Rodrigues, 2019). BIM use in construction has demonstrated significant cost and time savings, ranging from 5% to 52.36% (Sholeh, 2020; Lu, 2014; Chelson, 2010; Shin, 2018). HBIM further reduces costs by minimizing rework, automating quantity take-offs, and improving estimate preparation (Li, 2014), leading to quicker turnaround times and better cost control (Ismail, 2015). Effective teamwork during the design phase also contributes to cost reduction, as collaborative discussions allow stakeholders to agree on cost-saving measures (Egbelakin, 2008). Since HBIM enhances early-stage collaboration and coordination, it plays a vital role in this aspect (Sacks, 2017). Studies show that HBIM retrofits can combine cost and energy efficiencies through reduced operational costs, lower energy consumption, optimized capital investment, emission reductions, and improved comfort (Mehndi, 2020; Chaves, 2014).

3.4.3 Quality

In HBIM-retrofit, quality performance is defined by eight criteria: documentation accuracy, heritage value preservation, energy efficiency, structural performance, indoor comfort, certification compliance, asset management, and collaborative workflow (Alexa & David, 2019). Heritage value preservation is critical, while documentation accuracy, structural performance, and certification compliance carry high impact (Mora et al., 2021). HBIM-retrofit enhances quality through non-invasive 3D documentation (Bruno & Roncella, 2019), reversible interventions (Piselli et al., 2020), real-time monitoring (Meoni et al., 2022), and automated conservation standard verification (Žurić et al., 2022). A case study by Piselli et al. (2020b) showed 70% energy savings, 3,032 kg CO_2 reduction, and Gold Certification in the GBC Historic Building™ protocol, all while preserving architectural integrity. HBIM-retrofit also improves quality through better interdisciplinary interoperability, centralized data management, and integration with conservation practices (Mora et al., 2021). Despite challenges like data management and technical expertise needs, it offers a promising balance between modern performance and heritage preservation (Nieto-Julián et al., 2023).

3.4.4 Design

HBIM revolutionized the retrofitting of heritage structures by enhancing design performance through data-driven approaches. It integrates historical data, 3D laser scanning, and parametric modeling to reconstruct as-built conditions, facilitating accurate and precise retrofit design (Dore & Murphy et al., 2015). This allows architects and engineers to simulate various interventions, reducing errors while preserving heritage authenticity. HBIM improves decision-making by consolidating datasets on structural integrity, material decay, and environmental impact (Baik et al., 2020). It enables real-time simulations of structural performance, including seismic resilience and load distribution (Oreni et al., 2014). This ensures the selection of retrofit strategies that enhance stability while maintaining architectural integrity. Furthermore, HBIM aids in preserving heritage authenticity by documenting intricate historical details and ensuring design modifications remain faithful to the original aesthetic (Bruno & Roncella, 2019). It supports minimally invasive interventions aligned with international conservation guidelines. Despite its advantages, HBIM adoption faces challenges such as high costs, interoperability issues, and the complexity of modeling irregular historic features. Future research focuses on AI-driven automation and real-time monitoring for adaptive reuse strategies (Banfi et al., 2021).

3.4.5 Environmental

Research highlights BIM's influence on environmental aspects of design performance (Arayici, 2017; Hull & Bryan, 2019; Gigriarelli, 2019). Van Eldik (2020) proposed a BIM-based framework to evaluate environmental impacts, enabling automated, accurate, and integrated environmental impact assessments (EIA) to support better design decisions. BIM also serves as a tool for sustainability metrics, allowing early-stage sustainability analysis and promoting energy efficiency and resource optimization relative to project cost (Bank, 2010; Wong, 2015). Environmental performance is increasingly important in retrofit projects, influencing stakeholders' participation (Alberini, 2013; Achtnicht, 2014) and serving as a key performance indicator. McGinley (2022) defines environmental performance by outcomes in energy use, carbon emissions, and achieving targeted Energy Performance Certificate (EPC) ratings. Lizana (2016) applied a multi-criteria assessment from stakeholders' perspectives, measuring energy savings and emissions reductions. As sustainability becomes central to design delivery, managing and minimizing energy consumption and carbon emissions over a building's lifecycle has become critical for design performance (Wong, 2015).

3.4.6 Relationship

Effective relationship performance in HB conservation projects relies on clear collaboration, efficient communication, and seamless data integration among stakeholders. Several studies highlight these criteria through the role of shared digital environments. Bruno and Roncella (2019) note that incorporating non-geometric information enhances documentation, while García-Valldecabres et al. (2021) and Thravalou et al. (2023) report that collaborative platforms and centralized data repositories improve interaction and reduce conflict. HBIM-retrofit tools influence these relationships distinctly. Liu et al. (2023) show that Level of Detail guidelines clarify stakeholder priorities and standardize processes. Jouan and Hallot (2019) demonstrate that integrating cultural significance into digital models supports informed conservation decision-making. Khan et al. (2022) highlight improved communication through converged building data, and Piselli et al. (2020) show that user-centric facility management and energy retrofit applications enhance operations. Prizeman et al. (2020) further stress that HBIM applications facilitate knowledge sharing, supporting cost-efficient and sustainable strategies. Across ten studies, data sharing, stakeholder involvement, and knowledge transfer are consistently emphasized. In practice, HBIM-retrofit supports preventive maintenance, energy upgrades, and asset management, enhancing data integrity and aligning conservation goals. Overall, the evidence confirms that well-implemented HBIM-retrofit solutions significantly improve stakeholder interactions and project performance.

3.4.7 Communication

Historic Building Information Modeling (HBIM) enhances communication performance in the retrofitting of heritage structures by fostering collaboration among stakeholders. By integrating historical data, 3D models, and real-time updates, HBIM facilitates clear and effective communication between architects, engineers, conservators, and policymakers (Fai et al., 2011; Murphy et al., 2013). This reduces misunderstandings and ensures alignment in retrofit decision-making. One of HBIM's key contributions is its ability to centralize information in a digital environment, enabling interdisciplinary teams to access accurate and updated data remotely. Studies show that HBIM improves coordination by visualizing structural conditions, material decay, and intervention plans in a shared platform (Baik et al., 2019). This enhances stakeholder engagement and accelerates approval processes for retrofitting interventions. Moreover, HBIM strengthens public and institutional communication by providing interactive visualizations and simulations of conservation strategies (Bruno & Roncella, 2019). These tools help communicate retrofit impacts to non-technical stakeholders, fostering informed decision-making and greater public support for conservation projects.

4. METHODOLOGY

This review follows a systematic approach to collect and analyze relevant studies on HBIM and retrofitting practices for HB. A thorough literature search was conducted across various academic databases, focusing on research papers, industry reports, and grey documents that address HBIM in the context of heritage conservation and retrofitting. International best practices were identified to draw comparisons with the Malaysian context, while challenges and gaps in the local adoption of HBIM were highlighted in the earlier discussion. As this research is to develop a practical procedural guidance, the scope of the study will focus on keystone in HBIM-retrofit which focuses on workflows, process, strategies, methods and approaches. Through the content and thematic analysis, several key themes relevant to HBIM-retrofit are proposed based on the imposed research questions. postulate the need to create a theoretical knowledge classifying several themes namely the historic retrofit characteristics, HBIM-retrofit functionality and its relationship to the enhancement in design performance of heritage building conservation projects in Malaysia. This review paper also aims to establish a theoretical framework which further leads to the development of conceptual framework for this study. This framework is necessary to aid the researcher in structuring the research hypothesis and strategies towards the future directions of research.

5. RESULTS AND DISCUSSION

5.1 Historic Building Retrofitting Characteristics

Historic retrofit involves a strategic plan which includes the organizational, process and technical. Such criteria are being identified by providing discussions on the stakeholders' role and key phases involved in retrofit, the role of energy audit and its activities to support the process and the technical needs and methods in evaluating assessment of historic retrofit. In exploring the classifications needed in a retrofit process, the study has identified 8 grey documents, which consists of historic retrofit guidance from various countries. This selection was taken as it resembles the best practice in retrofit and is being used as reference in many projects. As there are coverage on energy retrofit is not available in the Malaysian context (Ismail, 2021; Azis, 2021). A document review of such international standards was conducted to further investigate the process and classifications of retrofitting of HB, the technical guidelines were identified which highlight in detail the scope and key stages of retrofitting for energy efficiency of HB. This enables the study to understand the historic HB retrofitting taxonomy, which covers the concept and principles, role of stakeholders, key development stages/phases, activities and techniques and methods applied in a historic retrofit assessment project which can be formulated to develop the framework as depicted in Table 1 below.

Table 1. Summary of selected guidance documents on energy retrofits in historic and traditional buildings

Guidelines	Date	Country	Organization	Reference
PAS 2038:2021 - Retrofitting non-domestic buildings for improved energy efficiency	2021	UK	PAS	(PAS, 2021)
ASHRAE Guideline 34-2019 - Energy Guideline for Historic Buildings	2019	US	ASHRAE	ASHRAE (2019)
Historic Environment Scotland 2021 - Historic Environment Scotland Guide to Energy Retrofit of Traditional Buildings	2021	UK	Historic Environment Scotland	Historic Scotland (2021)
EN 16883 Conservation of cultural heritage - Guidelines for improving the energy performance of historic buildings	2015	EU	CEN/TC 346	BSI (2017)
Planning Responsible Retrofit of Traditional Buildings	2015	UK	STBA	May & Griffiths (2015)
Energy Efficiency and Historic Buildings: How to Improve Energy Efficiency	2018	UK	Historic England	McCaig (2018)

continued on following page

Table 1. Continued

Guidelines	Date	Country	Organization	Reference
Energy Efficiency Solutions for Historic Buildings: A Handbook	2014	EU	3ENCULT	Troi et al. (2015)
Energy Heritage: A Guide to Improving Energy Efficiency in Traditional and Historic Homes	2008	UK	Changeworks	Changeworks (2008)
The Secretary of the Interior's Standards for Rehabilitation & Illustrated Guidelines on Sustainability for Rehabilitating Historic Buildings.	2013	US	US. National Park Service	National Park Service (2013)

Upon investigation of the documents, the literature has classified five important key themes that needed to be explored further, this includes the heritage aspects, development stages, energy audit criteria, performance measures and method of approach. Table 2 highlights the keystones according to respective authors and further discussions on the 4 characteristics are provided.

Table 2. The HBIM-retrofit characteristics (Developed by Author)

No.	Author/Year	Historic Retrofit Keystones			
		Heritage Aspects	Development Stages	Design Criteria	Method of Approach
1.	McCaig et al. (2018)	/	/	/	/
2.	ASHRAE (2019)	/	/	/	/
3.	May & Griffiths (2015)	/	/	/	/
4.	Troi et al. (2015)	/	/	/	/
5.	Historic Scotland (2021)	/		/	/
6.	BSI (2017)	/	/	/	/
7.	Changeworks (2008)	/	/	/	/
8.	National Park Service (2013)	/		/	/

5.2 HBIM-Retrofit Characteristics

Ever since the widespread adoption of BIM introduced the early years and the successful adoption, has influenced its further applications in the existing and HB context (Arayici, 2017). This has further emerged many studies related to energy efficiency or in other terminologies, such as BIM-Building Energy Performance (BIM-BEP), Energy-Efficiency BIM (EE-BIM) and HBIM (Di Bicari, 2022). As

BIM systems are complex, the ICT system adoption tend to fail if they are developed too prescriptively narrow to support the full range of practice, because then users have to developed and deploy approaches beyond that ICT system scope, Hence, it is important that a framework for HBIM that is sufficiently embracing the scope of a specific practice, even if not all modules are initially implemented (Counsell, 2017).

Moreover, the characteristics of HBIM for historic retrofit has different criteria and unique set of expectations which differs from BIM conventional practice, which requires deeper understanding on the social-environment, technical tools, process and the careful consideration of information, which not just factual, but also includes the site interpretation and the fuzzy qualities of historic places (Littlefield, 2017). Whilst it is generally agreed that BIM brings together Process, Technology and Policy (Succar, 2009), Dawood and Vukovic (2015) further extend this to add another 'pillar' to include people as a critical element in the process. However, currently, there are very limited studies on the HBIM associated with energy retrofit model that influences adoption into historic-built environment.

Hence, a total of 6 publications were identified that highlights the HBIM-Characteristics aspects and these contents were analyzed. Khoedier (2016), identifies in their study the framework of sustainable retrofit to be implemented in retrofitting of historic in Egypt, this can expedite the implementation of HBIM enabled retrofit, however the actual HBIM attributes (process) was unclear and ambiguous. Khaddaj et al. (2016), on the other hand, have brought further the knowledge and understanding of the criteria of HBIM application in existing buildings, by presenting four (4) key elements of HBIM-retrofit adoption which are the process, organizational, informational and technical needs. This are consistent with many previous studies in BIM, highlighting these factors as a key pillar in BIM implementation (Jamal, 2019).

Furthermore, Bruno (2018), developed the framework to provide more definition and classification through systematic review which leads towards the proposal of HBIM performance assessment for diagnosis-aided information modelling and management framework which includes the characteristics that defines the involved process, work organization and environment, BIM knowledge framework (informational) provides a methodology for automating acquisition, updating and management of knowledge in historic-retrofitting. Table 3 highlights previous studies on HBIM-retrofit categorized within these four key domains.

Table 3. The HBIM-Retrofit keystones (Developed by Author)

No.	Author/Year	HBIM-Retrofit Keystones			
		Organizational	Informational	Lifecycle	Technical
1.	Khalil (2021)	/	/	/	/
2.	Sanhudo (2018)		/	/	/
3.	Khaddaj (2016)	/	/	/	/
4.	Khan (2022)		/	/	/
5.	Gigliarelli (2022)	/	/	/	/
6.	Charlton (2021)	/	/	/	/

5.3 HBIM Retrofit Functionality

The key criteria for HBIM-enabled retrofit were discussed to form a fundamental knowledge area. Di Bicari (2022) explored and synthesis current available grey and document literature on the matter of HBIM-retrofit aspects, with a brief description of the main topics covered. Most of the publications deal in detail with data exchange (85%), methodologies and technical issues (70%). This paper was selected as it details the best modelling practice, with in-depth guidance on the BIM to BPS workflow, although they concentrate on different combinations of BIM and BPS tools. Other themes addressed are related to technical, organizational issues and informational development and management. A total of 17 publications have been identified outlining the guidance on the adoption framework of HBIM in various facets of conservation works, narrowed down into the scope of historic buildings conservation and energy efficiency. The methodologies were selected for their relevance and affinity with the proposed HBIM-enabled retrofit framework. This research model outlines important attributes and classification that could support the development of theoretical framework, especially on the specific implementation and guidance, key elements and categories related to HBIM-integration for energy efficiency assessment, this have been identified and discussed in the following subchapter. The list of authors using the HBIM-retrofit elements for their study is shown in Table 4.

Table 4. HBIM-Retrofit model elements from previous studies

	HBIM-Retrofit Elements	Gigliarelli (2022)	Antonopoulou (2017)	Messner et al. (2019)	Reeves (2015)	Ouellette et al. (2022)	Hull & Bryan (2019)	Nieto-Julián, (2021)	Baik (2021)	Banfi (2021)	Quattrini, (2018)	Capone (2019)	Bruno (2018)	Carvalho (2019)	Moyano (2022)	Khaddaj (2016)	Karlapudi (2018)	Stanga (2023)
	HBIM Organizational																	
1	Roles and responsibility	x	x	x		x	x	x	x			x	x			x		
2	Training and education			x			x									x		
3	BIM Protocol	x	x	x		x	x		x	x			x			x		x
	HBIM Informational																	
4	Geometrical	x	x		x	x	x		x	x	x	x	x	x	x		x	x
5	Historical and Archaeology	x	x				x	x	x	x	x	x	x		x			x
6	Pathology						x											x
7	Surveying	x	x			x	x	x	x	x		x	x		x			x
8	Energy Performance	x	x		x	x	x						x	x			x	
9	Material	x			x	x	x							x	x		x	
10	Structural	x			x	x	x			x					x			
	HBIM Project Lifecycle																	
11	Data acquisition/capture	x	x	x		x		x	x	x		x	x		x	x		x
12	Data modelling	x	x	x	x	x		x	x	x	x	x	x	x	x	x		x
13	Data simulation and analysis	x		x	x	x				x			x	x	x			
14	Data Visualization			x					x	x	x					x		x
	HBIM Technical																	
17	Software	x	x	x	x	x		x	x	x	x	x	x	x	x		x	x
18	Hardware and equipment	x	x	x	x	x		x	x	x	x	x	x		x	x		x
19	Technical support	x	x	x	x	x	x	x					x		x	x	x	x
20	Interoperability						x	x		x	x		x		x	x	x	

5.5 The Relationship Between HBIM-Retrofit Model and Performance Through Task Technology Fit

This section discusses the theory underpinning the development of the conceptual framework for this study. According to Iyamu (2013), an underpinning theory serves as the foundation for philosophical inquiry, supporting the rigorous pursuit of research and forming the basis for the study's assumptions. Underpinning theories articulate fundamental principles that guide the understanding of variables and their relationships (Iyamu, 2013). Thus, applying a theory, or a combination of theories, is crucial for guiding empirical data analysis. As discussed earlier, this study is situated within the context of Information Systems (IS) or Information and Communications Technology (ICT) adoption in Business Process Management (BPM) at the individual or organizational level.

The adoption of IS/ICT technologies has a significant impact on supporting business units in decision-making, design, and operational processes (Ahmad, 2019). Furthermore, the role of HBIM/BIM as an IS/ICT technology has driven innovation and fostered changes in business processes and functions, aiming to transform organizations towards greater efficiency, lean operations, and productivity (Jadhav, 2011; Heaton, 2019). Theories from the social sciences have been widely used to underpin IS/ICT-related studies in business process contexts (John, 2016). Socio-technical theories have made notable theoretical, practical, and methodological contributions to IS/ICT adoption research. In this study, the Task-Technology Fit (TTF) model is selected to underpin the conceptual framework.

In organizational contexts, technology is typically deployed to add value by improving or supporting specific tasks. However, adoption often demands considerable resources for acquisition, implementation, and usage, leading to a critical question: how much value does the technology create for the organization? (Melville, 2004). Due to the complex interplay between technology, users, systems, tasks, and processes, quantifying technology's value is highly challenging (Orlikowski, 2000; Goodhue, 1995). As an alternative, the TTF theory was proposed to measure technology's impact more effectively. Widely accepted in IS research (Lee, 2007), TTF has been called "one of the most important developments in information systems theory" (Melchor & Buendía, 2014), offering a way to quantify the effectiveness of technology in organizations (Spies & Botha, 2020; Goodhue, 1998).

Furneaux (2012) highlighted a notable increase in the application of TTF theory over the years, both in volume and across diverse environments. The basic TTF model, illustrated in Figure 1, provides a lens to understand how technology usage creates value for users (Goodhue & Thompson, 1995). The model's core premise is that technology improves task performance and utilization when there is a strong fit between the technology's functionalities and the users' task requirements (Goodhue, 1998; Goodhue et al., 2000; Larsen, 2009). Furthermore, TTF theory supports evaluating the positive effects of aligning task requirements and technology features on user outcomes such as satisfaction, performance, and productivity (Spies, 2020; Pal & Patra, 2021).

The theory posits that information systems enhance performance only when the technology's functionality matches the user's task needs (Larsen, 2009), a relationship that can be assessed through the perceived value from users' viewpoints. Similar applications of TTF can be seen in HBIM/BIM research. For example, Wong (2022) developed a conceptual model using TTF to examine the motivational and behavioural factors influencing facilities management users' adoption of digital twin technology. Similarly, Srećković (2021) designed a BIM-Blockchain workflow model, adapting configuration and task-technology fit theories to conceptualize emerging phenomena (Spies, 2020), aligning with the nature of this research.

The antecedents of task-technology fit involve the interaction between tasks, technologies, and individuals. Technology refers to computer systems (hardware, software, data) and support services (training, help desks), characterized by attributes such as availability, capability, and relevance (Goodhue & Thompson, 1995), as exemplified by HBIM platforms. Task characteristics involve the actions performed to transform inputs into outputs (Goodhue & Thompson, 1995); in this study, the task is historic building retrofit, with its own distinct process characteristics. Finally, individual characteristics refer to users or groups within an organization whose experience and motivation influence technology adoption.

Figure 1. Task-technology fit model

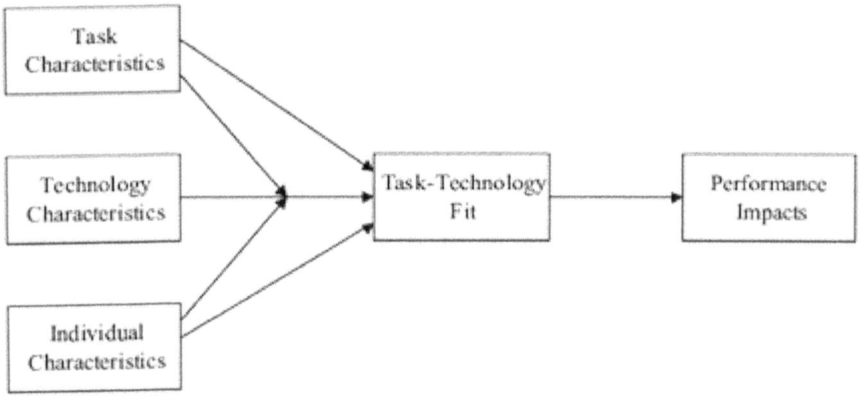

(Goodhue & Thompson, 1995)

Most of the extant literature that use TTF matches primarily the task and technology characteristics for evaluating how information technology impacts the users' performance (Wu & Chen, 2017), hence this justifies as the study is to investigate the influence of HBIM-retrofit model, to impact the design project performance (time, cost, quality, design, relationship, environmental and communication). By utilizing the TTF theory, it helps in providing a valuable insight that helps researchers to anticipate human behavior within the complex socio-technical systems, as such it can become a useful model for prediction in certain qualitative and quantitative research and practical applications (John, 2016; Spies, 2020).

Within this context, the theory is thus adopted in this study to provide a meta-theory that describes and explains the relevance of HBIM-retrofitting, which includes task (historic retrofitting), technology (HBIM) and individual (heritage building conservators) characteristics in a more consistent, systematic and theoretical grounded way, and also to identify impact and opportunities through the application of IS/ICT technology, in this case HBIM-retrofit. Moreover, the contribution of this study in relation to the literature review concerns investigating the relationship between HBIM-retrofit framework and it influence to motivate performance in HB conservation practice, and this can be under the light of the task-technology fit model.

5.6 The HBIM-Retrofit Conceptual Framework

A conceptual framework is crafted by researchers from a range of theoretical or conceptual viewpoints. According to Imenda (2014) and Grant & Osanloo (2016). The objectives of conceptual framework are to:

a) Help the researcher clearly view the core variables and concepts in a specific study.
b) Provide the researcher with a universal approach, methodology, research design, target population and research sample, data collection and analysis.
c) Guide the researcher in the analysis and explanation of the data since there is no prevailing theoretical perspective.
d) Guide forthcoming research, especially where the conceptual framework incorporates literature review and field data, synthesis of relevant aspects.

The conceptual framework for this study is developed based on Task Technology Fit (TTF) Theory and considered empirical evidence and gaps that have been identified from the literature review. The activity theory stated that the relationship between two variables is contingent or depending on the level of third variable (Deng & Smyth, 2013). In this study, the conceptual framework illustrates the relationship of HBIM-enabled retrofit model variables on the design performance of heritage building conservation project. A similar study was conducted by Van Tam (2023) and Cao & Wang (2017), which investigates the direct relationship of two independent and dependent variables between the BIM-implementation factors and the relationship on construction project performance, within the context of the construction industry. After studying the related literature, a new conceptual framework is designed as shown in Figure 2. The framework includes the indicators or items of each of the variables which have been discussed earlier in which the independent variables are the established Historic BIM-retrofitting framework keystones.

The framework consists of eight independent variables (exogenous variables) to establish the HBIM-retrofit attributes which consists of the heritage aspects, development stages, energy audit criteria, performance measures, retrofit method of approach, HBIM organizational, HBIM project lifecycle, HBIM technical and HBIM informational attributes which has three to seven indicators in the variables respectively. All the attributes postulate a direct relationship with the endogenous variables (DV) which is the design performance of HB conservation projects in which the indicators consist of seven (7) elements that influence the performance aspects indicators, namely time, cost, quality, design, environmental, relationship and communication as discussed previously. Therefore, a conceptual framework of an appropriate HBIM-retrofit for enhancing the heritage building conservation practice in Malaysia was further developed by testing the hypotheses in this study.

Figure 2. A conceptual framework of HBIM-retrofit relationship with the enhanced design performance of historic building conservation process in Malaysia.

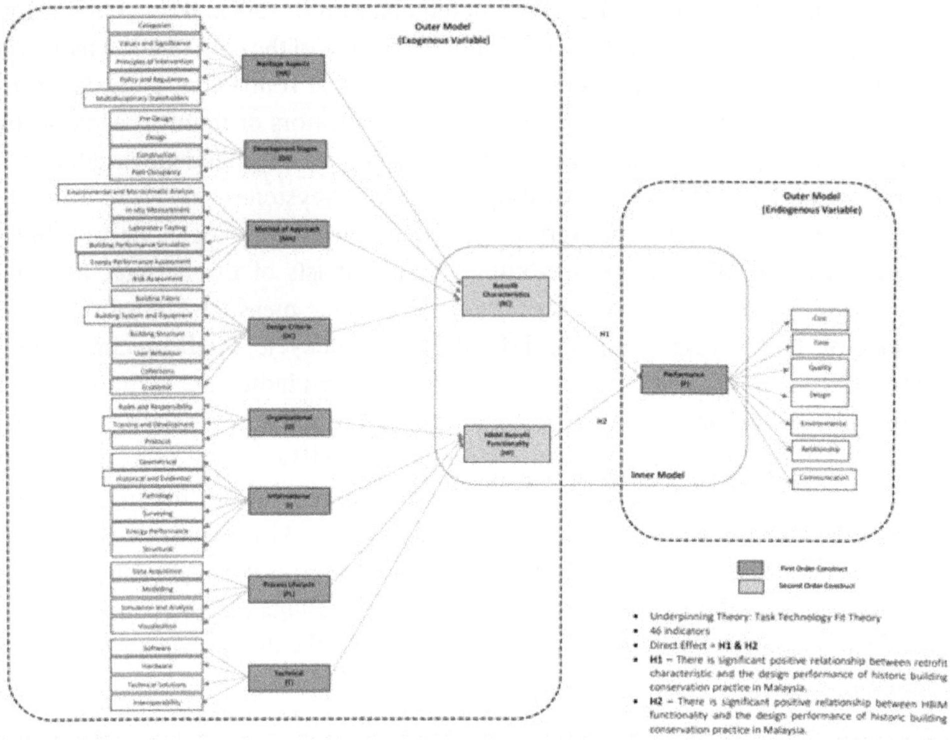

6. CONCLUSION AND FUTURE DIRECTION

Heritage Building Information Modelling (HBIM) and its application in retrofitting HB impose both the potential and challenges in improving the energy performance of Malaysia's HB while preserving their cultural significance. As Malaysia faces increasing environmental and developmental pressures, the integration of modern technologies like HBIM is essential for sustainable heritage conservation. However, the successful adoption of HBIM in the country remains limited due to fragmented conservation practices, unavailability of standardized practical guidelines, and the technical complexities associated with retrofitting HB projects. The HBIM-retrofit model, as outlined in this review, offers a systematic approach to addressing these challenges. It emphasizes the importance of combining digital technology solutions with a deep understanding of the unique process of heritage buildings retrofitting which involves the heritage aspects, retrofit development stages, method of approaches and retrofit design criteria. Furthermore, the HBIM-retrofit model keystones were identified in which the organizational, informational, lifecycle, and technical are crucial for guiding the retrofitting process, ensuring that energy efficiency improvements do not compromise the historical integrity of these structures. The HBIM-retrofit co-relates towards enhanced design performance in conservation projects in which the task technology fit (TTF) provides an underpinning theory that posits a relationship between the task and technology characteristics of HBIM-retrofit with performance impact. Moreover, this aids in the development of conceptual framework of the study. Future research is needed to further explore the application of HBIM-retrofitting in Malaysia, focusing on refining the conceptual framework with the industry player's experience and perspective. By leveraging international best practices and adapting them to local conditions, Malaysia can significantly improve the sustainability and longevity of its heritage buildings, contributing to both environmental goals and the preservation of its cultural heritage.

REFERENCES

Achtnicht, M., & Madlener, R. (2014). Factors influencing German house owners' preferences on energy retrofits. *Energy Policy*, *68*, 254–263. DOI: 10.1016/j.enpol.2014.01.006

Aikivuori, A. (2006). Periods and demand for private sector housing refurbishment. *Construction Management and Economics*, *14*(1), 3–12. DOI: 10.1080/01446199600000002

Alberini, A., Banfi, S., & Ramseier, C. (2013). Energy efficiency investments in the home: Swiss homeowners and expectations about future energy prices. *The Energy Journal (Cambridge, Mass.)*, *34*(1), 49–86. DOI: 10.5547/01956574.34.1.3

Alexa, W., & David, H. (2020). Implementing HBIM on conservation heritage projects: Lessons from renovation case studies.

Ali, M., Ismail, K. M., Hashim, K. S. H. Y., Suhaimi, S., & Mustafa, M. H. (2018). Historic Building Information Modelling (HBIM) for Malaysian Construction Industry. Planning Malaysia Journal, 16(7).

Allison, L. (2024, September 5). All projects worth RM10mil and above to use construction software BIM. *The Star*. https://www.thestar.com.my/news/nation/2024/09/04/all-projects-worth-rm10mil-and-above-to-use-construction-software-bim-says-ahmad-maslan

Almeida, S. L. G. D. (2014). Retrofitting and refurbishment processes of heritage buildings: application to three case studies (Doctoral dissertation).

Alshawabkeh, Y., Baik, A., & Miky, Y. (2024). HBIM for Conservation of Built Heritage. *ISPRS International Journal of Geo-Information*, *13*(7), 231. DOI: 10.3390/ijgi13070231

Antonopoulou, S., & Bryan, P. (2017). *BIM for heritage: developing a historic building information model*. Historic England.

Arayici, Y., Counsell, J., Mahdjoubi, L., Nagy, G. A., Hawas, S., & Dweidar, K. (Eds.). (2017). *Heritage building information modelling*. Taylor & Francis. DOI: 10.4324/9781315628011

Arplan. (2024). *Introduction to HBIM: Fundamentals and Applications*. Retrieved from. https://arplan.com.co/

ASHRAE. (2019). *ASHRAE terminology. ASHRAE searchable terminology tool. American Association of Heating, Refrigeration and Air-conditioning Engineers.* ASHRAE.

Azira, S., Rahim, A., Noor Hayati, I., Nurul Syala, A. L., Nurulhuda, A., Mohd Amir Shazwan, H., & Sangeetha, A. (2022). Historic Building Information Modelling (HBIM): The Application of Digitalization in Conserving Negeri Sembilan Traditional Malay House. *INTI JOURNAL, 2022*(10), 1–7.

Azis, I. A. A., Khaidzir, K. A. M., & Rashid, A. M. (2022). Identifying Guidelines Limitation Of Historic Interior In Adaptive Re-Use Of Heritage Building In Malaysia. *Journal of Tourism Hospitality and Environment Management, 7*(27), 69–76. DOI: 10.35631/JTHEM.727006

Azizi, N. Z. M., Razak, A. A., Din, M. A. M., & Nasir, N. M. (2016). Recurring Issues in Historic Building Conservation. *Procedia: Social and Behavioral Sciences, 222*, 587–595. DOI: 10.1016/j.sbspro.2016.05.217

Baharuddin, M. N., Bahardin, N. F., Harun, S. N., Abd Manap, M. D., & Ab Rashid, M. S. (2022). ASSESSING CRITICAL RISK FACTORS FOR HERITAGE CONSERVATION PROJECTS IN COMPLIANCE WITH NATIONAL HERITAGE ACT 2005 (ACT 645). *PLANNING MALAYSIA, 20.*

Baik, A. H. (2020). *Heritage building information modelling for implementing UNESCO procedures: Challenges, potentialities, and issues*. Routledge. DOI: 10.1201/9781003036548

Banfi, F. (2021). The evolution of interactivity, immersion and interoperability in HBIM: Digital model uses, VR and AR for built cultural heritage. *ISPRS International Journal of Geo-Information, 10*(10), 685. DOI: 10.3390/ijgi10100685

Biagini, C., Capone, P., Donato, V., & Facchini, N. (2016). Towards the BIM implementation for historical building restoration sites. *Automation in Construction, 71*, 74–86. DOI: 10.1016/j.autcon.2016.03.003

Blavier, C. L. S., Huerto-Cardenas, H. E., Aste, N., Del Pero, C., Leonforte, F., & Della Torre, S. (2023). Adaptive measures for preserving heritage buildings in the face of climate change: A review. *Building and Environment, 245*, 110832. DOI: 10.1016/j.buildenv.2023.110832

Bruno, N., & Roncella, R. (2019). HBIM for conservation: A new proposal for information modeling. *Remote Sensing (Basel), 11*(15), 1751. DOI: 10.3390/rs11151751

Bruno, S., De Fino, M., & Fatiguso, F. (2018). Historic Building Information Modelling: Performance assessment for diagnosis-aided information modelling and management. *Automation in Construction, 86*, 256–276. DOI: 10.1016/j.autcon.2017.11.009

Cao, D., Li, H., & Wang, G. (2017). Impacts of building information modeling (BIM) implementation on design and construction performance: A resource dependence theory perspective. *Frontiers of Engineering Management, 4*(1), 20–34. DOI: 10.15302/J-FEM-2017010

Capone, M., & Lanzara, E. (2019). Scan-to-BIM vs 3D ideal model HBIM: Parametric tools to study domes geometry. *The International Archives of the Photogrammetry, Remote Sensing and Spatial Information Sciences, 42*(W9), 219–226. DOI: 10.5194/isprs-archives-XLII-2-W9-219-2019

Carvalho, J. P., Bragança, L., & Mateus, R. (2019). Optimising building sustainability assessment using BIM. *Automation in Construction, 102*, 170–182. DOI: 10.1016/j.autcon.2019.02.021

Celoza, A., Leite, F., & de Oliveira, D. P. (2021). Impact of BIM-related contract factors on project performance. *Journal of Legal Affairs and Dispute Resolution in Engineering and Construction, 13*(3), 04521011. DOI: 10.1061/(ASCE)LA.1943-4170.0000478

Changeworks, E. W. H. (2008). Energy Heritage. A Guide to Improving Energy Efficiency in Traditional and Historic Homes.

Charlton, J., Kelly, K., Greenwood, D., & Moreton, L. (2021). The complexities of managing historic buildings with BIM. *Engineering, Construction, and Architectural Management, 28*(2), 570–583. DOI: 10.1108/ECAM-11-2019-0621

CIDB. (2015). *Construction Industry Transformation Programme 2016-2020. Construction Industry Development Board.* CIDB.

CIDB. (2021). *Construction 4.0 Strategic Plan 2021-2025. Construction Industry Development Board.* CIDB.

Counsell, J., & Taylor, T. (2017). What are the goals of HBIM? In *Heritage building information modelling* (pp. 15–31). Routledge. DOI: 10.4324/9781315628011-3

Deng, F., & Smyth, H. (2014). Nature of firm performance in construction. *Journal of Construction Engineering and Management, 140*(2), 04013040. DOI: 10.1061/(ASCE)CO.1943-7862.0000778

Di Biccari, C., Calcerano, F., D'Uffizi, F., Esposito, A., Campari, M., & Gigliarelli, E. (2022). Building information modeling and building performance simulation interoperability: State-of-the-art and trends in current literature. *Advanced Engineering Informatics, 54*, 101753. DOI: 10.1016/j.aei.2022.101753

Diara, F. (2022). HBIM open source: A review. *ISPRS International Journal of Geo-Information, 11*(9), 472. DOI: 10.3390/ijgi11090472

Dore, C., & Murphy, M. (2012, September). Integration of Historic Building Information Modeling (HBIM) and 3D GIS for recording and managing cultural heritage sites. In 2012 18th International conference on virtual systems and multimedia (pp. 369-376). IEEE.

Dore, C., & Murphy, M. (2015). Historic building information modelling (HBIM). In *Handbook of research on emerging digital tools for architectural surveying, modeling, and representation* (pp. 233-273). IGI Global Scientific Publishing.

Eames, M., Dixon, T., Lannon, S., Hunt, M., Laurentis, C., Marvin, S., Hodson, M., Guthrie, P., Georgiadou, M. (2014). Retrofit 2050: Critical Challenges for Urban Transitions.

Egbelakin, T., & Wilkinson, S. (2008). Factors affecting motivation for improved seismic retrofit implementation. In *Australian Earthquake Engineering Conference (AEES).*

García-Valldecabres, J., Galiano-Garrigós, A., Meseguer, L. C., & González, M. C. L. (2021). HBIM work methodology applied to preventive maintenance: A state-of-the-art review. *Building Information Modelling (BIM) in Design. Construction and Operations IV, 205*, 157.

Gigliarelli, E., Calcerano, F., & Cessari, L. (2017). Heritage Bim, numerical simulation and decision support systems: An integrated approach for historical buildings retrofit. *Energy Procedia, 133*, 135–144. DOI: 10.1016/j.egypro.2017.09.379

Gigliarelli, E., Calcerano, F., Martinelli, L., Artopoulos, G., Thravalou, S., & Alexandrou, K. (2022). *Methodology for the energy renovation of heritage buildings using BIM. Guidelines for the development of an Energy Efficient Heritage Building Information Model (EE-HBIM)* (Gigliarelli, E., Calcerano, F., & Martinelli, L., Eds.). Consiglio Nazionale delle Ricerche.

Goodhue, D. L. (1998). Development and measurement validity of a task-technology fit instrument for user evaluations of information system. *Decision Sciences, 29*(1), 105–138. DOI: 10.1111/j.1540-5915.1998.tb01346.x

Goodhue, D. L., Klein, B. D., & March, S. T. (2000). User evaluations of IS as surrogates for objective performance. *Information & Management*, *38*(2), 87–101. DOI: 10.1016/S0378-7206(00)00057-4

Goodhue, D. L., & Thompson, R. L. (1995). Task-technology fit and individual performance. *Management Information Systems Quarterly*, *19*(2), 213–236. DOI: 10.2307/249689

Grant, C., & Osanloo, A. (2014). Understanding, selecting, and integrating a theoretical framework in dissertation research: Creating the blueprint for your "house". *Administrative issues journal, 4*(2), 4.

Halim, U. L., & Tambi, N. (2021). Awareness of community on the conservation of heritage buildings in George Town, Penang. *Planning Malaysia, 19*.

Harun, S. N. (2021). Pemuliharaan Bangunan Warisan. Dewan Bahasa dan Pustaka, ISBN 978-983-49-1955-9.

Heaton, J., Parlikad, A. K., Owens, D., & Pawsey, N. (2019). BIM as an Enabler for Digital Transformation. In International Conference on Smart Infrastructure and Construction 2019 (ICSIC) Driving data-informed decision-making (pp. 49-54). ICE Publishing. DOI: 10.1680/icsic.64669.049

Historic England. (2015). Planning responsible retrofit of traditional buildings.

Historic Scotland. (2021). Guide to Energy Retrofit of Traditional Buildings. https://www.historicenvironment.scot/archives-and research/publications/publication/?publicationId=47c9f2eb-1ade-4a76-a775-add0008972f3

Hull, J., & Bryan, P. (2019). *Bim for heritage: Developing the asset information model*. Historic England.

Idris, M. L., Rahman, A. R. A., Ishiyaku, B., & Alias, A. R. (2024). Preserving heritage buildings with building information modelling: Addressing the challenges and opportunities. *Smart and Green Materials*, *1*(1), 33–43. DOI: 10.70028/sgm.v1i1.5

IEA. (2017). Towards a Zero-Emission, Efficient, and Resilient Buildings and Construction Sector [Global Status Report.]. *UN Environment and International Energy Agency*, *2017*, 2017.

Imenda, S. (2014). Is there a conceptual difference between theoretical and conceptual frameworks. *Journal of Social Sciences*, *38*(2), 185–195. DOI: 10.1080/09718923.2014.11893249

Ismail, E. D., Said, S. Y., Jalil, M. K. A., & Ismail, N. A. A. (2021). Benefits and Challenges of Heritage Building Information Modelling Application in Malaysia. *Environment-Behaviour Proceedings Journal, 6*(SI4), 179–184. DOI: 10.21834/ebpj.v6iSI4.2917

Iyamu, T. (2013). Underpinning theories: Order-of-use in information systems research. *Journal of Systems and Information Technology, 15*(3), 224–238. DOI: 10.1108/JSIT-11-2012-0064

Jamal, K. A. A., Harun, S. N., & Said, S. Y. (2024). A Review and Comparative Study of Retrofit for the Energy Efficiency of Malaysian Heritage Buildings. *Insights on Resiliency and Urban Development*, 42-58.

Jamal, K. A. A., Mohammad, M. F., Hashim, N., Mohamed, M. R., & Ramli, M. A. (2019). Challenges of Building Information Modelling (BIM) from the Malaysian Architect's Perspective. In MATEC Web of Conferences (Vol. 266, p. 05003). EDP Sciences.

Jiancheng, S., Saar, C. C., & Aminudin, E. (2023). Digital Twinning in the Malaysian Construction Industry. In *Digitalization in Construction* (pp. 271–293). Routledge. DOI: 10.1201/9781003408949-15

Jouan, P., & Hallot, P. (2019). Digital twin: A HBIM-based methodology to support preventive conservation of historic assets through heritage significance awareness. *The International Archives of the Photogrammetry, Remote Sensing and Spatial Information Sciences, 42*(W15), 609–615. DOI: 10.5194/isprs-archives-XLII-2-W15-609-2019

Kermanshahi, E. K., Tahir, M. B. M., Lim, N. H. A. S., Balasbaneh, A. T., & Roshanghalb, S. (2020, April). Awareness of Building Information Modeling Implementation in Retrofitting the Existing Residential Building in Malaysia. [). IOP Publishing.]. *IOP Conference Series. Earth and Environmental Science, 476*(1), 012007. DOI: 10.1088/1755-1315/476/1/012007

Khaddaj, M., & Srour, I. (2016). Using BIM to retrofit existing buildings. *Procedia Engineering, 145*, 1526–1533. DOI: 10.1016/j.proeng.2016.04.192

Khalaf, R. W. (2021). World Heritage on the Move: Abandoning the Assessment of Authenticity to Meet the Challenges of the Twenty-First Century. *Heritage, 4*(1), 371–386. DOI: 10.3390/heritage4010023

Khalil, A., Stravoravdis, S., & Backes, D. (2021). Categorisation of building data in the digital documentation of heritage buildings. *Applied Geomatics, 13*(1), 29–54. DOI: 10.1007/s12518-020-00322-7

Khan, M. S., Khan, M., Bughio, M., Talpur, B. D., Kim, I. S., & Seo, J. (2022). An Integrated HBIM Framework for the Management of Heritage Buildings. *Buildings (Basel, Switzerland)*, *12*(7), 964. DOI: 10.3390/buildings12070964

Khodeir, L. M., Aly, D., & Tarek, S. (2016). Integrating HBIM (Heritage Building Information Modeling) tools in the application of sustainable retrofitting of heritage buildings in Egypt. *Procedia Environmental Sciences*, *34*, 258–270. DOI: 10.1016/j.proenv.2016.04.024

Larsen, T. J., Sørebø, A. M., & Sørebø, Ø. (2009). The role of task-technology fit as users' motivation to continue information system use. *Computers in Human Behavior*, *25*(3), 778–784. DOI: 10.1016/j.chb.2009.02.006

LCCF. (2017). Low Carbon Cities Framework Version 2. Ministry of Energy, Green Technology and Water Malaysia (KeTTHA), ISBN 978-967-5893-10-0.

Lee, C. C., Cheng, H. K., & Cheng, H. H. (2007). An empirical study of mobile commerce in insurance industry: Task–technology fit and individual differences. *Decision Support Systems*, *43*(1), 95–110. DOI: 10.1016/j.dss.2005.05.008

Lim, Y. W., Chong, H. Y., Ling, P. C., & Tan, C. S. (2021). Greening existing buildings through Building Information Modelling: A review of the recent development. *Building and Environment*, *200*, 107924. DOI: 10.1016/j.buildenv.2021.107924

Littlefield, D. (2017). Heritage and time: mapping what is not there. In *Heritage Building Information Modelling* (pp. 32–45). Routledge. DOI: 10.4324/9781315628011-4

Liu, J., Foreman, G., Sattineni, A., & Li, B. (2023). Integrating Stakeholders' Priorities into Level of Development Supplemental Guidelines for HBIM Implementation. *Buildings (Basel, Switzerland)*, *13*(2), 530. DOI: 10.3390/buildings13020530

Liu, T., Ma, G., & Wang, D. (2022). Pathways to successful building green retrofit projects: Causality analysis of factors affecting decision making. *Energy and Building*, *276*, 112486. DOI: 10.1016/j.enbuild.2022.112486

Lizana, J., Molina-Huelva, M., & Chacartegui, R. (2016). Multi-criteria assessment for the effective decision management in residential energy retrofitting. *Energy and Building*, *129*, 284–307. DOI: 10.1016/j.enbuild.2016.07.043

Lovell, L. J., Davies, R. J., & Hunt, D. V. (2023). The application of historic building information modelling (HBIM) to cultural heritage: A review. *Heritage*, *6*(10), 6691–6717. DOI: 10.3390/heritage6100350

Ma, Z., Cooper, P., Daly, D., & Ledo, L. (2012). Existing building retrofits: Methodology and state-of-the-art. *Energy and Building*, *55*, 889–902. DOI: 10.1016/j.enbuild.2012.08.018

Mari, T., Chiat, C. P., Ng, V., Gunasagaran, S., & Kuppusamy, S. (2024). STRATEGIES FOR NET ZERO CARBON IN SUSTAINABLE CONSERVATION PRACTICES FOR HISTORICAL BUILDINGS IN MALAYSIA. *PLANNING MALAYSIA, 22.*

Martinelli, L., Calcerano, F., & Gigliarelli, E. (2022). Methodology for an HBIM workflow focused on the representation of construction systems of built heritage. *Journal of Cultural Heritage*, *55*, 277–289. DOI: 10.1016/j.culher.2022.03.016

McCaig, I., Pender, R., & Pickles, D. (2018). *Energy Efficiency and Historic Buildings: How to Improve Energy Efficiency*. Historic England.

McGinley, O., Moran, P., & Goggins, J. (2022). An Assessment of the Key Performance Indicators (KPIs) of Energy Efficient Retrofits to Existing Residential Buildings. *Energies*, *15*(1), 334. DOI: 10.3390/en15010334

Mehndi, S. M., & Chakraborty, I. (2020, February). Simulation for a cost-effective and energy efficient retrofits of the existing building stock in India using BIM. In 2020 International conference on contemporary computing and applications (IC3A) (pp. 237-243). IEEE. DOI: 10.1109/IC3A48958.2020.233305

Melchor-Ferrer, E., & Buendía-Carrillo, D. (2014). Financial information management for university departments, using open-source software. *International Journal of Information Management*, *34*(2), 191–199. DOI: 10.1016/j.ijinfomgt.2013.12.009

Melville, N., Kraemer, K., & Gurbaxani, V. (2004). Information technology and organizational performance: An integrative model of IT business value. *Management Information Systems Quarterly*, *28*(2), 283–322. DOI: 10.2307/25148636

Meoni, A., Vittori, F., Piselli, C., D'Alessandro, A., Pisello, A. L., & Ubertini, F. (2022). Integration of structural performance and human-centric comfort monitoring in historical building information modeling. *Automation in Construction*, *138*, 104220. DOI: 10.1016/j.autcon.2022.104220

Messner, J., Anumba, C., Dubler, C., Goodman, S., Kasprzak, C., Kreider, R., Leicht, R., Saluja, C., & Zikic, N. (2019). BIM Project Execution Planning Guide, Version 2.2. Computer Integrated Construction Research Program, The Pennsylvania State University, University Park, PA, USA, August, Retrieved from http://bim.psu.edu

Mora, R., Sánchez-Aparicio, L. J., Maté-González, M. Á., García-Álvarez, J., Sánchez-Aparicio, M., & González-Aguilera, D. (2021). An historical building information modelling approach for the preventive conservation of historical constructions: Application to the Historical Library of Salamanca. *Automation in Construction, 121*, 103449. DOI: 10.1016/j.autcon.2020.103449

Moran, F., Blight, T., Natarajan, S., & Shea, A. (2013). The use of passive house planning package to reduce energy use and CO2 emissions in historic dwellings. *Energy and Building, 75*, 216–227. DOI: 10.1016/j.enbuild.2013.12.043

Moyano, J., Carreño, E., Nieto-Julian, J. E., Gil-Arizon, I., & Bruno, S. (2022). Systematic approach to generate Historical Building Information Modelling (HBIM) in architectural restoration project. *Automation in Construction, 143*, 104551. DOI: 10.1016/j.autcon.2022.104551

Murphy, M., McGovern, E., & Pavia, S. (2009). Historic building information modelling (HBIM). *Structural Survey, 27*(4), 311–327. DOI: 10.1108/02630800910985108

Nawi, N. H. M., Haron, R. C., & Kamarudin, Z. (2020). Risk cost analysis in Malay heritage conservation project. Planning Malaysia, 18.

Nieto-Julián, J. E., Farratell, J., Bouzas Cavada, M., & Moyano, J. (2023). Collaborative workflow in an HBIM project for the restoration and conservation of cultural heritage. *International Journal of Architectural Heritage, 17*(11), 1813–1832. DOI: 10.1080/15583058.2022.2073294

Nieto-Julián, J. E., Lara, L., & Moyano, J. (2021). Implementation of a TeamWork-HBIM for the Management and Sustainability of Architectural Heritage. *Sustainability (Basel), 13*(4), 2161. DOI: 10.3390/su13042161

Oreni, D., Brumana, R., Della Torre, S., Banfi, F., Barazzetti, L., & Previtali, M. (2014). Survey turned into HBIM: The restoration and the work involved concerning the Basilica di Collemaggio after the earthquake (L'Aquila). *ISPRS Annals of the Photogrammetry, Remote Sensing and Spatial Information Sciences, 2*(5), 267–273. DOI: 10.5194/isprsannals-II-5-267-2014

Orihuela, P., Pacheco, S., & Orihuela, J. (2017). Proposal of performance indicators for the design of housing projects. *Procedia Engineering, 196*, 498–505. DOI: 10.1016/j.proeng.2017.07.230

Orlikowski, W. J. (2000). Using technology and constituting structures: A practice lens for studying technology in organizations. *Organization Science, 11*(4), 404–428. DOI: 10.1287/orsc.11.4.404.14600

Ouellette, J. W., Woodard, P., & Bekboliev, M. (2022). Information Delivery Manual (IDM) Development for Building Information Modelling (BIM) and Building Energy Modelling (BEM) Workflows. 112.

Pal, D., & Patra, S. (2021). University students' perception of video-based learning in times of COVID-19: A TAM/TTF perspective. *International Journal of Human-Computer Interaction, 37*(10), 903–921. DOI: 10.1080/10447318.2020.1848164

Penjor, T., Banihashemi, S., Hajirasouli, A., & Golzad, H. (2024). Heritage Building Information Modelling (HBIM) for Heritage Conservation: Framework of Challenges, Gaps, and Existing Limitations of HBIM. *Digital Applications in Archaeology and Cultural Heritage*, •••, 00366.

Piselli, C., Guastaveglia, A., Romanelli, J., Cotana, F., & Pisello, A. L. (2020). Facility energy management application of HBIM for historical low-carbon communities: Design, modelling and operation control of geothermal energy retrofit in a real Italian case study. *Energies, 13*(23), 6338. DOI: 10.3390/en13236338

Quattrini, R., Battini, C., & Mammoli, R. (2018). HBIM to VR. Semantic awareness and data enrichment interoperability for parametric libraries of historical architecture. *The International Archives of the Photogrammetry, Remote Sensing and Spatial Information Sciences, 42*(2), 937–943. DOI: 10.5194/isprs-archives-XLII-2-937-2018

Ramli, M. A., Kamarazaly, M. A., Shi, L. X., Ibrahim, F. I., & Rizal, M. (2021). Building Defects and Restoration Techniques of Heritage Buildings in Malaysia. *Malaysian Construction Research Journal (MCRJ)*, 76.

Ramli, N. N., Ab Wahab, L., & Mohd Isa, H. (2021). A pilot survey on the assessment technique in accessing defects on heritage buildings in Malaysia. In *Virtual Go-Green: Conference and Publication (V-GoGreen 2020)* (pp. 37-42).

Reeves, T., Olbina, S., & Issa, R. R. (2015). Guidelines for using building information modeling for energy analysis of buildings. *Buildings (Basel, Switzerland), 5*(4), 1361–1388. DOI: 10.3390/buildings5041361

Rodrigues, F., Teixeira, J., Matos, R., & Rodrigues, H. (2019). Development of a web application for historical building management through BIM technology. *Advances in Civil Engineering, 2019*(1), 2019. DOI: 10.1155/2019/9872736

Samarghandi, H., Mousavi, S., Taabayan, P., Mir Hashemi, A., & Willoughby, K. (2016). Studying the Reasons for Delay and Cost Overrun in Construction Projects: The Case of Iran.

Sanhudo, L., Ramos, N. M., Martins, J. P., Almeida, R. M., Barreira, E., Simões, M. L., & Cardoso, V. (2018). Building information modeling for energy retrofitting–A review. *Renewable & Sustainable Energy Reviews*, *89*, 249–260. DOI: 10.1016/j.rser.2018.03.064

Shan, M., Chen, Y. F., Zhai, Z., & Du, J. (2022). Investigating the critical issues in the conservation of heritage building: The case of China. *Journal of Building Engineering*, *51*, 104319. DOI: 10.1016/j.jobe.2022.104319

Shehu, Z., Endut, I. R., Akintoye, A., & Holt, G. D. (2014). Cost overrun in the Malaysian construction industry projects: A deeper insight. *International Journal of Project Management*, *32*(8), 1471–1480. DOI: 10.1016/j.ijproman.2014.04.004

Sholeh, M. N., Fauziyah, S., & Khasani, R. R. (2020). Effect of Building Information Modeling (BIM) on reduced construction time-costs: a case study. In E3S Web of Conferences (Vol. 202, p. 02012). EDP Sciences.

Sopery, L. H. M., & Ayob, M. F. (2023). An Investigation on Sustainable Refurbishment Practice of Building Works in Malaysia. Journal of Architecture [JAPCM]. *Planning and Construction Management*, *13*(1), 29–39.

Spies, R., Grobbelaar, S., & Botha, A. (2020, April). A scoping review of the application of the task-technology fit theory. In *Conference on e-Business, e-Services and e-Society* (pp. 397-408). Cham: Springer International Publishing. DOI: 10.1007/978-3-030-44999-5_33

Srećković, M., Šibenik, G., Breitfuß, D., Preindl, T., & Kastner, W. (2021). Analysis of design phase processes with BIM for blockchain implementation. In *ECPPM 2021-eWork and eBusiness in Architecture, Engineering and Construction* (pp. 125–131). CRC Press. DOI: 10.1201/9781003191476-17

Stanga, C., Banfi, F., & Roascio, S. (2023). Enhancing building archaeology: Drawing, UAV photogrammetry and scan-to-BIM-to-VR process of ancient Roman ruins. *Drones (Basel)*, *7*(8), 521. DOI: 10.3390/drones7080521

Syed Zainol Abidin, I. (1995). *Pemeliharaan Warisan Rupa Bandar*. Badan Warisan Malaysia.

Tahir, M. M., Haron, N. A., Alias, A. H., Harun, A. N., Muhammad, I. B., & Baba, D. L. (2018). Improving Cost and Time Control in Construction Using Building Information Model (BIM): A Review. *Pertanika Journal of Science & Technology*, *26*(1).

Thravalou, S., Michopoulos, A., Alexandrou, K., & Artopoulos, G. (2023, June). Energy retrofit strategies of built heritage: Using Building Information Modelling tools for streamlined energy and economic analysis. [). IOP Publishing.]. *IOP Conference Series. Earth and Environmental Science, 1196*(1), 012115. DOI: 10.1088/1755-1315/1196/1/012115

Tobias, M. (2004, March 18). Elements of Sustainable architecture. Retrieved from https://www.ny-engineers.com/blog/elements-of-sustainable-architecture

UN. (2022). Transforming our World: The 2030 Agenda for Sustainable Development. A/RES/70/1. 2015. Available online: https://sustainabledevelopment.un.org/post2015/transformingourworld (accessed on 3 August 2022).

UNSDG. (2015). UN General Assembly, Transforming our world: the 2030 Agenda for Sustainable Development, 21 October 2015, A/RES/70/1, available at: https://www.refworld.org/docid/57b6e3e44.html [accessed 1 December 2021]

Van Eldik, M. A., Vahdatikhaki, F., dos Santos, J. M. O., Visser, M., & Doree, A. (2020). BIM-based environmental impact assessment for infrastructure design projects. *Automation in Construction, 120*, 103379. DOI: 10.1016/j.autcon.2020.103379

Van Tam, N., Quoc Toan, N., Phong, V. V., & Durdyev, S. (2023). Impact of BIM-related factors affecting construction project performance. *International Journal of Building Pathology and Adaptation, 41*(2), 454–475. DOI: 10.1108/IJBPA-05-2021-0068

Wilkinson, S. (2011). *Sustainable retrofit potential in lower quality office stock in the central business district.* CIB Management and Innovation in the Sustainable Built Environment.

Wise, F., Moncaster, A., & Jones, D. (2021). Rethinking retrofit of residential heritage buildings. *Buildings & Cities, 2*(1), 495–517. DOI: 10.5334/bc.94

Wong, J., Hoong, P., Teo, E., & Lin, A. (2022, November). Digital twin: A conceptualization of the task-technology fit for individual users in the building maintenance sector. [). IOP Publishing.]. *IOP Conference Series. Earth and Environmental Science, 1101*(9), 092041. DOI: 10.1088/1755-1315/1101/9/092041

Wong, J. H., Rashidi, A., & Arashpour, M. (2020). Evaluating the impact of building information modeling on the labor productivity of construction projects in Malaysia. *Buildings (Basel, Switzerland), 10*(4), 66. DOI: 10.3390/buildings10040066

Wong, J. K. W., & Zhou, J. (2015). Enhancing environmental sustainability over building life cycles through green BIM: A review. *Automation in Construction, 57*, 156–165. DOI: 10.1016/j.autcon.2015.06.003

Wu, B., & Chen, X. (2017). Continuance intention to use MOOCs: Integrating the technology acceptance model (TAM) and task technology fit (TTF) model. *Computers in Human Behavior*, *67*, 221–232. DOI: 10.1016/j.chb.2016.10.028

Yau, Y. H., & Hasbi, S. J. E. P. (2017). A comprehensive case study of climate change impacts on the cooling load in an air-conditioned office building in Malaysia. *Energy Procedia*, *143*, 295–300. DOI: 10.1016/j.egypro.2017.12.687

Yusoff, S. N. S., & Brahim, J. (2021). Implementation of building information modeling (Bim) for social heritage buildings in kuala lumpur. *International Journal of Sustainable Construction Engineering and Technology*, *12*(1), 88–99. DOI: 10.30880/ijscet.2021.12.01.009

Zaid, N. S. M., Ghazali, M. I., Azmi, N. F., Kamarulzaman, P. N., & Akashah, F. W. (2021). Environmental sustainability in Malaysian Heritage Building Conservation: Review of Heritage Legistation and Policy. Malaysian Construction Research Journal (MCRJ), 203.

Žurić, J., Zichi, A., & Azenha, M. (2023). Integrating HBIM and sustainability certification: A pilot study using GBC historic building certification. *International Journal of Architectural Heritage*, *17*(9), 1464–1483. DOI: 10.1080/15583058.2022.2042623

Chapter 4
Analytic Hierarchy Progress (AHP) for Construction Technologies Adoption in the Malaysian Construction Industry

Changsaar Chai
https://orcid.org/0000-0001-7733-1690
Taylor's University, Malaysia

Ming Gai
https://orcid.org/0009-0007-1920-9911
Taylor's University, Malaysia

Haw Yang Ang
https://orcid.org/0009-0008-7170-401X
Taylor's University, Malaysia

Chia Kuang Lee
https://orcid.org/0000-0001-6063-8071
Universiti Malaysia Pahang Al-Sultan Abdullah, Malaysia

Wan Siang Chong
https://orcid.org/0009-0007-8891-2643
Universiti Malaysia Sarawak, Malaysia

Mervyn Hsin Jyi Wong
https://orcid.org/0009-0002-0057-8657
University of Malaya, Malaysia

ABSTRACT

Digital technologies are vital for boosting productivity and efficiency in Malaysia's construction industry. Despite a 74% awareness rate of Building Information Modeling (BIM) reported in the Malaysia BIM Report 2019, adoption remains limited. This study employs a quantitative methodology, using questionnaire surveys to assess the need for technology adoption among industry players. The findings

DOI: 10.4018/979-8-3373-5278-7.ch004

aim to inform the development of a decision-making tool for selecting suitable technologies for construction projects. By identifying key challenges arising from limited digital adoption, the research proposes strategies to enhance technology integration. Expected outcomes include improved project outcomes, reduced cost overruns, minimized delays, and enhanced industry productivity. Ultimately, these advancements will contribute to Malaysia's GDP growth while addressing critical inefficiencies in the construction sector. Promoting digital adoption is essential for elevating the industry's performance and ensuring its long-term sustainability.

INTRODUCTION

The Construction 4.0 Strategic Plan (2021–2025), developed by CIDB through the Construction Research Institute of Malaysia (CREAM), highlights twelve key technologies capable of transforming Malaysia's construction companies into technologically advanced entities. These technologies have demonstrated effectiveness in addressing major industry challenges and delivering tangible benefits to stakeholders (Subramaniam et al., 2020; CIDB Malaysia, 2022).

Traditional construction practices in the industry are often characterized by inefficiencies such as cost overruns, schedule delays, and quality issues (Zimmermann, 2023). Fragmented processes and a heavy reliance on manual methods contribute to low productivity and waste (CIDB, 2023). In contrast, the integrated use of advanced technologies offers a path to overcome these challenges by streamlining workflows and improving accuracy. For instance, Building Information Modeling (BIM) enables better coordination and clash detection, prefabrication techniques reduce material waste and speed up assembly, and digital project management platforms facilitate real-time communication and decision-making (ProjectTeam, 2023). By combining such tools, construction projects can achieve higher efficiency, improved safety, and greater sustainability.

Despite the proven effectiveness of these technologies in addressing industry challenges (Subramaniam et al., 2020), adoption rates among Malaysian AEC (Architects, Engineers, and Consultants) firms remain suboptimal, with fewer than 40% implementing digital tools. This reluctance stems from uncertainties such as resistance to changing conventional methods, difficulties in integrating new systems, and a lack of decision-support frameworks (Meor Gheda et al., 2025; Nnaji & Karakhan, 2020). The gap between the proven benefits of Construction 4.0 technologies and their limited uptake highlights the need for effective decision-making aids to guide practitioners in selecting and implementing innovations.

The Analytic Hierarchy Process (AHP) offers a robust decision-making framework that can help bridge this gap. AHP enables construction practitioners to evaluate and select appropriate technologies based on multiple criteria, thus supporting more rational adoption choices. Research indicates that adopting suitable technologies can ensure project completion within budget and on schedule (Tan, 2023). Increasingly, stakeholders—including clients and project managers—are utilizing AHP to optimize project outcomes by integrating relevant digital tools (Aladayleh & Aladaileh, 2024). This study proposes AHP as a strategic decision-making tool to enhance technology adoption among AEC companies, ultimately improving efficiency, cost-effectiveness, and overall productivity in Malaysia's construction sector.

LITERATURE REVIEW

The adoption of construction technologies has been widely promoted, with studies emphasizing their benefits, including increased productivity, cost-effectiveness, and reduced construction time and costs (Ibrahim et al., 2021). These tools also enhance building quality by enabling precise measurements and reducing errors, thus ensuring the delivery of high-standard structures (GoCodes, 2022). Nevertheless, significant barriers persist, impeding the widespread integration of these technologies within the construction sector (Subramaniam et al., 2020).

Despite advancements in digital tools, the construction industry has been slow to embrace digitalization. Only a small fraction of companies fully utilize advanced planning and management technologies, leading to productivity levels that lag behind other sectors (CIDB, 2023). Key challenges include budget constraints, as many organizations hesitate to invest heavily in construction technologies (Omar et al., 2022). Additionally, low adoption rates are exacerbated by limited staff knowledge, inadequate training, and management resistance to change. The industry's fragmented nature and continued reliance on manual processes further hinder progress (LetsBuild, 2023). Overcoming these challenges requires comprehensive training initiatives, cultural shifts within organizations, and strategic investments by stakeholders to ensure effective implementation and maximize the potential of digital tools.

VOSviewer analysis visualized the landscape of construction technologies in the Malaysian construction industry, categorizing 2,000 articles into five clusters: blockchain, construction industry, developing countries, technology adoption, and the analytic hierarchy process (AHP). The bibliometric mapping revealed relationships among these articles, with the AHP cluster emerging as the largest. Within this cluster, associations were identified with concepts such as remote sensing, risk assessment, and structural equation modeling. In contrast, the technology adoption cluster is linked to topics like technology acceptance, DEMATEL, construction barriers, and sustainability. Notably, the connection between the AHP and technology adoption clusters is relatively weak, highlighting a research gap. This suggests limited integration of AHP in facilitating the adoption of construction technologies. Bridging this gap could offer valuable insights into overcoming adoption barriers and promoting efficient decision-making processes in the construction sector.

Figure 1. Keyword co-occurrence analysis

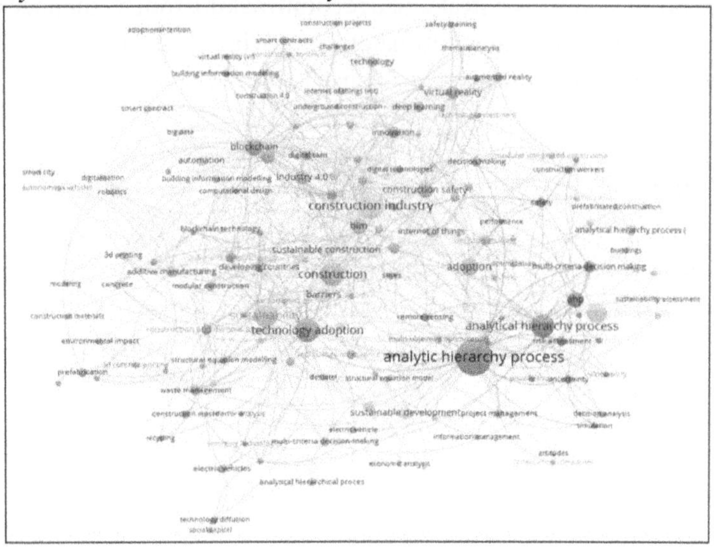

Current research provides limited guidance on implementing construction technologies in Malaysia's construction sector. To bridge this gap, the Analytical Hierarchy Process (AHP) is proposed as an effective decision-making tool for AEC companies to facilitate the adoption of advanced technologies.

As noted by CIDB Malaysia (2021), the construction sector plays a vital role in contributing to the country's GDP, with over 90% of firms comprising small and medium enterprises (SMEs). This sector also has a significant multiplier effect of 2.03, underscoring its economic impact. Consequently, construction stakeholders must explore advanced technologies and evaluate their projects' "health"—defined as their functionality and progress toward successful completion. The integration of digital tools is essential for optimizing project performance, enabling stakeholders to achieve and potentially surpass their project objectives while advancing the overall productivity and efficiency of Malaysia's construction industry.

Construction Technologies

The Malaysian construction industry has outlined twelve key construction technologies in the Construction 4.0 Strategy Plan, shown in Table 1, developed by CIDB, which aligns with the principles of the Fourth Industrial Revolution (IR 4.0). This five-year plan aims to enhance the industry's competitiveness, productivity, and efficiency while reducing delays by implementing smart construction practices (Senior Minister of Infrastructure, 2020).

Table 1. Articles on twelve (12) construction technologies outlined by CIDB in the construction 4.0 strategy plan.

Causes	Authors
T1 Building Information Modelling (BIM).	Chai et al., 2023; Roslan et al., 2022; Tan, 2023.
T2 Prefabrication & Modular Construction.	Zahrizan et al., 2013; Ismail et al., 2023.
T3 Autonomous Construction.	Technology Cards, 2019; Xu & Borja, 2020.
T4 Augmented Reality & Virtualization.	Orihuela et al., 2019; Chai et al., 2023; Nassereddine et al., 2022; Henze, 2022.
T5 Cloud and Realtime Collaboration.	Watson. J, 2020; Bello et al., 2021; CIDB Malaysia, 2023.
T6 3D Scanning and Photogrammetry.	Cousins, 2018; DJMCADCoordination, 2021.
T7 Big Data and Predictive Analysis.	Yousif et al., 2021; Shanthi, 2023.
T8 Internet of Things (IoT). **T9** 3D Printing and Additive Manufacturing.	Gbadamosi et al., 2019; Dinesh, 2022. Steenhuis et al., 2015; Michell, 2016
T10 Advanced Building Materials.	Asite, 2021; Aung, 2022.
T11 Blockchain. **T12** Artificial Intelligence (AI).	Kim et al., 2020; Farhadi, 2023. Abioye et al., 2021; Holzmann & Lechiara, 2022

Building Information Modeling (BIM)

Building Information Modeling (BIM) is a transformative process that significantly enhances construction management. One of its key advantages is its ability to create a model that remains functional from the initial design phase through construction and building occupancy, making BIM a fundamental tool in construction management (Chai et al., 2023).

Effective and systematic communication is crucial in the construction industry, where multiple stakeholders—such as clients, contractors, engineers, architects, and consultants—are involved. BIM provides an excellent platform for ensuring efficient communication among these parties, reducing misunderstandings, information gaps, and disputes. Additionally, BIM facilitates the digitalization of the construction industry, streamlining workflows and minimizing delays and cost overruns. Its success in improving collaboration and communication among project stakeholders is well-documented (Roslan et al., 2022). In line with this, Prime Minister Datuk Seri Anwar Ibrahim announced that the government will implement BIM across both public and private sector projects (BERNAMA, 2024).

Prefabrication and Modular Construction

Prefabrication and modular construction, collectively known as the Industrialized Building System (IBS), involve manufacturing building components off-site before transporting them to the construction site for assembly. This approach shows significant potential for residential projects, such as flats, condominiums, and terrace houses, due to their repetitive nature. The private sector, with its higher volume of projects compared to the public sector, stands to benefit substantially from adopting IBS (Zahrizan et al., 2013). Modular construction, also known as volumetric construction, modular building systems, or modular systems, is a form of off-site prefabrication and a modern construction method that streamlines the building process (Musa et al., 2018).

Autonomous Construction

Autonomous construction involves the use of automated machinery on-site to enhance operator safety and improve construction productivity. Autonomous equipment, including excavators, bulldozers, haul trucks, and loaders, can perform tasks such as soil excavation and grading. These machines can be monitored by a single operator overseeing multiple units working simultaneously. Smaller autonomous equipment, such as rovers, can assist workers by transporting heavy construction materials across the site (Technology Cards, 2019). The integration of robotic systems in construction significantly reduces project timelines and increases safety by replacing workers in repetitive or hazardous tasks (Xu & Borja, 2020).

Augmented Reality and Virtualization

Parties involved in a construction project can use Augmented Reality (AR) to visualize construction plan elements, leading to a clearer understanding of the project (Henze, 2022). AR-BIM enables users to experience virtual representations of reality, addressing current obstacles, seamlessly integrating with on-site activities, and overlaying planned data onto the existing environment (Chai et al., 2023). AR is primarily employed during the pre-construction phase, especially in the design stage, allowing stakeholders to visualize the project's quality before construction begins (Nassereddine et al., 2022). In contrast, Virtual Reality (VR) offers virtual simulations of environments, enabling users to engage with realistic three-dimensional scenarios. It is a computer-generated space where users can experience events that are not occurring in real-time or at their current location (Orihuela et al., 2019).

Cloud and Realtime Collaboration

The construction sector generates vast amounts of data throughout project development, with diverse data produced at various stages and often stored separately on team servers, individual desktops, laptops, and smartphones. To ensure effective project coordination, it is essential to integrate this data, as poor integration can lead to incorrect decisions that negatively impact overall performance, cause project delays, and reduce profitability (Bello et al., 2021). Cloud technology and real-time collaboration have revolutionized construction projects by offering cost-effectiveness, flexibility, and scalability. These tools enhance collaboration among key stakeholders by enabling seamless data sharing (CIDB Malaysia, 2023). Additionally, Watson (2020) highlights that cloud and real-time collaboration play a crucial role in ensuring efficient project scheduling and coordination.

3D Scanning and Photogrammetry

Laser scanning is a technique used to acquire digital information about an object through a non-intrusive laser beam. The scanner rotates 360 degrees, capturing data from various surfaces across the scanned area (DJMCADCoordination, 2021). This technology provides highly accurate and regularly updated records, offering benefits in cost management, site development, project delivery, and build quality (Cousins, 2018). One key feature of 3D scanning is its ability to perform a virtual walk-through of the project site, allowing stakeholders to gain a comprehensive understanding of the site's condition and progress.

Big Data and Predictive Analysis

Big data refers to large datasets characterized by high volume, diverse variety, and rapid velocity, resulting from continuous data generation across various sources (Yousif et al., 2021). The construction industry holds significant potential to leverage big data and predictive analytics, enhancing project management by utilizing historical data to predict and mitigate risks (Shanthi, 2023). By analyzing these datasets, industry stakeholders can make informed decisions, improve efficiency, and proactively address challenges that may arise during project execution.

Internet of Thing (IoT)

The Internet of Things (IoT) is a technological advancement that is transforming the interaction between devices, the internet, and humans. It enables the collection of data on performance, conditions, and activities at construction sites, which is then transmitted to a central dashboard for analysis and decision-making (Dinesh, 2022). In construction, IoT is applied in various areas, including infrared and heat sensors, temperature, humidity, and pressure sensors, track circuit monitoring, and motion detectors. By integrating IoT, construction projects can effectively assess and monitor risks related to safety and quality, enhancing overall project management (Gbadamosi et al., 2019).

3D Printing and Additive Manufacturing

Additive manufacturing differs from traditional processes such as milling and grinding, which shape products by removing material. Instead, it builds products layer by layer, enabling three-dimensional manufacturing with minimal waste (Steenhuis et al., 2015). 3D printers convert CAD models into a series of layers that are printed sequentially, allowing for precise and customized fabrication. A critical factor for success in the 3D printing business is minimizing the costs and complexity of producing parts while ensuring seamless integration with consumer software (Michell, 2016). This technology offers significant advantages in terms of efficiency, material conservation, and the ability to create complex geometries that would be difficult or impossible with traditional manufacturing methods.

Advanced Building Materials

A new class of construction materials has emerged, exhibiting exceptional properties that enhance the intelligence, energy efficiency, strength, and sustainability of infrastructure and buildings (Asite, 2021). Developed through extensive research and analysis, these advanced materials aim to address various challenges facing the construction industry today. Although many of these materials have yet to see widespread adoption, their increasing affordability and accessibility are expected to drive broader integration into construction practices in the near future.

Blockchain

In construction, blockchain functions as a decentralized, immutable digital ledger that securely and transparently records transactions and other data. Its structure consists of several key components working together to ensure the accuracy and integrity of the stored information (Farhadi, 2023). Kim et al. (2020) highlighted the diverse applications of blockchain in construction, including smart contracts, construction coins, blockchain-based Building Information Modeling (BIM), securities, digital currency, and record-keeping. Additionally, blockchain offers significant advantages such as enhanced security, reduced construction costs, and increased transparency, making it a valuable tool for improving efficiency and accountability in the construction industry.

Artificial Intelligence (AI)

AI was first introduced in the 1950s with the goal of replicating human intelligence through computer programs (Holzmann & Lechiara, 2022). In the construction industry, AI has the potential to streamline numerous tasks, improving decision-making, automating repetitive processes, and enhancing overall efficiency (Abioye et al., 2021). By leveraging machine learning and data analytics, AI can optimize various aspects of construction, from design to project management, reducing costs and improving productivity.

METHODOLOGY

Primary Data Collection: Interviews and Criteria Identification

The primary data collection involved semi-structured interviews, where ten respondents (industry experts and contractors with varied experience in technology use) provided insights into factors affecting technology adoption. From the qualitative data, eight (8) adoption criteria were initially identified, as shown in Table 2.

Table 2. Adoption criteria for construction technologies.

Adoption Criteria	
A1 Individual Attributes	
A2 Organization Characteristics	
A3 Project Characteristics	
A4 Technology Attributes	
A5 Vendor's Attributes	
A6 Reduce Printing Fee	
A7 More time for training	
A8 Improving safety	

During the initial data collection phase, eight adoption criteria were identified (see Table 2). However, after expert interviews, "Reducing Printing Fee," "More Time for Training," and "Improving Safety" were deemed to be encapsulated within "Individual Attributes," "Organizational Characteristics," and "Technology Attributes," respectively. Consequently, five core criteria were retained: Individual Attributes, Organizational Characteristics, Project Characteristics, Technology Attributes, and Vendor's Attributes. This selection process ensures a clear and efficient evaluation model.

During the interviews, these criteria were presented to the respondents, who were asked to select or prioritize the factors they deemed most critical for technology adoption. As a result of this exercise, five criteria were ultimately shortlisted (A1 through A5), which correspond to broader categories of influence, while three criteria (A6, A7, A8) were dropped from further analysis.

In subsequent data collection and analysis, all questionnaire designs and the construction of the AHP model were based on these five core criteria. This approach not only ensures consistency in the research framework but also guarantees that the ranking results of various technologies, evaluated under multiple criteria, are both representative and meaningful.

Thematic Analysis

A thematic analysis was conducted on the interview data to validate and organize the criteria. Thematic analysis is a qualitative research method that systematically identifies, analyzes, and presents patterns (themes) within data. As illustrated in Figure 2, the process begins with familiarization, where researchers immerse themselves in the data by reading and re-reading to fully understand its depth. Initial codes are then generated by highlighting interesting segments of the data. These codes are grouped into potential themes, which are reviewed and refined to ensure they accurately reflect the data. Each theme is then defined and named to capture its core meaning. Finally, the themes are integrated into a coherent narrative, supported by relevant data excerpts and contextualized within existing research. This method enables researchers to extract meaningful insights and identify underlying patterns in qualitative data.

Figure 2. Thematic analysis process.

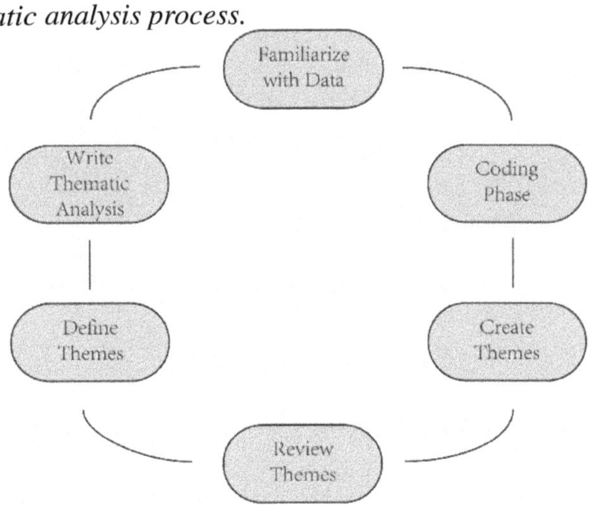

Analytical Hierarchy Process (AHP)

The adoption of the Analytic Hierarchy Process (AHP) has significantly enhanced and streamlined the selection and decision-making processes (Jahanfar et al.,2021). This study employs AHP as a decision-making tool due to its ability to effectively address complex problems. Key advantages of AHP include its simplicity and flexibility in solving intricate issues, as well as its capacity to assess both qualitative and quantitative data from subjective and objective perspectives. By decomposing complex problems into hierarchical levels, AHP facilitates the analysis of issues at varying levels of granularity. Additionally, it enables the evaluation of the consistency of judgments made by decision-makers (Krstić et al., 2022).

Expert opinions, obtained through AHP, provide a robust framework for decision-making, integrating both rational and intuitive elements to select the optimal solution based on multiple criteria, ultimately aligning outcomes with user satisfaction. As shown in Figure 3, AHP involves a three-level structure: the top level (Goal), the second level (Attributes or Criteria), and the third level (Alternatives). During this process, decision-makers perform pairwise comparisons to establish priorities and rankings for the alternatives. AHP offers a comprehensive measurement theory that derives ratio scales from paired comparisons within multilevel hierarchical structures, ensuring consistency while accounting for dependencies within the structure.

Figure 3. Example of hierarchy of objectives.

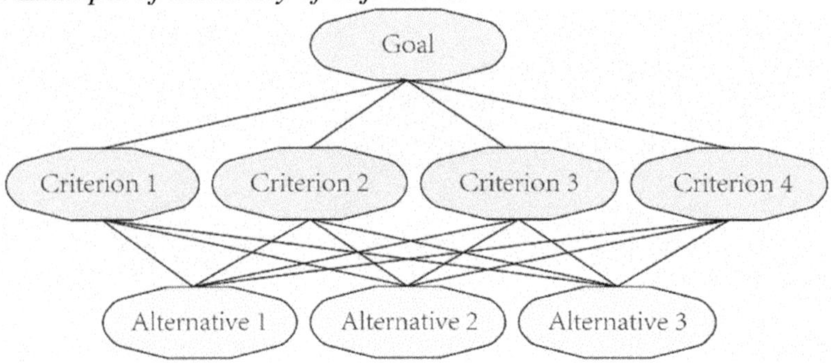

Pairwise Comparison Analysis

Pairwise comparison analysis is a widely used technique in decision-making that assesses the relative importance or preference among different elements or criteria. This method involves systematically comparing each element against every other element in pairs and assigning a numerical value to reflect the degree of preference or significance. It aids in task prioritization, evaluates the importance of criteria, and supports decision-making when selecting between alternatives.

A common application of pairwise comparison analysis is in the Analytic Hierarchy Process (AHP), where decision-makers utilize this method to construct a pairwise comparison matrix. This matrix serves as the basis for calculating weights and priorities for each element, leading to a more informed and rational decision-making process. As depicted in Figure 3.2, the numerical scale used in pairwise comparison analysis typically ranges from 1 (indicating equal importance) to 9 (indicating absolute importance), with intermediate values representing varying degrees of preference or significance.

Application of Superdecision

An alternative approach employed in this study is the use of Superdecision software, as shown in Figure 3.9. Superdecision facilitates the administration of pairwise questionnaires and comparisons, which are then used to generate the pairwise matrix. The use of Superdecision significantly simplifies the process compared to manual calculations and helps reduce human errors, thereby improving the validity and reliability of the data. Once the data collection process was completed, the researcher ensured that all inconsistency ratio (IR) values among the selected expert panelists were equal to or less than 0.10.

Superdecision software is specifically designed for multi-criteria decision-making, aimed at achieving specific goals. It assigns numerical values to represent the relative preference or importance between criteria, which in turn allows for the generation of ranked criteria.

Respondents Background

The selected respondents comprise skilled experts from contractor firms within the Malaysian construction sector, specifically those engaged in the adoption of construction technologies. To ensure expertise, the chosen respondents are required to have a minimum of 25 years of experience in the construction industry. A total of 15 contractors were selected from 15 different firms. However, only 10 respondents agreed to participate in the interview and provide their expert insights on construction technology adoption in the industry, as shown in Table 3. With their extensive experience, these experts are well-equipped to offer reliable responses to the pairwise questionnaires addressing various construction technologies in Malaysia, ensuring that the inconsistency ratio remains at or below 0.1.

Table 3. Respondent's background.

Respondents	**Years of Experience**
R1 Respondent 1 – Contractor	>25
R2 Respondent 2 – Contractor	
R3 Respondent 3 – Contractor	>25
R4 Respondent 4 – Contractor	>28
R5 Respondent 5 – Consultant	>26
R6 Respondent 6 – Consultant	>27
R7 Respondent 7 – Consultant	>30
R8 Respondent 8 – Engineer	>29
R9 Respondent 9 – Engineer	>26
R10 Respondent 10 – Engineer	>32

Inconsistency Value

In qualitative research methodology, the term "inconsistency value" denotes a measure of the coherence level in data or information. Within the Analytical Hierarchy Process (AHP), inconsistency values reflect the extent to which the preferences or judgments given by respondents are conflicting or inconsistent. These values in AHP assist in evaluating the reliability and consistency of pairwise comparisons. Lower inconsistency values indicate greater reliability of the data or information. The inconsistency values of the respondents are illustrated in Table 4.

Table 4. Inconsistency value

Respondents	**Adoption Criterions**				
	Individual Attributes	**Organization Characteristics**	**Project Characteristics**	**Technology Attributes**	**Vendor's Attributes**
Respondent 7	0.128	0.114	0.161	0.115	0.151
Respondent 8	0.136	0.123	0.126	0.198	0.153
Respondent 9	0.138	0.126	0.131	0.230	0.230
Respondent 10	0.140	0.134	0.139	0.230	0.163

Process Of Achieving Criteria For Construction Technologies In The Malaysian Construction Industry

To establish the criteria for the adoption of construction technologies in Malaysia, several studies were conducted for primary data collection. The collected data were then refined to identify five key criteria for adoption, as presented in Table 5. These criteria were highlighted by the interviewees in the transcript from the previous section.

Table 5. Shortlisted adoption criterion from respondents.

Shortlisted Adoption Criteria
S1 Individual Attributes
S2 Organization Characteristics
S3 Project Characteristics
S4 Technology Attributes
S5 Vendor's Attributes

ANALYSIS AND DISCUSSION

Examine The Adoption Criteria For The Construction Technologies In Malaysia

The categories outlined in Table 6 showcase the diversity of adoption criteria, each indicating the sub-cluster influencing adoption decisions in the Malaysian construction sector.

Project Characteristic cover project-specific contexts like project size, complexity, and delivery timelines that can facilitate or hinder the use of new tools on a case-by-case basis. Technology Attributes refer to inherent qualities of the technology itself—such as usability, compatibility, perceived usefulness, and reliability—that influence user acceptance. Organizational Characteristics encompass firm-level conditions including management support, financial capacity, and organizational culture, aligning with an organization's readiness to implement technology. Individual Attributes represent personal factors such as a decision-maker's experience, skills, and openness to innovation, which affect the willingness to adopt new technology. Finally, Vendor's Attributes capture external influences including the reliability, support services, and reputation of technology providers, which can impact a contractor's confidence in adopting a solution. Together, these five criteria holistically reflect the multifaceted theme of construction technology adoption: they address individual and organizational readiness, project-specific demands, technological factors, and external support. Ensuring clarity on these categories provides a comprehensive lens for analyzing adoption decisions.

Table 6. Categories of variables (CIDB Malaysia (2022)

Project characteristics	**Technology attributes**	**Organization characteristics**	**Individual attributes**	**Vendors' attributes**
Task	Maturity	Need		After sales service
Productivity	Brand	Goal	Relationship with a vendor	Guaranty
Quality	Reliability	Cost reduction	Response to risk	Warranty
Project cost	Versatility	Technology policy	Culture	Training service
Time	Capacity	Formalization/ centralization	Innovativeness	Maintenance service
Safety	Power	Company culture	Subjective norms	Financial offer

continued on following page

Table 6. Continued

Project characteristics	**Technology attributes**	**Organization characteristics**	**Individual attributes**	**Vendors' attributes**
Project condition	Durability	Company size	Personal response to motivation	Relationship & Motivation program
Geography and location of the project	Site performance			Vendor: size of organization

A comprehensive survey and study were conducted, involving contractors with varying levels of experience in adopting construction technologies, with the aim of exploring the adoption criteria among Malaysian contractors. Through this analysis, several adoption criteria and challenges in the adoption process were identified and organized thematically. The analysis revealed that the most significant factor contributing to the lack of adoption is "knowledge acquisition."

Knowledge Acquisition

Several participants emphasized the ongoing challenges related to acquiring the necessary knowledge and skills to effectively utilize construction technologies. The complexity of digital construction technologies requires a deep understanding of the technology itself, its integration into various construction processes, its impact on project management, and its potential to foster innovation and efficiency.

(i) Training and Educational Programs:

Accessible and comprehensive training programs on construction technologies play a significant role in facilitating knowledge acquisition. The availability and effectiveness of such programs can greatly enhance contractors' understanding and skills, enabling them to implement these technologies more effectively.

(ii) Access to Information and Resources:

Access to relevant information, resources, and educational materials is essential for acquiring knowledge. The availability of up-to-date manuals, tutorials, case studies, and best practices can streamline the learning process, enhancing contractors' competence in construction technologies.

(iii) Organizational Culture and Support:

The organizational culture within construction firms, including management support for learning and skill development, has a significant influence on knowledge acquisition. A culture that encourages continuous learning and provides opportunities for skill enhancement fosters effective knowledge acquisition among employees.

(iv) Hands-on Experience and Exposure:

Practical experience and exposure to real-world applications of construction technologies are critical for knowledge acquisition. Participation in hands-on training, pilot projects, and exposure to innovative construction methods all contribute to enhancing contractors' competence in utilizing these technologies.

(v) Collaboration and Knowledge Sharing:

Engagement with colleagues, industry peers, and experts plays an important role in acquiring knowledge. Networking events, knowledge-sharing sessions, and collaborative projects provide opportunities for individuals to learn from others' experiences, exchange ideas, and discover effective strategies for implementing construction technologies.

(vi) Technological Infrastructure and Tools:

Access to the necessary technological infrastructure, software tools, and equipment is pivotal for implementing construction technologies. The availability of advanced tools and technologies enables individuals to experiment, practice, and gain familiarity with various aspects of these technologies.

(vii) Motivation and Interest:

Personal motivation and a strong interest in acquiring knowledge about construction technologies are crucial factors. Internal drive, curiosity, and a genuine eagerness to explore emerging technologies can inspire individuals to actively seek opportunities for learning and skill development in this field.

(viii) Feedback and Continuous Improvement:

Regular feedback, performance evaluations, and opportunities for continuous improvement are vital for effective knowledge acquisition. Constructive feedback helps contractors identify areas for growth, refine their learning methods, and improve their proficiency in utilizing construction technologies over time.

These factors are interrelated and collectively shape the knowledge acquisition process in the context of adopting construction technologies. In the current study, insufficient knowledge acquisition clearly emerged as a critical barrier to adoption. Many contractors do not receive enough training or accessible information, resulting in a limited understanding of how to use new tools effectively. Organizational support for learning is also often lacking, which means employees may not be fully prepared or confident to implement novel technologies even when they are available. These shortcomings indicate that knowledge gaps directly hinder technology uptake.

To address this issue, targeted improvements are necessary. Construction firms and industry bodies should invest in comprehensive training programs and professional development courses on digital construction technologies. Establishing mentorship or knowledge-sharing platforms (for example, regular workshops or an online knowledge base) can help disseminate best practices and lessons learned. It is also important to foster a corporate culture that values continuous learning—management should encourage staff to attend training and recognize those who upskill. On a practical level, providing more hands-on opportunities, such as pilot projects or on-site demonstrations of new technologies, will allow practitioners to build experience and confidence. Ensuring easier access to updated resources (technical guides, case studies, etc.) can further empower individuals to self-educate. These improvements in the knowledge acquisition process would directly mitigate one of the main barriers identified. Indeed, recent research underscores the need for comprehensive training and structured change management to accelerate digital integration in construction (Meor Gheda et al., 2025). By closing the knowledge and skills gap, companies and policymakers can significantly enhance the readiness of the workforce to adopt and utilize sustainable construction technologies.

Developing A Decision-Making Tool For Digital Technologies Adoption In Malaysia

An alternative approach employed in this study involves the use of Superdecision software, as depicted in Figure 4. Superdecision is utilized to conduct pairwise questionnaires and comparisons, which generate the pairwise matrix. The application of Superdecision simplifies the process compared to manual calculations and reduces the risk of human error, thereby improving the validity and reliability of the data. After the researcher completed the data collection process, they ensured that all inconsistency ratio (IR) values were equal to or less than 0.10 among the selected expert panelists.

Figure 4. Superdecision questionnaires and collected data

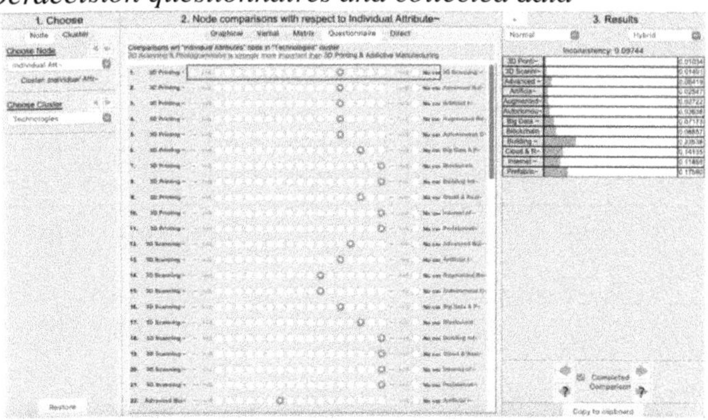

Superdecision software is designed for decision-making based on multiple criteria to achieve specific objectives. It assigns numerical values to represent the level of preference or importance between criteria, thereby generating rankings. The software was employed for face-to-face data collection with the targeted respondents. The data collection process is simple and easily comprehensible for potential experts, with immediate guidance provided throughout. The primary focus was on the inconsistency value, which was used to assess the accuracy and reliability of the data.

Table 7. Ranking of construction technologies

Rank	Individual Attributes	Organization Characteristics	Project Characteristics	Technology Attributes	Vendor's Attributes
1	Building Information Modelling (BIM)	Prefabrication & Modular Construction	Prefabrication & Modular Construction	Advanced Building Material	Prefabrication & Modular Construction
2	Prefabrication & Modular Construction	Advanced Building Material	Advanced Building Material	Prefabrication & Modular Construction	Building Information Modelling (BIM)
3	Advanced Building Material	Building Information Modelling (BIM)	Building Information Modelling (BIM)	Building Information Modelling (BIM)	Advanced Building Material

continued on following page

Table 7. Continued

Rank	Individual Attributes	Organization Characteristics	Project Characteristics	Technology Attributes	Vendor's Attributes
4	Cloud & Realtime Collaboration	Internet of Thing (IoT)	Internet of Thing (IoT)	Big Data & Predictive Analysis	Internet of Thing (IoT)
5	Internet of Thing (IoT)	Autonomous Construction	Autonomous Construction	Internet of Thing (IoT)	Cloud & Realtime Collaboration
6	Blockchain	Cloud & Realtime Collaboration	Cloud & Realtime Collaboration	Cloud & Realtime Collaboration	Big Data & Predictive Analysis

Table 8. Ranking of construction technologies

Rank	Individual Attributes	Organization Characteristics	Project Characteristics	Technology Attributes	Vendor's Attributes
7	Big Data & Predictive Analysis	Blockchain	Blockchain	Blockchain	Autonomous Construction
8	Autonomous Construction	Big Data & Predictive Analysis	Big Data & Predictive Analysis	3D Printing & Addictive Manufacturing	Augmented Reality & Virtualization
9	Augmented Reality & Virtualization	Augmented Reality & Virtualization	Augmented Reality & Virtualization	Artificial Intelligence (AI)	Artificial Intelligence (AI)
10	Artificial Intelligence (AI)	3D Printing & Addictive Manufacturing	3D Printing & Addictive Manufacturing	3D Scanning & Photogrammetry	3D Scanning & Photogrammetry
11	3D Scanning & Photogrammetry	Artificial Intelligence (AI)	Artificial Intelligence (AI)	Augmented Reality & Virtualization	3D Printing & Addictive Manufacturing
12	3D Printing & Addictive Manufacturing	3D Scanning & Photogrammetry	3D Scanning & Photogrammetry	Autonomous Construction	Autonomous Construction

The Analytical Hierarchy Process (AHP) is utilized to provide a systematic analysis of the complex interrelationships and priority levels among various digital construction technologies and the criteria for preferred adoption in the Malaysian construction industry.

Individual Attribute

Table 7 indicates that Building Information Modelling (BIM) ranks highest. According to the Technology Acceptance Model (TAM), individuals' willingness to adopt a new technology is influenced by perceived usefulness, which refers to the belief that using the system will enhance job performance, and perceived ease of use, which refers to the belief that using the system will be effortless. TAM proposes that the impact of external variables on the intention to use is moderated by these two perceptions (Zhao et al., 2022), as highlighted by the respondents in the previous section.

The second ranking is attributed to prefabrication and modular construction. Musa et al. (2018) introduced the term Industrialized Building System (IBS) to describe the concept of prefabrication in the Malaysian construction sector. IBS refers to a construction process in which standardized building components are manufactured either in a factory or on-site, then transported and assembled into a structure, requiring minimal on-site labor. Key factors contributing to the increased adoption of prefabrication and modular construction include the use of high-quality, uniform three-dimensional units, which have the potential to reduce waste.

Following this is advanced building materials. Asite (2021) emphasized that the integration of advanced building materials represents a commitment to adopting innovative, economically viable sustainable solutions that aim to enhance profitability, reduce construction time, and benefit the environment. Despite skepticism within the construction industry regarding investments in technologies or resources lacking a proven track record, advanced building materials offer the potential for significantly lower lifecycle costs.

Organization and Project Characteristics

Prefabrication and modular construction rank highest in this criterion. Previous studies (Wu et al., 2021) have highlighted several benefits of this technology for the construction industry. According to Wu et al. (2021), the advantages include the establishment of a fixed design early in the process, which facilitates better prefabrication adoption; improved supervision leading to higher quality of prefabricated products; reduced overall construction costs; shortened construction time; enhanced environmental performance through waste minimization; and increased integrity in building design.

Following this, advanced building materials are ranked second. Factors supporting this ranking include their contribution to improved energy efficiency. Advanced building materials optimize energy consumption by enhancing insulation, reducing heat gain, and decreasing reliance on artificial lighting for certain projects, thereby increasing sustainability. These materials promote sustainable construction by reducing resource consumption and lowering the carbon footprint, contributing to a more environmentally friendly future. With their unique attributes improving energy efficiency, durability, safety, and occupant comfort, advanced materials are positioned to revolutionize the future of the built environment, fostering sustainable, intelligent, and innovative structures.

Building Information Modelling (BIM) occupies the third ranking in these criteria. Olbina and Elliott (2019) suggest that the successful implementation of BIM in a project is influenced by several factors. Human, financial, technological, and technical resources are primary enablers of BIM adoption, while project, policy, and industry-related factors have a moderate impact. Key elements contributing to BIM implementation include the presence of effective leadership, qualified staff, and access to information technology. Additional factors, such as project type, size, and complexity, also influence the decision to adopt BIM.

Technology Attributes

Advanced building materials hold the top ranking. Figure 9 below illustrates the outcomes of utilizing these materials. Moreover, the use of advanced building materials increases the likelihood of the Malaysian construction industry adopting these innovations to maximize the benefits of their construction projects. These benefits include reducing construction and material costs, accelerating the construction process, lowering life-cycle costs, improving energy efficiency, and enhancing health and well-being by minimizing carbon dioxide emissions from conventional construction materials.

Figure 5. Illustration of consequences of employing advanced building material (Rodrigues de Almeida et al., 2016)

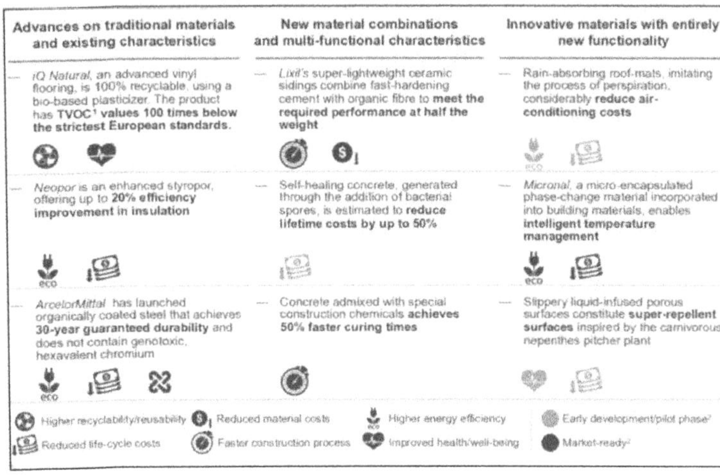

Prefabrication and modular construction hold the second ranking. Unlike traditional construction methods, prefabrication and modular construction involve the manufacturing of unit modules in factories. This approach enhances efficiency and productivity by ensuring quality and reducing waste, which has garnered global attention. The factory-based production process allows for shorter construction project timelines. Site work, including substructure tasks such as excavation and foundation work, can proceed concurrently with module production in factories, unlike the sequential processes in conventional construction (Shin et al., 2022). Additionally, respondents highlighted that modular construction offers sustainability benefits by reducing waste generation through effective material control, reuse, and recycling, as demonstrated in a study by Wu et al. (2021) in Bantul and Bandung cities within the Badan region.

Building Information Modelling (BIM) is ranked third in this criterion. BIM serves as a dynamic tool for decision-making, facilitating the creation and management of parametric digital representations of buildings or infrastructure projects (Chougan et al., 2023). It enables the integration and sharing of information throughout the building's life cycle, addressing information disparities and providing a reliable foundation for decision-making throughout the project. This approach effectively mitigates traditional project performance challenges, such as schedule delays, quality issues, and cost overruns, as highlighted by the interview respondents.

Vendor's Attributes

Based on the interviews with the respondents, survey results were collected through verbal discussions. After analyzing the data, prefabrication and modular construction emerged as the top-ranking criterion, followed by Building Information Modelling (BIM) and advanced building materials. Several factors were considered during the ranking process, including after-sales services, warranties, training services, maintenance services, financial offers, vendor relationships, and motivation programs. As a result, prefabrication and modular construction ranked highest, particularly in terms of maintenance services and vendor relationships. BIM ranked second, supported by factors such as reliability, customer relationship management, and a commitment to excellence. Advanced building materials ranked third, driven by factors including reliability, supplier relationships, sustainability, and environmental considerations. Notably, the most appealing aspects of advanced building materials include their reduced carbon footprint and lower maintenance costs.

Ranking of Technologies (AHP Results)

With the five major criteria established and weighted through the AHP, the study then ranked various construction technologies in terms of their priority for adoption. The analysis (summarized in Table 7 and Table 8) revealed the following order of preference:

Prefabrication and Modular Construction: This option emerged as the top-ranked technology. Respondents gave high importance to aspects like maintenance services and vendor relationships associated with prefabrication, indicating confidence in the support and after-sales service ecosystems of modular construction solutions. In the Malaysian context, prefabrication and modular methods are often discussed under the term Industrialized Building System (IBS). CIDB defines IBS as a construction process in which standardized building components are manufactured in a controlled environment (onsite or offsite) and then transported and assembled into a structure with minimal on-site labor (Ismail et al., 2023). The strong ranking for prefabrication/IBS is supported by its known benefits: it allows key design decisions to be finalized early, improves quality through better supervision of factory-made components, reduces overall construction costs and time, and minimizes on-site waste. These advantages not only enhance efficiency but also contribute to sustainability (through waste reduction and improved environmental performance), which resonated with many respondents (Rocha et al., 2023).

Building Information Modeling (BIM): BIM was the next highly ranked technology. Factors such as reliability, customer relationship management by BIM vendors, and a demonstrated commitment to excellence in implementation contributed to its strong position. According to the Technology Acceptance Model (TAM), an individual's willingness to adopt a new technology is heavily influenced by perceived usefulness (the belief that using the technology will improve job performance) and perceived ease of use (the belief that the technology will be easy to use). These perceptions mediate the impact of external variables on the intention to use a technology. In the context of BIM, respondents highlighted its usefulness in enhancing collaboration and reducing errors, and many found modern BIM software increasingly user-friendly. This aligns with TAM, which has been applied in construction to explain BIM adoption (Zhao et al., 2022). In our findings, respondents who perceived clear benefits of BIM (such as easier coordination of plans and clash detection) and found it relatively accessible were more inclined to prioritize it.

Advanced Building Materials: The third-ranking technology category encompassed advanced or innovative building materials (for example, self-healing concrete, high-performance composites, or sustainable materials). Respondents valued factors such as material reliability, supplier relationships, and especially the sustainability and environmental benefits of these materials. Notably, many participants were drawn to advanced materials due to their potential for a reduced carbon footprint and lower long-term maintenance costs. These materials often contribute to energy efficiency and durability, which align with industry goals for sustainable construction. The strong interest here suggests that when the cost and performance of new materials are proven, contractors are willing to adopt them for the sake of long-term project value and compliance with emerging green building standards.

Overall, the AHP results provide a clear hierarchy of technology preferences when multiple criteria are considered. It is evident that prefabrication/modular construction and BIM stand out as top priorities for Malaysian contractors, with advanced materials also highly regarded. These preferences reflect a balance of practical benefits and strategic value: prefabrication addresses efficiency and quality, BIM improves process management and decision-making, and new materials offer sustainability benefits.

The AHP model results indicate that prefabrication/modular construction and BIM are the most promising technologies. This finding provides clear guidance for policymakers: governments could introduce tax breaks, subsidies, or other incentives to encourage the adoption of these key technologies. Additionally, industry associations may launch specialized training programs to elevate the overall digital capability of the construction workforce. Furthermore, promoting advanced building materials through green building standards and environmental incentives could further support sustainable construction practices.

Based on the above analysis, several conclusions can be drawn. Decisions regarding the adoption of digital construction technologies are influenced by the challenges and issues within the Malaysian construction industry, as reflected in the key categories of consideration (S1–S5). The rankings of technologies relative to these criteria illustrate how well each innovation addresses the prevailing challenges identified (such as the need for better knowledge acquisition, organizational support, etc.). Embracing appropriate construction technologies can clearly enhance project productivity, lead to cost savings, and improve overall effectiveness and efficiency in the construction process.

CONCLUSION

The primary aim of this study is to elucidate how construction technologies can enhance productivity, leading to cost-effectiveness and increased efficiency within the Malaysian construction industry. The study investigates the prioritization of digital technologies for adoption based on the considerations of Malaysian contractors, revealing their hierarchical significance and interconnections within decision-making processes. This comprehensive examination unequivocally confirms that construction technologies play a critical role in improving collaboration, communication, and decision-making in the construction sector.

The findings of this study present a significant opportunity for contractors in Malaysia. By utilizing the Analytic Hierarchy Process (AHP) technique for analysis, contractors can identify the most appropriate construction technologies to adopt for their projects, thereby ensuring project success.

This study began with expert interviews that identified eight adoption criteria, which were then refined and consolidated into five core criteria. An AHP model was subsequently constructed to systematically evaluate and rank construction technologies based on these criteria, forming a structured and integrated decision-support framework. This methodology not only strengthens the credibility of the findings but also offers a replicable model for future research in technology adoption.

Addressing the challenge of insufficient understanding of construction technologies is vital. Providing training programs and courses on digital construction technologies for the Malaysian construction sector can help overcome this barrier, encouraging staff to embrace the new era of construction innovation.

One of the most significant advantages of construction technologies is the potential for cost savings. To fully capitalize on these benefits, contractors must take proactive steps. These technologies can generate substantial savings and promote sustainability in construction projects by reducing rework, enabling informed lifecycle costing, optimizing resource allocation, and providing predictive analysis capabilities.

Finally, the AHP methodology proved to be a valuable framework for decision-making in this context. Its structured approach to evaluating construction technology options against multiple criteria allowed for a nuanced understanding of priorities and trade-offs. We recommend that contractors and industry stakeholders apply this method more broadly, as it can help systematically address the multifaceted challenges of technology adoption and establish clear priorities for action. Policymakers and firms could use an AHP-based approach to develop roadmaps for technology implementation that align with both business objectives and national development targets.

In summary, the research design of this study successfully integrated the five key adoption criteria throughout its methodology. The initial qualitative phase (expert interviews and thematic analysis) distilled the broad range of adoption factors into the five core categories of individual, organizational, project, technology, and vendor-related considerations. The subsequent quantitative phase then incorporated these categories into a unified AHP decision-making model. By aligning data collection and analysis with these five criteria, the study ensured a holistic evaluation of technology adoption decisions. This integrative approach demonstrates how diverse factors can be concurrently addressed in assessing new construction technologies, yielding a comprehensive model that mirrors the multifaceted nature of technology adoption in construction. Future research and practical initiatives can adopt a similar integrated design to tackle complex adoption issues in a structured, inclusive manner, thereby further bridging the gap between technology potential and on-the-ground practice in the construction industry.

REFERENCES

Abioye, S. O., Oyedele, L. O., Akanbi, L., Ajayi, A., Davila Delgado, J. M., Bilal, M., Akinade, O. O., & Ahmed, A. (2021). Artificial intelligence in the construction industry: A review of present status, opportunities and future challenges. *Journal of Building Engineering*, *44*, 103299. DOI: 10.1016/j.jobe.2021.103299

Aladayleh, K. J., & Aladaileh, M. J. (2024). Applying Analytical Hierarchy Process (AHP) to BIM-Based Risk Management for Optimal Performance in Construction Projects. *Buildings (Basel, Switzerland)*, *14*(11), 3632. DOI: 10.3390/buildings14113632

Asite. (2021, January, 20). Why we should be paying more attention to advanced building materials. [Online blog post]. https://www.asite.com/blogs/why-we-should-be-paying-more-attention-to-advanced-building-materials

Aung, T. H. S. (2022). 10 - Trendy advanced construction materials to know (2022). [Online blog post]. https://www.novatr.com/blog/advanced-construction-materials

Bello, S. A., Oyedele, L. O., Akinade, O. O., Bilal, M., Delgado, J. M. D., Akanbi, L. A., Ajayi, A. O., & Owolabi, H. A. (2021). Cloud computing in construction industry: Use cases, benefits and challenges. *Automation in Construction*, *122*, 103441. DOI: 10.1016/j.autcon.2020.103441

BERNAMA. (2024, June, 07). *Govt ready to adopt BIM for development projects - PM Anwar.* https://thesun.my/malaysia-news/govt-ready-to-adopt-bim-for-development-projects-pm-anwar-IC12546781

Chai, C., Chan, S., Xiong, L., Lim, B., & Shan, J. (2023) Digital tools adoption towards construction industry revolution. *Journal of Engineering Science and Technology, 11(2022),* 231-243. https://jestec.taylors.edu.my/Special%20Issue%20THINK%20SPACE%202022/STAAUH%202022_19.pdf

Chougan, M., Al-Kheetan, M. J., & Ghaffar, S. H. (2023). Additive Manufacturing and the Construction Industry. In *Disrupting Buildings* (pp. 97–109). Palgrave Macmillan., DOI: 10.1007/978-3-031-32309-6_7

Cousins, S. (2018, May, 8). *How 3D building scanning transforms construction projects.* https://www.navvis.com/blog/54-how-3d-scanning-is-being-used-to-transform-construction-projects

Dinesh, B. (2022, August, 9). *IoT is changing the construction industry.* https://www.datasciencecentral.com/iot-in-construction-changing-the-construction-industry/

DJMCADCoordination. (2021, August, 26). The ROI of 3D scanning: Is it worth it? [Online blog post]. https://caddjm.com/is-3d-scanning-worth-it/

Farhadi, F. (2023, July, 13). *Block chain in Construction: Ultimate Guide for 2024*. https://neuroject.com/blockchain-in-construction/

Gbadamosi, A. Q., Oyedele, L., Mahamadu, A. M., Kusimo, H., & Olawale, O. (2019). *The role of internet of things in* delivering *smart construction* [Paper presentation]. CIB world building congress. Hong Kong, China.

GoCodes. (2022). *7 Positive impacts of construction technology on businesses*. https://gocodes.com/positive-impacts-construction-technology/?srsltid=AfmBOoqED4nehztCrB2DXHk8mZoCIqYfKe3hlZlZlo1pdOZGXXcYQRDb

hanthi R. (2023, July, 20). *Leveraging big data analytics for construction success*. https://www.contractormag.com/technology/article/21269911/leveraging-big-data-analytics-for-construction-success

Henze, A. (2022, February, 8). *Augmented reality in construction: Benefits and limitations*. https://www.teamviewer.com/en/insights/augmented-reality-in-construction/

Holzmann, V., & Lechiara, M. (2022). Artificial intelligence in construction Projects: An explorative study of professionals' expectations. *European Journal of Business Management and Research*, *7*(3), 151–162. DOI: 10.24018/ejbmr.2022.7.3.1432

Ibrahim, F. S., Esa, M., & Rahman, R. A. (2021). The adoption of IOT in the Malaysian construction industry: Towards construction 4.0. *International Journal of Sustainable Construction Engineering Technology*, *12*(1). Advance online publication. DOI: 10.30880/ijscet.2021.12.01.006

Ismail, S., Hon, C. K. H., Crowther, P., Skitmore, M., & Lamari, F. (2023). The drivers and challenges of adopting the Malaysia industrialised building system for sustainable infrastructure development. *Construction Innovation*, *23*(5), 1054–1074. DOI: 10.1108/CI-05-2021-0088

Jahanfar, A., Mohamed, O., & Zayed, T. (2021). Multi-criteria decision-making in construction: Analytic Hierarchy Process applications and perspectives. *Journal of Construction Engineering and Management*, *147*(9), 04021105. DOI: 10.1061/(ASCE)CO.1943-7862.0002117

Kim, K., Lee, G., & Kim, S. (2020). A study on the application of blockchain technology in the construction industry. *KSCE Journal of Civil Engineering*, *24*(9), 2561–2571. DOI: 10.1007/s12205-020-0188-x

Krstić, M., Agnusdei, G. P., Miglietta, P. P., Tadić, S., & Roso, V. (2022). Applicability of industry 4.0 technologies in the reverse logistics: A circular economy approach based on comprehensive distance based ranking (COBRA) method. *Sustainability (Basel), 14*(9), 5632. DOI: 10.3390/su14095632

LetsBuild. (2023, October, 18). *Why the slow uptake of technology in construction is holding the industry back.* https://www.letsbuild.com/blog/why-the-slow-uptake-of-technology-in-construction-is-holding-the-industry-back

Malaysia, C. I. D. B. (2021, August, 30). *MCO and Its Impact on the Construction Industry.* https://cidbibs.com.my/mco-and-its-impact-on-the-construction-industry/

Malaysia, C. I. D. B. (2022, March, 31). *The adoption of technology in the construction industry.* https://www.cidb.gov.my/eng/the-adoption-of-technology-in-the-construction-industry/

Malaysia, C. I. D. B. (2023, August, 17). *Unlocking the potential of cloud computing.* https://www.cidb.gov.my/eng/unlocking-the-potential-of-cloud-computing/

Meor Gheda, M. L., Chung, H. W., Abdul Aziz, F. F., & Mohamad Nusran, N. F. (2025). Bridging the digital divide: Overcoming challenges in technology adoption in Malaysia's construction industry. *Malaysian Journal of Social Sciences and Humanities, 10*(1), 1–15. DOI: 10.47405/mjssh.v10i1.3137

Michell, V. (2016). 3D Printing and Additive Manufacturing Capability Modelling. *Sixth International Symposium on Business Modeling and Software Design.* SCITEPRESS. DOI: 10.5220/0006222400730083

Musa, M. F., Mohammad, M. F., Mahbub, R., & Yusof, M. R. (2018). Adopting modular construction in the Malaysian construction industry. *Asian Journal of Environment-Behaviour Studies, 3*(10), 1–9. DOI: 10.21834/aje-bs.v3i10.307

Nassereddine, H., Hanna, A. S., Veeramani, D., & Lotfallah, W. (2022). Augmented Reality in the Construction Industry: Use-Cases, Benefits, Obstacles, and Future Trends. *Frontiers in Built Environment, 8*, 730094. DOI: 10.3389/fbuil.2022.730094

Nnaji, C., & Karakhan, A. A. (2020). Technologies for safety and health management in construction: Current use, implementation benefits and limitations, and adoption barriers. *Journal of Building Engineering, 29*, 101212. DOI: 10.1016/j.jobe.2020.101212

Olbina, S., & Elliott, J. W. (2019). Contributing project characteristics and realized benefits of successful BIM implementation: A comparison of complex and simple buildings. *Buildings (Basel, Switzerland), 9*(8), 175. DOI: 10.3390/buildings9080175

Omar, A. F., Zainordin, N., Khoo, S. L., & Peng, J. O. H. (2022, November). Barriers in implementing artificial intelligence (AI) and Internet of things (IoTs) among Malaysian construction industry. *AIP Conference Proceedings, 2644*(1), 030002. Advance online publication. DOI: 10.1063/5.0111461

Orihuela, P., Noel, M., Pacheco, S., Orihuela, J., Yaya, C., & Aguilar, R. (2019). Application of virtual and augmented reality techniques during design and construction process of building projects. *Annual Conference of the International Group for Lean Construction*. https://doi.org/DOI: 10.24928/2019/0220

ProjectTeam. (2023). *Revolutionizing construction: The impact of technology on industry practices* [Blog post]. ProjectTeam.com. https://blog.projectteam.com/revolutionizing-construction-the-impact-of-technology-on-industry-practices

Rocha, P. F., Ferreira, N. O., Pimenta, F., & Pereira, N. B. (2023). Impacts of prefabrication in the building construction industry. *Encyclopedia, 3*(1), 28–45. DOI: 10.3390/encyclopedia3010003

Rodrigues de Almeida, P., Zafra Solas, M., Renz, A., Bühler, M. M., Gerbert, P., Castagnino, S., & Rothballer, C. (2016). Shaping the future of construction a breakthrough in mindset and technology. *World Economic Forum*. https://doi.org/DOI: 10.13140/RG.2.2.21381.37605

Roslan, A. F., Aminudin, E., Lau, S. E. N., & Abidin, N. I. (2022). Construction 4.0 to transform the Malaysian construction industry. *The Ingenieur magazine, 88*, 6-23. https://bem.org.my/documents/20181/268613/5.1.pdf/ea3269cb-ec2e-4d89-884a-33e24d2faa2a

Senior Minister of Infrastructure. (2020). *Construction 4.0 Strategic Plan (2021-2025)-Strategic Plan for Construction Industry: Gearing Up for the Fourth Industrial Revolution*. CIDB. https://www.cidb.gov.my/wp-content/uploads/2022/12/CR4_web_compressed.pdf

Shin, J., Moon, S., Cho, B., Hwang, S., & Choi, B. (2022). Extended technology acceptance model to explain the mechanism of modular construction adoption. *Journal of Cleaner Production, 342*, 130963. DOI: 10.1016/j.jclepro.2022.130963

Steenhuis, H. J., & Pretorius, L. (2015). Additive manufacturing or 3D printing and its adoption [Paper presentation]. *International Association for Management of Technology 2015 Conference Proceedings. University of Pretoria, Pretoria, South Africa.*

Subramaniam, R., & Ramachandiran, C. R. (2020). Emergence of advanced digital technology to increase productivity in civil and infrastructure industry in Malaysia. *PalArch's Journal of Archaeology of Egypt/Egyptology, 17*(7), 5309-5318.

Tan, K. W. (2023). *Examine digital transformation in the Malaysian construction industry* [Unpublished doctoral dissertation]. UTAR, Perak, Malaysia.

Tan, K. W. (2023, September, 27). *Integrating technology in construction*. https://thesun.my/opinion-news/integrating-technology-in-construction-PB11543936

Technology Cards. (2019, August, 16). *Autonomous construction vehicles*. https://www.technologycards.net/the-technologies/autonomous-construction-vehicles

Watson, J. (2020, October, 30). How cloud can enable collaboration in construction industry. https://www.acecloudhosting.com/blog/cloud-collaboration-in-construction-industry/

Wu, Z., Luo, L., Li, H., Wang, Y., Bi, G., & Antwi-Afari, M. F. (2021). An Analysis on Promoting Prefabrication Implementation in Construction Industry towards Sustainability. *International Journal of Environmental Research and Public Health, 18*(21), 11493. DOI: 10.3390/ijerph182111493 PMID: 34770008

Xu, X., & De Soto, B. G. (2020). On-site autonomous construction robots: A review of research areas, technologies, and suggestions for advancement. *Proceedings of the International Symposium on Automation and Robotics in Construction, 37*, 385-392. https://doi.org/DOI: 10.22260/ISARC2020/0055

Yousif, O. S., Zakaria, R., Aminudin, E., Yahya, K., Mohd Sam, A. R., Singaram, L., Munikanan, V., & Shamsuddin, S. M. (2021). Review of Big Data Integration in Construction Industry Digitalization. *Frontiers in Built Environment, 7*, 770496. DOI: 10.3389/fbuil.2021.770496

Zahrizan, Z., Ali, N. M., Haron, A. T., Marshall-Ponting, A., & Hamid, Z. A. (2013). Exploring the adoption of Building Information Modelling (BIM) in the Malaysian construction industry: A qualitative approach. *International Journal of Research in Engineering and Technology, 2*(8), 384–395. https://citeseerx.ist.psu.edu/document?repid=rep1&type=pdf&doi=c2dde67f06be045aadb766e739241ca02c996b70. DOI: 10.15623/ijret.2013.0208060

Zhao, Y., Sun, Y., Zhou, Q., Cui, C. Y., & Liu, Y. (2022). How A/E/C professionals accept BIM technologies in China: A TAM perspective. *Engineering, Construction, and Architectural Management, 30*(10), 4569–4589. DOI: 10.1108/ECAM-04-2022-0308

Zimmermann, M. (2023, September 26). Industrialized construction (IC), the key to reducing the environmental impact of industry [Blog post]. *Cemex Ventures*. https://www.cemexventures.com/industrialized-construction/

Chapter 5
Challenges in Adopting Internet of Thing (IoT)-Based Real-Time Monitoring Technologies in Malaysian Construction Industry

Nur Syaimasyaza Mansor
https://orcid.org/0000-0002-5699-8848
School of Housing, Building, and Planning, Gelugor, Malaysia

Hong Cing Cing
https://orcid.org/0009-0009-7142-4621
PMT Enterprise, Malaysia

ABSTRACT

Internet of Things (IoT) provides real-time monitoring function for the construction players to assess accurate data for decision making. The technologies being used in Malaysia construction industry include sensors, drones, RFIDs and others. These technologies improve project's efficiency in several areas with its real-time report, yet adoption still low. Most studies focused on IoT implementation during the design phase, with few studies concentrating on the construction phase, particularly from the standpoint of contractors. Considering contractors make many decisions during the construction phase, IoT assist them for quick and precise decisions. This paper focus on the challenges in adopting IoT-based real-time monitoring technologies in the Malaysian construction industry. The research adopted qualitative method via

DOI: 10.4018/979-8-3373-5278-7.ch005

semi-structured interviews with three G7 contractors in using those IoT and result in ten challenges. The findings are to increase industry participants' understanding and adoption of IoT in construction projects, hence enhancing project efficiency.

INTRODUCTION

Internet of Things (IoT) is said to be an advanced digital technology that is leading a new revolution, Industrial Revolution 4.0 (IR4.0) in today's world (Mahmud et al., 2018). Significant contributions of IoT includes smart cities, smart homes, smart transportation, pollution control and energy savings (Aleksandrova et al., 2019; Beniwal & Singhrova, 2022; Chen et al., 2020; Elghaish, Hosseini, et al., 2021; Hassan et al., 2020; Hou & Li, 2020; Kumar et al., 2019). In construction industry, IoT is being introducing to smoothen the construction operations and also overcome the challenges that might be faced (Mahmud et al., 2018). In relation to Construction 4.0, nine (9) pillars have been introduced for the construction players to monitor and control the whole project lifecycle. They are IoT, advanced robotics / autonomous robots, additive manufacturing, augmented reality, simulation, system integration, cloud computing, cybersecurity and big-data analytics (Ibrahim et al., 2021; Mahmud et al., 2018). Among all these, IoT has been highlighted as one of the core elements to improve productivity and bring better performance (Ibrahim et al., 2021; Oke et al., 2022). The development of the IoT has made a huge impact and offers various benefits to the construction projects and the construction participants. IoT systems are capable of gathering and sending data through the hardware and software by connecting to the Internet to enhance the process in various stages of construction (Gbadamosi et al., 2019). IoT has the potential to shift the construction industry up to a higher level by the usage of the innovative solutions as an assistance for the construction players to adapt the IoT technologies and make changes to the construction industry (Halim et al., 2021).

The IoT provided with real-time monitoring function can strengthen the information collection, storage and analysis for improved decision-making (Al-Amleh, 2020). The growth of IoT in the Dubai construction industry has shown effective waste management, safety management and cost reduction (Al-Amleh, 2020). IoT has greatly enhanced sustainability smart buildings or smart homes (Beniwal & Singhrova, 2022; Fernandez, 2019; Jia et al., 2019; Oke et al., 2022). Besides, IoT has also developed some intelligent applications in some of the built environment domain such as smart healthcare facilities, smart transportations and traffic systems, smart cities, control of logistic supply chain, fleet tracking systems, industrial automation, energy efficiency, waste management and environmental monitoring (Beniwal & Singhrova, 2022; Chen et al., 2020; Ghosh et al., 2020; Hassan et al.,

2020; Kang, 2022). Some examples of projects that have used IoT in the construction industry are mostly the large projects such as bridges, railways, tunnels, offshore and onshore facilities (Gamil et al., 2020; Mohammed et al., 2022).

Although the IoT-based real-time monitoring technologies bring a lot of benefits, especially in terms of decision making, there are still a small amount of construction firm that implement the IoT technologies (Hong & Mansor, 2023). Yap et al. (2019) stated that the critical construction industry problems in Malaysian construction projects are poor project performance, low productivity and reluctance toward implementing innovative solutions. The construction industry is facing with some significant challenges that inhibit the widespread adoption of these IoT technologies (Ibrahim et al., 2021). The construction firms have to struggle in solving these challenges so that they can enjoy the broader benefits that IoT adoption could bring to the construction sector (Al-Amleh, 2020). Project delays, increased costs and all the common project failures are all can be reduced to minimum if the IoT is brought to the construction industry and well implemented (Hong & Mansor, 2023).

This paper intends to find out the challenges in adopting the IoT-based real-time monitoring technologies in the Malaysian construction industry. The findings of this research are noteworthy to increase industry participants' understanding and adoption of those IoT technologies in construction projects, hence enhancing the project efficiency and performance. The construction players can review the challenges and develop the alternatives to overcome the challenges in order to implement the IoT to unlock the benefits that the IoT could bring to the construction project, ranging from improved decision-making to heightened operational efficiency.

LITERATURE REVIEW

Internet of Things (IoT)

IoT consists wide range of technologies because no official definition is given for IoT (Ghosh et al., 2020). Gbadamosi et al. (2019) stated that IoT is a network of things that equipped with a unique identification (UID) or Internet Protocol (IP) address which used to transfer and collect data. Arslan et al. (2019) also added that IoT is a worldwide network of things that are interconnected together and uniquely addressable based on its standard communication protocols. According to Chanal & Kakkasageri (2020), IoT is a network that links the physical and virtual objects globally to convert the information into digital form via the Internet. Gamil et al. (2020) explained that IoT is the possibility of the things that are connected to the Internet to form a platform to execute the activities. Al-Amleh (2020) and Oke et al. (2021) both defined that IoT is the network of the physical and virtual objects to connect and exchange data within each other by using the Internet with the help of the sensors, software and other technologies. Kevin Ashton, a British technologist from the United Kingdom, who introduced the term IoT, give a brief definition that IoT is the combination of various objects that are physically connected to the Internet via sensors (Arowoiya et al., 2020; Noura et al., 2019).

In this paper, IoT is defined as the worldwide things consist of interconnected smart devices that require the Internet for sharing and collecting data among each other. IoT in construction refers to the application of Internet-connected equipment or software in transferring data to enhance the construction project efficiency (Al-Amleh, 2020).

Real-Time Monitoring

Real-time monitoring is the live streaming of continuously updated data (Aravind et al., 2021). It is a process to collect and store data from devices, software, tools or even human actions so that the individual can detect the performance of the activities, applications and services in real time (Aceyus Team, 2021; Tunis, 2021). The data will be collected from time to time so that actions can be taken as soon as possible if any problems arise (Aravind et al., 2021; Tunis, 2021). The data will be collected at the exact time and be available at that moment, meaning that the data will be processed not even within one second. This is why it is named 'Real-Time' (Aceyus Team, 2021).

IoT provides real-time monitoring in several ways. IoT can facilitate the real-time communication between the items in the environment of the electrical appliances (Arslan et al., 2019). In addition, IoT has the availability to control the progress and gather the data at real time (Gillis, 2022).

Adoption of IoT in Malaysian Construction Industry

Based on the Construction 4.0 Strategic Plan (2021 – 2025) introduced by CIDB, IoT is one of the steps for this industry revolution (Mahmud et al., 2018). However, there are still a major of the construction industry players have yet to bring the IoT into practical (Gamil et al., 2020; Ibrahim et al., 2021; Tze Qing, 2019). Research shows that it is because there is a lack of experts to operate the IoT systems and lack of awareness of the construction players towards the IoT functions (Gamil et al., 2020; Ibrahim et al., 2021). However, JKR and CIDB is developing some strategies to persuade the construction players to accept IoT in their projects, such as organising the seminars, workshops and trainings. Moreover, some of the tertiary education institutions such as universities, polytechnics and vocational colleges have also included IoT-related courses to let the students to be IoT-literate as they are the future leaders in the industry. Malaysian construction industry needs continuous improvement for it to stand still in the worldwide competition (Ibrahim et al., 2021). IoT system is an efficient tool for gathering and transmitting data through interconnected hardware and software. This connectivity, facilitated by internet integration, enhances the construction processes at various stages, providing a technological edge (Gbadamosi et al., 2019).

The growth of IoT signifies a transformative era, enabling innovative solutions to assist construction players in adapting to evolving technologies and reshaping the construction industry (Halim et al., 2021). By implementing the IoT in the Malaysian construction industry, the construction players are transforming into the technological era. IoT has played a significant role in enhancing sustainability, particularly in the realm of smart buildings and smart homes (Beniwal & Singhrova, 2022; Fernandez, 2019; Jia et al., 2019; Oke et al., 2022). One of the well-known smart cities that have in Malaysia is the Forest City that are located in Johor Bahru. This Forest City has incorporated the IoT technologies which are the sensors like smart metres and building monitoring sensoring devices (Onag, 2019).

Types of IoT-Based Real-Time Monitoring Technologies used by Malaysian Contractors

Even though IoT has not been developed well in Malaysia, there are still some IoT technologies being used by a very small group (Ibrahim et al., 2021). According to Ibrahim et al (2021), the IoT devices that are widely used in the construction industry in Malaysia are the smart communication tools such as WhatsApp, Telegram, E-mail, Google Meet and Zoom Meeting. This medium gives the construction players an information sharing platform to update progress, exchange information and report problems (Assan, 2018; Ibrahim et al., 2021; Mahmud et al., 2018; Tze Qing, 2019). This type of communication can trigger a fast-decision-making process and actions may be taken immediately (Ibrahim et al., 2021).

In addition, other IoT tools that are mostly used by the construction players are the location services software. Project delays are sometimes happened due to the delays of the materials to the site. With IoT, the construction players may control the delivery of the construction materials and resources by using services like Global Positioning System (GPS), Waze and Google Map to ensure that the materials can be delivered on time. To use all these software, a smartphone equipped with the network connection will do where this will not cause any burdens for the contractors on the cost. They can select for the fastest travel route and escape from the road congestions (Ibrahim et al., 2021; Kang, 2022; Mahmud et al., 2018).

Besides, some large construction companies are using drones to monitor the construction site (Ibrahim et al., 2021; Tze Qing, 2019). It is mostly used to take photos for updating progress without going to the site to take the photos personally (Al-Amleh, 2020; Elghaish, Matarneh, et al., 2021; Mahmud et al., 2018). It is also be used by the engineers to survey the construction site topography for them to analyse the geotechnical elements. The small and medium-sized construction companies found it difficult to adopt this remote-operated device due to its high cost which will be a burden for the company. The high cost not only means the cost of the drone devices, but also the specialist's cost and perhaps the training fees. It is because a specialist is needed for the operation of drone so that the information that is shown is accurate and clear (Ibrahim et al., 2021).

Moreover, the Malaysian construction industry also started to adopt the sensor devices at the construction sites. The sensors are used to detect and gather the information on the usage of the construction equipment and machineries (Gamil et al., 2020; Ibrahim et al., 2021). It is also important for the construction site safety by monitoring and detecting the construction hazards at the construction site (Ibrahim et al., 2021).

Furthermore, radio frequency identification (RFID) is also being used in the Malaysian construction industry. Its main purpose is to track the materials on site (Kavinesh Radhakrishna et al., 2023). However, its adoption is still very limited because most of the construction players prefer to use their manual method to track the materials at the site rather to learn this new technology which may ease their work in the future (Ibrahim et al., 2021).

There are also some types of IoT-based real-time monitoring technologies that are introduced to the Malaysian construction industry but still have limited used by the contractors. They are 1) wearable devices which can detect the worker's movement and health conditions (Al-Amleh, 2020; Awolusi et al., 2018; Ibrahim et al., 2021; Kumar et al., 2019; Oke & Arowoiya, 2021; Tze Qing, 2019), 2) security cameras to ensure the safety and security at the construction site (Berawi et al., 2019), 3) conservation data analytical model (CDAM) used to predict the weather condition and dynamic climate changes (Ma et al., 2020), 4) document management system such as Oracle Aconex which mainly for the communication and data sharing (Hong & Mansor, 2023), 5) Building Information Modelling (BIM) software that is familiar for modelling, quantification and clash detection analysis (Hong & Mansor, 2023), 6) augmented reality (AR) to superimpose the images for the entire building (Hong & Mansor, 2023) and 7) cyber physical system (CPS) like ModianGo application which is applicable to site monitoring (Hong & Mansor, 2023).

Potential Improvements for Decision Making of a Contractor

A contractor has to make a lot of decisions at the construction site. In order for the contractors to decide and plan, they not only need the knowledge and previous experiences, but also sufficient data for them to decide for the next taken actions. IoT allows the contractors to make good decisions and optimise the performance at the construction site by providing real-time monitoring function (Gbadamosi et al., 2019; Ibrahim et al., 2021). Quick thinking and good decision making are very important for them to deliver a good project.

The manual means for monitoring and controlling the construction site are time-consuming, error prone, subjective and costly. This will result in too much information being slowed down in receiving and transferring. The decisions that are made are also ineffectively (Louis & Dunston, 2018). By using the IoT-based monitoring function, a large volume of data and field information during the whole project lifecycle can be generated automatically (Ibrahim et al., 2021; Louis & Dunston, 2018). This massive amount of data may be collected for the contractors to control the performance of the activities (Louis & Dunston, 2018).

The real-time monitoring of IoT allows the contractors to access the real-time data and reports at any time. The agentless real-time monitoring helps the contractors to save time to go to the site and reduce the numbers of site supervisors. With the sudden change notifications, alerts and warnings from the IoT, the contractors can be alerts to the sudden changes and react quickly. The IoT provides accurate predictions and root causes analysis for the contractors to plan for their work and solve the problems during the project (Aravind et al., 2021; Hong & Mansor, 2023). The functions of the real-time monitoring of IoT that can benefit the contractor have been summarised in Table 1.

Table 1. Functions of real-time monitoring of IoT

Functions	Description
Produce real-time data and reports automatically	Contractor can always access the up-to-date data to plan for their next action to minimise any fault occur at the construction site.
Allow agentless real-time monitoring	Contractor can reduce the frequency going to the site but at the same time knowing the conditions and status of the work progress at the construction site.
Provide alerts, warnings and notifications	When there are any things happen beyond the schedule or proposed programme or any abnormal things happen, the IoT can notify the contractor at the first time so that the contractor can take action immediately.
Potential to make accurate prediction	The contractor can use IoT such as RFID to predict the peaks and valleys in operation to ensure that the supply can meet the demand and also identify the potential delays.
Generate root cause analysis	The contractor can find out the main cause of the problem and solve them accordingly by using the IoT such as sensor devices.

The IoT benefits can be greatly utilised by the contractors on their decision-making in several areas, such as machineries and equipment, materials, labours, project sequences, waste management, risk management, productivity and quality of works, cost control and environmental protection (Al-Amleh, 2020; Awolusi et al., 2018; Du et al., 2020; Górecki & Núñez-Cacho, 2022; Mahmud et al., 2018; Yusof et al., 2021; Oke et al., 2022; Sandanayake et al., 2019). The application of these IoT will be explained in Table 2. A contractor has to make sure all the resources such as machineries, equipment, materials are in sufficient amount and of good condition, labours in good health and high productivity and also have a well-managed project flow to ensure a smooth project delivery. With the help of IoT, the contractors can minimise the waste, reduce the risks, increase the productivity and quality of works, prevent overspent project and produce a project harmless to environment (Hong & Mansor, 2023).

Table 2. Applications of IoT

Areas	Applications
Machineries and equipment	Contractor can know the schedule for the maintenance of the machineries and equipment. Information such as fuel consumption, temperature and the working-hours of the equipment can be monitored by the contractor. With IoT, predictive maintenance can be practiced.
Materials	Contractor can track the transportation and the delivery of the materials to ensure that the material can be sent to the site on time. Besides, the contractor can check the utilisation and leftover of the materials so that they can make sure that the supply of the materials on site are always enough. Y using RFID, the contractor can monitor on the quality of the materials.
Labours	The contractor can detect the location and the motion of the labours. The contractor can also monitor the number of workers at the site to arrange the manpower needed. Besides, the contractors can detect the health and safety of the workers to ensure they work in their best conditions.
Project sequences	With various data collected such as the physical work progress and financial progress, the contractor can know the progress and the status of the project. This can help them to well organise the project to make sure all the activities are done following the schedule.
Waste management	The data given by the IoT can help the contractor to analyse the source and the generation of the garbage in order to formulate best garbage collection route and time for the garbage collection to improve garbage collection efficiency.
Risk management	With the IoT, the contractor can reduce some of the projects risks especially the machineries and equipment failure, theft, etc.
Productivity and quality of works	By tracking the activities, the contractor can control the output of the project to make sure that the project achieves high productivity and high quality.
Cost control	The contractor can trace the incoming and outgoing of the cost of the project.
Environmental protection	The IoT can help to reduce the resource consumption, by monitoring the energy usage at the construction sites such as lighting, heating and cooling systems. The IoT can be integrated with the modular construction to lessen the carbon footprints, thus meeting the concepts and requirements of the green buildings.

Challenges in Adopting IoT

Although there are a lot of benefits of IoT towards the decision making of a contractor, Malaysia is still having the low adoption rate of IoT in construction industry. The low adoption rate of IoT not only because it is new to the Malaysian construction industry, but there are also other factors that cause the construction players to reject the adoption of IoT technologies on the execution of their projects (Ibrahim et al., 2021). Due to the challenges that can foresee and currently facing, the construction players are averse to change because they are fear to be fail provided that they have put a lot of efforts into the implementation especially their investment (Maqbool et al., 2023; Waqar et al., 2023). The challenges that prohibit the growth of IoT in the construction industry have been stated as follows: -

Security

Technical security or privacy issues are one of the significant challenges in adopting the IoT technologies (Aamer et al., 2021; Tze Qing, 2019). Due to the connectivity to the Internet, it is difficult to safeguard the connected devices and information on the IoT from the external attacks (Gamil et al., 2020). Security concerns arise if unauthorized access occurs, with the potential exploitation of personal information (Waqar et al., 2023). The security concern comes from the sensitivity of the information and the assets. The data from the sensors to the base station faces privacy and security issues because of the wireless platform (Jennifer S. Raj, 2020). There are many new nodes and IoT added to the network. This increases the opportunity and target for attacks (Ku, 2019; Selvaraj & Sundaravaradhan, 2020). Duncan Jefferies, a London journalist said that "the greater number of smart devices connected to the Internet, the more potential for a hacker to find a way to get into the sensitive systems" (Lee, n.d.). Since IoT has no standards and specific regulations, many small or medium scales and start-ups will invent into the IoT business without any security measures (Janssen et al., 2019).

Connectivity

The IoT system has a centralised server to authorise all the devices in a network (Jennifer S. Raj, 2020; Selvaraj & Sundaravaradhan, 2020). If there are billions of devices, the network will not be able to support and become a bottleneck (Ku, 2019).

Storage system

Due to the fact that many IoT devices are connected to a main server, the server needs to have a single storage system that has massive space to store and manage the large amount of data (Jennifer S. Raj, 2020; Selvaraj & Sundaravaradhan, 2020). This storage is also important for backup and recovery of necessary information in the future (Jennifer S. Raj, 2020).

Data Complexity

Most IoT integration will generate unstructured data (Ku, 2019; Weinberg, 2021). The IoT devices will collect as much information as possible and create complex databases, which is also known as big data. The required data is hard to be extracted due to the massive volume (Gamil et al., 2020). The data has to be carefully analysed for effective decision making (Weinberg, 2021). If the team members are not data literate, they will not be able to analyse the data correctly. The wrong data that are extracted will cause the inaccurate information to execute the certain function (Gamil et al., 2020). Hence, intelligent analysis is important to make use of all the information collected. Diomedes Kastanis, the head of IoT at Singtel, said that "the real challenge of IoT is to make sense of all the information that has been collected". It is pointless if the data is only for display purposes without any use for the decision making (Ku, 2019).

Internet Availability

Network is the key to monitor the IoT devices. The Internet is needed for the devices to transfer the data to the server (Tze Qing, 2019). Weak Internet connectivity will create the technical difficulties in incorporating the smart devices into the industry (Aamer et al., 2021; Sharma et al., 2020). The poor Internet connection will result in the interruption of the interoperability of the devices and the operator cannot receive the information (Ibrahim et al., 2021). In most of the construction sites, the Internet availability remains a challenge (Gamil et al., 2020).

Knowledge

IoT implementation is still at its early stages and there are some potential benefits that seem undiscovered by some companies (Gamil et al., 2020). Inadequacy of knowledge on the IoT technologies may inhibit the companies from trying to adopt the respective technologies. Besides, different opinions about the IoT from the stakeholders may lead to unsuccessful adoption of IoT due to the misperception and wrong expectations on the respective IoT technologies (Aamer et al., 2021). In addition, there are insufficient skills and expertise on the IoT implementation (Ibrahim et al., 2021; Janssen et al., 2019).

Cost

The implementation of the IoT can be expensive (Tze Qing, 2019). The creation of the network between the smart devices requires significant investment of the companies, especially the long-term investment (de Vass et al., 2021; Janssen et al., 2019). Besides, there are operational and maintenance costs for the IoT implementation (Aamer et al., 2021; Tze Qing, 2019). In some cases, the professional has to be hired to implement the IoT. Not only the professional fees, the training to impart knowledge to the workers is also needed and the training consumes lots of money. Furthermore, the implementation of IoT would increase the cost in terms of the energy where the energy consumption is high (Sharma et al., 2020). Hence, the company that intends to adopt the IoT is required to have extra funds because the implementation really burn money (Gamil et al., 2020).

Energy Consumption

IoT implementation has high power consumption. The growing demands of IoT technologies require higher energy to process (Selvaraj & Sundaravaradhan, 2020). It will increase the energy cost of the value chains (Sharma et al., 2020). IoT devices mostly use batteries to operate because there is no direct access to the power supply (Biňas, 2019). It is impractical to frequently change the batteries for thousands of IoT devices (Tze Qing, 2019).

METHODOLOGY

The research flowchart is illustrated in Figure 1. Literature review was carried out to provide background information on the IoT and to determine the challenges of IoT adoption among the Malaysian Contractors. By searching the online documents (journal articles, thesis, articles, book, etc.) with the keywords such as "Internet of Things", "IoT", "Construction Industry", "Malaysian Construction Industry", "Contractor", "Real-Time Monitoring", "Challenges", "Limitations" and "Barriers", information can be obtained as the literature review.

Qualitative research method had been chosen for conducting this study. Semi-structured interview was conducted because most of the contractors are not familiar with IoT, so the target respondents are small. Hence, this research is good to understand in depth instead of getting the data based on the quantity. Besides, IoT is still new to the Malaysian construction industry. In order to understand more about IoT, the researcher has to get more in-depth details. G7 contractors has been set as the target respondents because they have higher potential to implement the IoT technologies for the projects compared to other grades of contractors. Non-probability sampling with snowball method was selected. The respondents were asked to recommend someone who meet the criteria if any. This method helps the researcher to identify the respondents easily.

Before sending out the invitation or conducting the interviews, the questions have been developed as the preparation. Since this research are straightforward to the challenges, the questions are divided into two (2) sections. They are Section A: Demographic Information of the respondents, Section B: Challenges in Adopting the IoT-based Real-Time Monitoring Technologies from the perspective of contractors. Section A is aimed to gather the background information of the respondents to ensure that the respondents fulfil the respondents' requirements. Section B are the core section where it is used to achieve the objective of this research.

The invitation letters had been sent to a total of 750 G7 companies via email, LinkedIn, Facebook and WhatsApp from 13th March 2023 to 15th May 2023 to briefly explain the titles, objectives and requirements of the respondents. However, only 15 of them replied and they met the requirements. A formal email was sent to them to explain the details and objectives of the interview. There were only 4 respondents replied that they were willing to join the interview for the data collection. The semi-structured interviews were conducted where some questions were predetermined based on the findings from literature review. At last, only 3 respondents (coded as R1, R2 and R3 in this paper) provided information that are relevant whereas the other respondent stated that their company has familiar by using the IoT and can overcome the challenges given, so they did not face any challenges. The respondents were also asked to recommend other potential respondent but the respondents did not give any recommendations and some recommended those that the researcher has approached but failed to get their approval to be interviewed.

To analyse the data collected, narrative analysis was used. First, the verbal words and sentences from the interviews were transcribed into structured text with proper grammar written in English. Next, the data were sorted based on the required data to achieve the objectives of the interview. Codes were given to conclude the messages from the selected text. After the coding was done, all the codes were reviewed and group into a theme. The theme then be given a clear name to reflect the meaning of the text. Then, the data are reviewed to make sure the connection between the given data. The data are de-contextualised and re-contextualised and the process is called as data deciphering. The data are distinguished into smaller pieces and consolidated into larger pieces to show whether the connection of the data exists. After all the data were analysed, the results were presented in tables and graphical forms.

Figure 1. Research flowchart

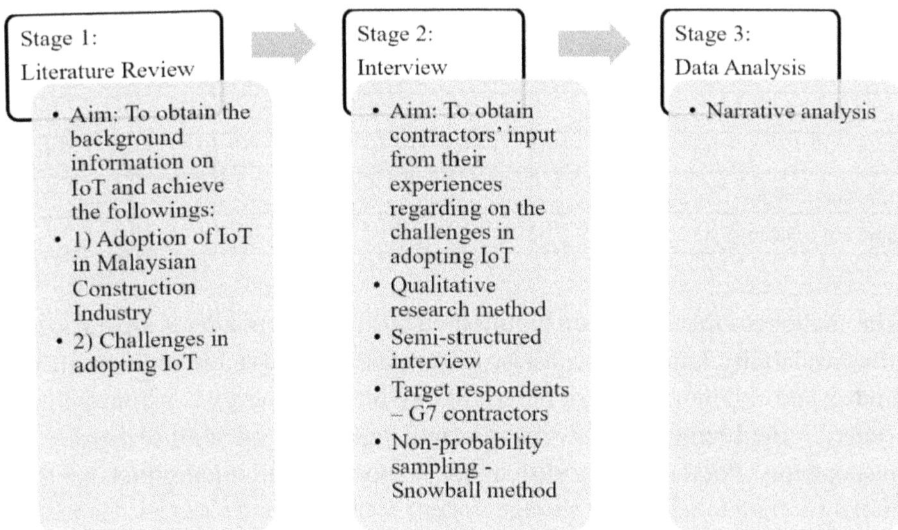

Fig. 1. Research Flowchart

FINDINGS AND DISCUSSION

According to Ibrahim et al. (2021), the IoT technologies have started to be implemented by the Malaysian construction industry but at a low adoption rate. On the route to find the respondents for the data collection, it can be noted that there is lack of contractor companies who know about the IoT in construction. From this research, if assuming that those do not reply to the email are those that are not adopting the IoT technologies at all, there are only 15 out of 750 companies adopting IoT, which is just about 20% of the G7 construction companies adopt the IoT technologies. It is just a small percentage and further increments are needed. According to both the primary and secondary data shown in Table 3, the challenges on the adoption of IoT-based real-time monitoring technologies among the Malaysian G7 contractors are the security concerns, connectivity problem, data complexity, Internet availability, knowledge towards IoT, cost, mindset and collaboration.

Table 3. Challenges of adoption of IoT-based real-time monitoring technologies

Challenges	Literature Review	R1	R2	R3
Security	/	/	/	/
Connectivity	/	/		
Storage system	/			
Data complexity	/			/
Internet availability	/	/	/	/
Knowledge	/	/	/	/
Cost	/	/	/	/
Energy consumption	/			
Mindset and collaboration		/	/	

The challenges that have been highlighted by all the respondents are the security, internet availability, knowledge, cost issue and another point stated additionally which is mindset and collaboration. For the storage system and energy consumption which are stated in the literature review as the challenges have been all disagreed by all the respondents. For the storage system, the respondents all highlighted that the IoT are mostly complete with a big storage system or using cloud services. Hence, there is no any issue regarding the storage. For the energy consumption, the respondents mentioned that the IoT can used renewable energy and rechargeable battery where this is something that can reduce their concerns on the energy consumption.

The first challenge is security concern. Two of the respondents (R1 & R2) strengthened that the contracts need security because the contracts are private and confidential. The issue that commonly happen by using IoT devices is the data breaches on the contract commercial. The contract documents for every project are deemed to be private and confidential unless it is shared publicly by the clients. All the data need to be protected where a really strong mobile or devices security protection has to be placed. Ku (2019) and Selvaraj & Sundaravaradhan (2020) highlighted that the implementation of IoT increases the risk of cyber attacks, making robust security measures essential. The other respondent (R3) also stated that some data are not so confident to give to the other people. There maybe lack of confidence between the parties. The implementation of IoT requires trust among the construction players that involved in the same project. It is because all the information is shared simultaneously with all the parties involved. All the parties should be able to access all the information through the IoT platform. Due to this fact, some parties are reluctant to the IoT because some of the information they are not dared to share. The respondents' opinions on this security concerns have been stated in Table 4.

Table 4. Respondents' opinions on security concern as challenge of IoT adoption

	Opinions on Security Concern
R1	"Security concern is like our contract commercial issue. It's all private and confidential. We have to protect all the data to prevent data breach by using IoT technologies."
R2	"Supposed that you have the digital network, basically you tend to attach to the cybersecurity issue, especially some contracts. It needs security. We need a really strong mobile security protection placed to prevent the data breach."
R3	"Most of these are cloud service. Meaning to begin with, the data security we are not so confident to give it to people, to give our data in hands of others, for example, knowing that they have access to 10 other construction firms' data. I think that's one of the major concerns that we need to get over."

All respondents (R1, R2 & R3) agreed that the Internet availability is a challenge when using the IoT-based real-time monitoring technologies. Tze Qing (2019) has also mentioned the need to have internet coverage for the device to transfer the data to server. Construction projects often take place in remote or challenging environments where reliable internet connectivity is not readily available. The respondent (R1) said that this is a big issue when comes to an infrastructure work like opening up a new road. There is usually no Internet connectivity over there. He also gave his opinions on the building projects. For the building projects, if it is in the very mature or developed and even developing township, then the Internet is available. However, if it is in a new or underdeveloped township, there might be no Internet coverage to that new area. They usually need to wait for the Telekom to put in a new lease line to pull for the high-speed Internet connectivity. At those rural sites, e.g. in the jungle, and some infrastructure works, a satellite GSM might be needed. Gamil et al. (2020) in their research also found internet availability remains a challenge in most of the construction site. Table 5 shows further opinion from the respondents on Internet availability problem, especially at the site.

Table 5. Respondents' opinions on internet availability issue as challenge of IoT adoption

	Opinions on Internet Availability Issue
R1	"If you are doing infrastructure, opening up a new road or new creature, there's no Internet connectivity, unless you are using a satellite GSM. So, we have to think about this problem." *Real Case Given: Waiting for the Internet line for one year.* *"I have waited for 1 year for the Telekom to put in a new lease line to pull inside my project for high-speed Internet connectivity."*
R2	"All these sound goods, provided we have all the supports. For example, we have the good Internet, with a coverage with a good internet speed and availability. Especially in some of those vibrable site, meaning sometimes the site require the satellite. We have projects also in the jungle, that's away from the GSM network, things like that. Even though the project is totally like domain, we require a satellite, especially in the infrastructure works in order to use all these IoT devices. Otherwise, we can't really stick to these."
R3	"Second, Internet connectivity is definitely an issue because in Malaysia construction most of the time happen only in developing new areas."

From the information given by the respondent (R3), in Singapore, the whole island has Internet connectivity, but in Malaysia, most of the time the Internet connectivity only happened in the developing areas. In the areas that do not support Internet connectivity or have the network connectivity issue, in order to install the sensors, additional boosters and receptors may be needed. The boosters and receptors might need to lay throughout the whole site to make sure the Internet connectivity can be reached. This will then lead to the high-cost implementation, which is also another challenge in adopting the IoT technologies. The respondents stressed on the cost issue a few times along the interview as shown in Table 6. It is because there are a lot of things need money when implementing a new IoT technology. For example, attending the relevant courses, purchasing the technologies, buying the licenses, hiring the specialists or consultants for the installation and operation, spending on the Internet and the lifecycle cost which includes the ongoing maintenance, updating to latest version, etc. (Sharma et al., 2020; Tze Qing, 2019). From the literature review, the cost also included the energy costs which is referred to the cost of the battery because most of the IoT are consuming the batteries (Sharma et al., 2020). However, this view of point is less agreed by the respondents where one of them stated that the energy is cheap and will not be a burden for them to adopt the IoT. He also stated that the energy used by the IoT can used the rechargeable batteries or the renewable energy. Some IoT that use the renewable energy are also available in the market.

Table 6. Respondents' opinions on cost issue as challenge of IoT adoption

	Opinions on Cost Issue
R1	"As a contractor, cost is very important to us...... If the cost is low, then of course we can consider it."
R2	"The cost is very important. The contractors are service providers. We have always been reminded on cost."
R3	"It's all high-cost implementation which needs a lot of business...... What needs to be done and how it needs to be done still requires an external consultant and in that sense, not only the complexity of hiring an external consultant, but the cost will also be high."

The next challenge that the respondents always strengthen is the knowledge and skills on the implementation of the IoT, which can be seen from Table 7. Knowledge is important for everything. Only with knowledge then we can proceed. A person has to be equipped with certain knowledge on the IoT in order to operate the IoT technologies. Without the knowledge, it is dangerous and useless for the person to utilise that IoT technologies (Aamer et al., 2021). IoT is risky for the external attacks, if it is utilised without certain knowledge, it will be easily to be exposed to attacks. The knowledge we can get it through reading, however we still need practical to develop the hard skills. The hard skills are also important for the IoT operation. If the company decided to implement IoT in the project, the parties that are involved are required to get the education and trainings on the relevant IoT technologies so that they can well implement and troubleshoot the IoT. According to Ibrahim et al. (2021), there are lack of workers that are equipped with the relevant skills and knowledge on the IoT. In other words, they are not aware of the IoT in construction and cause that the IoT technologies are hardly to be developed in the construction market due to the lack of the practical implementation by the construction players.

Table 7. Respondents' opinions on knowledge and skills issue as challenge of IoT adoption

	Opinions on Knowledge Issue
R1	"Inadequacy of knowledge is also a challenge. Contractors, we are engineers, we are laymen, sometimes we are not trained to these new types of technologies now. So, we have to educate everyone to use all these technologies to adapt and to accept all these technologies to help us in the future. Education is very important."
R2	"The industry needs the experienced people to compliment. We also need their knowledge."
R3	"As a human being, we can actually use our own instinct and knowledge to judge as in why is it that the system is proposing this method, and how alternatively could be done and if it done, what are the differences compared to the proposed system? Then we are able to make fact-based decisions if we are equipped with the knowledge, especially in the data analysis."

Besides the above-mentioned challenges, there are two respondents (R1 & R2) pointed out the same opinion on the mindset and collaboration as an additional challenge in adopting the IoT technologies. Table 8 shows their opinions on this challenge. In order to implement the new technologies like IoT, the firm or company needs people that have the mindset to appreciate the technologies and use the data to make decisions so that they can adapt and adopt those data driven IoT technologies. The construction players have to change their mindsets on whether they want to admit the technologies to the projects, thinking whether it is worth to spend the money to adopt the technologies, e.g. does it help to reduce the resources and speed up the progress, and whether they want to learn new things. From the respondent (R2), the older people usually will not use all these data, they preferred to go to the site to feel and look themselves. They have used to the traditional method and believe their experience are more reliable than the technologies. Due to their traditional mindset, they may see all these IoT technologies not so useful for them, so they are not willing to learn. Here comes to the importance of the collaboration. As we all know, there are different groups of people involved in one construction project, the management have to be collaborated in order to have good project delivery. If there are different groups of people using the different databases, then it will be a mess. Certain information may not be shared together. The respondent (R1) reviewed that in today's construction industry, most of the construction firms still operate in their traditional way because of their mindsets. They do not like to make this change since they have familiar with the traditional ways where they have worked for many years.

Table 8. Respondents' opinions on mindset and collaboration as challenge of IoT adoption

	Opinions on Knowledge Issue
R1	"To me, to adapt a new IoT depends on your mindset whether you want to admit it, whether you think it's worth to spend money to adapt the technology to help, to reduce your resources, to help to speed up your progress... Your mindset or whether you want to learn new things. Some of the people see that all these things are not useful, so no point for them on learning all these things. You have to change the mindset."
R2	"In order to appreciate the data, we need people to have a mindset to use data to make decision... suddenly it will have certain information that may not got to share. This grand of collaboration is an obstacle actually, meaning that the digital transformation, you have the data, you have the IoT, but this is the challenge because the collaboration is not there. This is still cooperated with the human mindset."

CONCLUSIONS

The objective, to investigate the challenges on the adoption of IoT-based real-time monitoring technologies among the Malaysian contractors, have been achieved through literature review and interview. Although there are a lot of benefits of IoT, especially towards the decision making of a contractor, Malaysia is still having the low adoption rate of IoT in construction industry because of the concerns on the security, connectivity problem of the IoT devices, the complexity of data from the IoT, the Internet availability for the IoT, knowledge towards IoT among the construction players, cost for the IoT implementation, the construction players' mindset and their collaboration. By successfully identifying the challenges on the IoT adoption, it is believed that the construction players can be aware of these challenges and find alternatives to solve it. This is not only to increase the awareness of the construction players, especially the contractors, but also to encourage the use of IoT in the Malaysian construction industry.

Throughout the process of conducting this research, some limitations have been faced. First, the researcher faces difficulty in finding the right respondents. Even though the research used the snowball method, the person suggested was not someone who fulfil the criteria of the respondents for this research. Besides, some respondents did not know anyone to recommend. As a result, the respondents for this research ended up with three. Hence, this little number of respondents cannot generalise the opinions of all the contractors in the Malaysian construction industry.

Some recommendations for future researchers are given below.

- Wider the target respondents. This research can be expanded to include other grades of contractors or the other construction players in Malaysian construction industry.

- Explore on the strategies to overcome the challenges faced in IoT adoption. In this research, the challenges on IoT adoption have been clearly identified. However, there are no strategies stated to overcome all these challenges. The future researchers can consider discovering the strategies to overcome these challenges so that the strategies can be well adopted by the construction players in order to increase the IoT adoption in the construction projects.
- Conduct case studies. The future researcher can conduct case studies to observe the practical implementation of IoT to better evaluate the challenges with its alternatives.

REFERENCES

Aamer, A. M., Al-Awlaqi, M. A., Affia, I., Arumsari, S., & Mandahawi, N. (2021). The internet of things in the food supply chain: Adoption challenges. *Benchmarking*, *28*(8), 2521–2541. DOI: 10.1108/BIJ-07-2020-0371

Aceyus Team. (2021, June 8). Four benefits of real-time monitoring in the modern contact center. Aceyus. https://www.aceyus.com/four-benefits-of-real-time-monitoring-in-the-modern-contact-center/

Al-Amleh, K. I. (2020). A study into the adoption of Internet of Things – IoT technologies within contractors in Dubai, United Arab Emirates [British University]. https://bspace.buid.ac.ae/bitstream/handle/1234/1811/20172453.pdf?sequence=3&isAllowed=y

Aleksandrova, E., Vinogradova, V., & Tokunova, G. (2019). Integration of digital technologies in the field of construction in the Russian Federation. *Engineering Management in Production and Services*, *11*(3), 40–49. DOI: 10.2478/emj-2019-0019

Aravind, R. Yadikiumarani, Meghna, K., R, D., & Naregowda, H. (2021). IoT based real time data monitoring for industry. Journal of Engineering Research & Technology (IJERT) NCCDS, 09(12). https://www.ijert.org/iot-based-real-time-data-monitoring-for-industry

Arowoiya, V. A., Oke, A. E., Aigbavboa, C. O., & Aliu, J. (2020). An appraisal of the adoption Internet of Things (IoT) elements for sustainable construction. *Journal of Engineering. Defence Technology*, *18*(5), 1193–1208. DOI: 10.1108/JEDT-10-2019-0270

Arslan, V., Ulubeyli, S., & Kazaz, A. (2019). The use of Internet of Things in the construction industry. https://www.researchgate.net/publication/337890965

Assan, L. (2018). *Potensi internet saling berhubung (internet of things) dalam industri pembinaan di Malaysia* [Ijazah Sarjana Muda Ukur Bahan, UNIVERSITI TEKNOLOGI MALAYSIA]. https://bic.utm.my/files/2018/08/LAROMI-ASSAN.pdf

Awolusi, I., Marks, E., & Hallowell, M. (2018). Wearable technology for personalized construction safety monitoring and trending: Review of applicable devices. *Automation in Construction*, *85*, 96–106. DOI: 10.1016/j.autcon.2017.10.010

Beniwal, G., & Singhrova, A. (2022). A systematic literature review on IoT gateways. *Journal of King Saud University. Computer and Information Sciences*, *34*(10), 9543–9563. DOI: 10.1016/j.jksuci.2021.11.007

Berawi, M. A., Sunardi, A., & Ichsan, M. (2019). Chief-Screen 1.0 as the Internet of Things platform in project monitoring & controlling to improve project schedule performance. *Procedia Computer Science, 161*, 1249–1257. DOI: 10.1016/j.procs.2019.11.239

Biňas, M. (2019). Advanced applications of IoT: low power. https://bletvaska.gitbooks.io/advanced-iot-applications/content/en/chapter-1/introduction.html

Chanal, P. M., & Kakkasageri, M. S. (2020). Security and privacy in IoT: A survey. *Wireless Personal Communications, 115*(2), 1667–1693. DOI: 10.1007/s11277-020-07649-9

Chen, J.-H., Ha, N. T. T., Tai, H.-W., & Chang, C.-A. (2020). The willingness to adopt the Internet of Things (IoT) conception in Taiwan's construction industry. *Journal of Civil Engineering and Management, 26*(6), 534–550. DOI: 10.3846/jcem.2020.12639

de Vass, T., Shee, H., & Miah, S. (2021). IoT in supply chain management: opportunities and challenges for businesses in early industry 4.0 context. *Operations and Supply Chain Management: An International Journal*, 148–161. DOI: 10.31387/oscm0450293

Du, L., Feng, Y., Lu, W., Kong, L., & Yang, Z. (2020). Evolutionary game analysis of stakeholders' decision-making behaviours in construction and demolition waste management. *Environmental Impact Assessment Review, 84*, 106428. DOI: 10.1016/j.eiar.2020.106408

Elghaish, F., Hosseini, M. R., Matarneh, S., Talebi, S., Wu, S., Martek, I., Poshdar, M., & Ghodrati, N. (2021). Blockchain and the 'Internet of Things' for the construction industry: Research trends and opportunities. *Automation in Construction, 132*, 104144. DOI: 10.1016/j.autcon.2021.103942

Elghaish, F., Matarneh, S., Talebi, S., Kagioglou, M., Hosseini, M. R., & Abrishami, S. (2021). Toward digitalization in the construction industry with immersive and drones technologies: A critical literature review. *Smart and Sustainable Built Environment, 10*(3), 347–373. DOI: 10.1108/SASBE-06-2020-0077

Fernandez, M. A. S. (2019). *Implementation of smart devices in the construction industry* [Faculty of Science and Engineering, University of Wolverhampton]. https://wlv.openrepository.com/bitstream/handle/2436/622965/Silverio_Fernandez_PhD_Thesis.pdf?sequence=1&isAllowed=y

Gamil, Y., Abdullah, A., Abd, M., Rahman, I., & Asad, M. M. (2020). Internet of things in construction industry revolution 4.0. *Journal of Engineering. Defence Technology*, *18*(5), 1091–1102. DOI: 10.1108/JEDT-06-2019-0164

Gbadamosi, A.-Q., Oyedele, L., Mahamadu, A.-M., Kusimo, H., & Olawale, O. (2019). The role of Internet of Things in delivering smart construction. In CIB World Building Congress. https://uwe-repository.worktribe.com/preview/1492592/Gbadamosi%20et

Ghosh, A., Edwards, D. J., & Hosseini, M. R. (2020). Patterns and trends in Internet of Things (IoT) research: Future applications in the construction industry. *Engineering, Construction, and Architectural Management*, *28*(2), 477–501. DOI: 10.1108/ECAM-04-2020-0271

Gillis, A. S. (2022, March). What is the Internet of Things (IoT)? Tech Target. https://www.techtarget.com/iotagenda/definition/Internet-of-Things-IoT

Górecki, J., & Núñez-Cacho, P. (2022). Decision-making problems in construction projects executed under the principles of sustainable development—Bridge construction case. *Applied Sciences (Basel, Switzerland)*, *12*(12), 6132. Advance online publication. DOI: 10.3390/app12126132

Halim, M. I., Rusuli, M. S. C., & Yaziz, M. F. A. (2021). Attitudes, awareness, readiness and barriers towards the Internet of Things adoption among the construction industry in East Coast Malaysia. *Journal of Sustainable Management Studies*, *2*(1), 7–15. http://myscholar.umk.edu.my/bitstream/123456789/2573/1/Che%20Rusuli%20et%20al._Published.pdf

Hassan, R., Qamar, F., Hasan, M. K., Aman, A. H. M., & Ahmed, A. S. (2020). Internet of Things and its applications: A comprehensive survey. *Symmetry*, *12*(10), 1674. DOI: 10.3390/sym12101674

Hong, C. C., & Mansor, N. S. (2023). Internet of Things (IoT): Real-time monitoring for decision making among the Malaysian contractors. *Journal of Advanced Research in Applied Sciences and Engineering Technology*, *32*(3), 455–470. DOI: 10.37934/araset.32.3.455470

Hou, J., & Li, B. (2020). The evolutionary game for collaborative innovation of the IoT industry under government leadership in China: An IoT infrastructure perspective. *Sustainability (Basel)*, *12*(9), 3750. DOI: 10.3390/su12093648

Ibrahim, F. S., Esa, M., & Rahman, A., R. (. (2021). The adoption of IoT in the Malaysian construction industry: Towards construction 4.0. *International Journal of Sustainable Construction Engineering and Technology*, *12*(1). Advance online publication. DOI: 10.30880/ijscet.2021.12.01.006

Janssen, M., Luthra, S., Mangla, S., Rana, N. P., & Dwivedi, Y. K. (2019). Challenges for adopting and implementing IoT in smart cities. *Internet Research*, *29*(6), 1589–1616. DOI: 10.1108/INTR-06-2018-0252

Jia, M., Komeily, A., Wang, Y., & Srinivasan, R. S. (2019). Adopting Internet of Things for the development of smart buildings: A review of enabling technologies and applications. *Automation in Construction*, *101*, 111–126. DOI: 10.1016/j.autcon.2019.01.023

Kang, K.-D. (2022). A review of efficient real-time decision making in the Internet of Things. *Technologies*, *10*(1), 12. DOI: 10.3390/technologies10010012

Kavinesh, R., Khairul Najmy, A. R., Alawiyah, A. W., Siti Julia, R., Hasliza, A. R., Lee, Y. S., Mohd Hafizi, O., & Khairul Affendi, R. (2023). Design of A 20-bit chipless RFID tag utilizing multiple resonators in UWB frequency range. *Journal of Advanced Research in Applied Sciences and Engineering Technology*, *29*(2), 25–38. DOI: 10.37934/araset.29.2.2537

Ku, M. (2019, December 27). Challenges and limitations of adopting IoT. IndSights Research. https://www.indsights.sg/industry-perspective/challenges-and-limitations-of-adopting-iot/

Kumar, S., Tiwari, P., & Zymbler, M. (2019). Internet of Things is a revolutionary approach for future technology enhancement: A review. *Journal of Big Data*, *6*(1), 111. DOI: 10.1186/s40537-019-0268-2

Lee. (n.d.). 3 challenges of IoT adoption and how enterprises can overcome them. Tech Genius. Retrieved January 14, 2023, from https://techgenies.com/3-challenges-of-iot-adoption-and-how-enterprises-can-overcome-them/

Louis, J., & Dunston, P. S. (2018). Integrating IoT into operational workflows for real-time and automated decision-making in repetitive construction operations. *Automation in Construction*, *94*, 317–327. DOI: 10.1016/j.autcon.2018.07.005

Ma, J., Yu, H., Xu, Y., & Deng, K. (2020). CDAM: Conservative data analytical model for dynamic climate information evaluation using intelligent IoT environment—An application perspective. *Computer Communications*, *150*, 177–184. DOI: 10.1016/j.comcom.2019.11.014

Mahmud, S. H., Assan, L., & Islam, R. (2018). Potentials of Internet of Things (IoT) in Malaysian Construction Industry. [AETiC]. *Annals of Emerging Technologies in Computing*, *2*(4), 46–52. DOI: 10.33166/AETiC.2018.04.004

Maqbool, R., Saiba, M. R., & Ashfaq, S. (2023). Emerging industry 4.0 and Internet of Things (IoT) technologies in the Ghanaian construction industry: Sustainability, implementation challenges, and benefits. *Environmental Science and Pollution Research International*, *30*(13), 37076–37091. DOI: 10.1007/s11356-022-24764-1 PMID: 36565423

Mohammed, B. H., Sallehuddin, H., Yadegaridehkordi, E., Safie Mohd Satar, N., Hussain, A. H., & Abdelghanymohamed, S. (2022). Nexus between Building Information Modeling and Internet of Things in the construction industries. *Applied Sciences (Basel, Switzerland)*, *12*(20), 10629. DOI: 10.3390/app122010629

Noura, M., Atiquzzaman, M., & Gaedke, M. (2019). Interoperability in Internet of Things: Taxonomies and open challenges. *Mobile Networks and Applications*, *24*(3), 796–809. DOI: 10.1007/s11036-018-1089-9

Oke, A. E., & Arowoiya, V. A. (2021). Evaluation of Internet of Things (IoT) application areas for sustainable construction. *Smart and Sustainable Built Environment*, *10*(3), 407–422. DOI: 10.1108/SASBE-11-2020-0167

Oke, A. E., Arowoiya, V. A., & Akomolafe, O. T. (2022). Influence of the Internet of Things' application on construction project performance. *International Journal of Construction Management*, *22*(13), 2517–2527. DOI: 10.1080/15623599.2020.1807731

Onag, G. (2019). Malaysia's Forest City bags IDC Smart City APAC Awards. https://futureiot.tech/malaysias-forest-city-bags-idc-smart-city-apac-awards

Raj, J. S.Jennifer S. Raj. (2020). A novel information processing in IoT based real time health care monitoring system. *Journal of Electronics and Informatics*, *2*(3), 188–196. DOI: 10.36548/jei.2020.3.006

Sandanayake, M., Zhang, G., & Setunge, S. (2019). Estimation of environmental emissions and impacts of building construction – A decision making tool for contractors. *Journal of Building Engineering*, *21*, 173–185. DOI: 10.1016/j.jobe.2018.10.023

Selvaraj, S., & Sundaravaradhan, S. (2020). Challenges and opportunities in IoT healthcare systems: A systematic review. *SN Applied Sciences*, *2*(1), 139. DOI: 10.1007/s42452-019-1925-y

Sharma, M., Joshi, S., Kannan, D., Govindan, K., Singh, R., & Purohit, H. C. (2020). Internet of Things (IoT) adoption barriers of smart cities' waste management: An Indian context. *Journal of Cleaner Production*, *270*, 122047. DOI: 10.1016/j.jclepro.2020.122047

Tunis, J. (2021, September 7). A detailed guide to real-time monitoring. Netreo. https://www.netreo.com/blog/real-time-monitoring/#:~:text=A%20Quick%20Definition%20of%20Real,applications%2C%20and%20services%20are%20performing

Tze Qing, A. (2019). Exploring the adoption of Internet of Things in Malaysian construction industry [Quantity Surveying, Universiti Tunku Abdul Rahman]. http://eprints.utar.edu.my/3384/1/QS-2019-1425152-1.pdf

Waqar, A., Khan, M. B., Shafiq, N., Skrzypkowski, K., Zagórski, K., & Zagórska, A. (2023). Assessment of challenges to the adoption of IOT for the safety management of small construction projects in Malaysia: Structural equation modeling approach. *Applied Sciences (Basel, Switzerland)*, *13*(5), 3340. DOI: 10.3390/app13053340

Weinberg, A. (2021, October 24). Top 4 challenges in IoT data collection and management. first point. https://www.firstpoint-mg.com/blog/top-4-challenges-in-iot-data-collection-and-management/

Yap, J. B. H., Chow, I. N., & Shavarebi, K. (2019). Criticality of construction industry problems in developing countries: Analyzing Malaysian projects. *Journal of Management Engineering*, *35*(5), 04019020. Advance online publication. DOI: 10.1061/(ASCE)ME.1943-5479.0000709

Yusof, A., Khoso, A. R., Sohu, S., Khahro, S. H., & Chai, C. S. (2021). Improving performance in construction projects: A case study of Malaysian public projects. *Pertanika Journal of Science & Technology*, *29*(4), 2579–2604. DOI: 10.47836/pjst.29.4.19

Chapter 6
Exploring New Frontiers: Design Thinking Guidelines for Virtual Built Environments Within the Metaverse

Erna Looi
https://orcid.org/0009-0007-8894-7387
Taylor's University, Malaysia

Sujatavani Gunasagaran
https://orcid.org/0000-0002-0377-2319
Taylor's University, Malaysia

TamilSalvi Mari
https://orcid.org/0000-0001-5442-3816
Taylor's University, Malaysia

ABSTRACT

With the emergence of the Metaverse, the concept of the built environment has transformed, presenting architects and designers with both new opportunities and challenges. Unlike physical spaces, the Metaverse offers freedom from physical constraints and provides immersive user experiences. This research explores how design thinking for virtual environments differs from established theories for physical and gaming spaces. It aims to develop a new set of design thinking guidelines for the Metaverse through literature review and one-on-one interviews with architects and designers working in virtual environments. The literature review informs the interview questions to ensure relevance, while the interviews uncover how practitioners approach designing in the Metaverse. Ultimately, this research seeks to bridge the gap between physical and virtual design thinking, ensuring architects remain relevant in the face of advancing technologies by equipping them with tools

DOI: 10.4018/979-8-3373-5278-7.ch006

tailored for designing in the Metaverse.

BACKGROUND

In the midst of the ever-evolving digital age, society is experiencing a profound transformation, transcending the boundaries of the physical world into uncharted territories of the digital realm. One of the most remarkable developments in this digital era is the meteoric rise of the Metaverse, a multifaceted, interconnected virtual universe where people interact, create, socialize, and even engage in economic activities. The development of the Metaverse, already underway before the COVID-19 pandemic, has accelerated exponentially in the wake of the global health crisis. With restrictions on physical interactions and a heightened reliance on digital platforms, the Metaverse has witnessed a surge in users and platforms. An intriguing aspect of this Metaverse development is the emergent trend of commissioning architects to design spaces within the virtual world (Finney. A., 2022).

Finney (2022) has mentioned a fascinating trend of architects being commissioned to design spaces within the virtual world. With the start of this, it could significantly alter how architects design as it challenges the traditional design thinking methods to be adapted towards the Metaverse. The exploration of this evolution of design thinking is paramount due to the diverse qualities between the physical built environment and virtual built environment. The Metaverse's unique qualities that differs from the physical world allows for boundless creative opportunities. Moneta (2020) agrees that the virtual environment requires and in-depth exploration for its own theory and rules due to its distinctive qualities. The influence of the Metaverse can't be avoided with the advancement of technology and time. In order to grow along with time, architects need to learn and design this unprecedented environment which requires a new set of design thinking.

As architects, we have been taught a relatively standardized set of design thinking procedures to follow when initiating the development of a design.Since the start of our architectural journey, we have been taught a fixed process when designing. Factors such as the site's physical characteristics, budgetary constraints, and compliance with building regulations and codes play pivotal roles in shaping the architectural design process. Bielefeld (2013) suggested that architecture design is portrayed as a reaction to the specific context in which it transforms into a tangible reality. It is shaped by ecstatic intuition, active spatial engagement and the transformation of an abstract concept into a tangible material form. In contrast to traditional design approaches for physical places, the ability to design the Metaverse require different methodologies due to their diverse and multifaceted methods of interactivity with the user within a virtual environment. Kim, Lee and Lee, (2017), discussed that unlike the physical reality, virtual landscapes in the context of digital games exhibit unique spatial qualities, such as distinct gravitational forces, varying dimensions and other unique environmental conditions.

While existing research demonstrates the prevalent trend of virtual world design mirroring the physical world. This belief is shared in a study by Kalay and Marx (2005) as they discussed that many 3D worlds often appear as caricatured imitations of real-world spaces, closely resembling physical environments. However, these digital realms frequently offer suboptimal conditions for human inhabitations. Most studies have been discussing on how to design for virtual environments that functions as a gaming experience but fail to acknowledge that the Metaverse is essentially a more expansive concept. The Metaverse aims to create a versatile and interconnected digital universe that serves various aspect of modern life besides gaming. In a study by Laeeq (2022), the Metaverse is understood as a concept of a fully immersive virtual world where individuals could come together to socialize, engage in recreational activities and conduct work-related tasks. It is crucial to understand the differences of user-interaction between gaming virtual environments versus the Metaverse as both presents different functions and problems.

The theories of "Design Thinking" proposed by Lawson (2006), explores the design process and methods that designers use when designing in reality or the physical world. From that, the gaming industry has also developed its personal set of design thinking rules that guides designers to design virtual environments that are engaging to gamers. This research aims to explore how designers and architects design within these two vastly different type of environments to create new knowledge of design thinking for built environments within the Metaverse which is a crossover of qualities between physical and gaming environments and more.

PROBLEM STATEMENT

Design Thinking has been long established in the built environments field as a tool for guiding the thought process in solving design problems. However, there is a notable gap seen for how the existing design thinking principles can be applied for designing built environments within the Metaverse as the qualities and challenges within the Metaverse greatly differ from physical built environments and digital gaming environments. The application of the existing principles to the context of built environments within the Metaverse remains underexplored due to it being recently developed, posing a need for research into this. The need to relook into how the traditional design thinking approaches can be altered to suit the unique qualities of the Metaverse is needed due to the many special qualities of the Metaverse that are newly being introduced to design like the need for user immersion, the lack of physical and structural constraints and more. The lack of physical and structural constraints within the Metaverse creates a whole new spectrum of design possibilities allowing architects to design spaces that go beyond things that are only achievable in the physical world, this understanding was also agreed in a study by Moneta (2020). This evolution in the built environment challenges the designers of the built environment to rethink how to design and how can they integrate user immersion and experience, spatial perception, and functionality into built environments within the Metaverse. Kim, Lee, and Lee (2017) mentioned that virtual environments are different from traditional environments due to how users interact as physical spaces are static whereas virtual environments are dynamic and interactive. The interactiveness of the Metaverse requires architects to pay greater attention to user interaction and immersion. The need to understand the users' behaviors and needs during design as emphasized by Lawson (2006) becomes even more critical in this context as the interaction and immersion of virtual spaces are the main drivers of how to retain users in order for the virtual environment to continue to stay relevant. In a recent study by Laeeq (2022), it was mentioned that the Metaverse has varied and evolving needs of users which justifies this research on how it requires a new approach to design thinking. The existing research of design thinking focuses more on the functional and aesthetic aspects of built environments without addressing the needs for user immersion, user interaction spatial perception and more which are topics that needs to be addressed in designing the Metaverse. The lack of a proper design thinking guideline for designing built environments within the Metaverse has led to these built environments not being designed to its full potential. This research aims to fill this existing gap by re-examining how traditional design thinking principles can be altered to suit the unique qualities of the Metaverse and also to generate new knowledge on new design thinking principles applicable to the Metaverse. With this exploration, valuable knowledge generated will provide architects and designers guidance on how to navigate the ever-changing nature of the virtual built environment within the Metaverse, ensuring that the design profession stays relevant along with the advancement of technology.

LITERATURE REVIEW

This literature review aims to explore the theoretical underpinnings and practical applications of design thinking in three distinct yet interconnected domains: physical built environments, digital gaming environments, and the Metaverse. This comprehensive review will identify key differences and similarities in design thinking across these contexts, providing a foundation for understanding how traditional architectural practices can be adapted for the Metaverse. By comparing and contrasting these areas, the review will highlight the unique challenges and opportunities presented by virtual spaces, offering valuable insights for architects and designers.

Metaverse

The covid-19 lockdown has evidently disrupted the physical interaction of human with spaces causing human activities to take a shift to the virtual realm. This has expanded the virtual environment market and introduced a whole new digital lifestyle for us humans. Shcumacher (2022) stated that this new digital lifestyle can't be erased and will continue to stay and develop. The development of the Metaverse has evolved from a mere concept to a real thing throughout the years. World building has seen in the early development of the Metaverse in the form of games such as The Sims and Second life. In the recent years, world building has been escalated to another level where users can actually purchase land and built their own buildings in virtual worlds such as Decentraland or The Sandbox. The figure below illustrates the evolution of the Metaverse throughout the years and how architecture can be seen being used in the Metaverse.

Figure 1. Timeline of evolution of the metaverse.

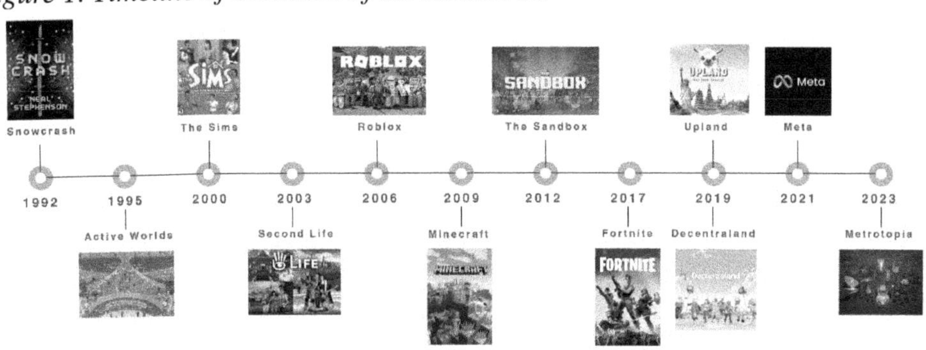

(Author, 2024)

Despite being in a completely different context from physical environments, virtual environments do share characteristics with physical environments as it is believed that fiction mimics reality. Shakeri and Ornek (2023) suggested that virtual place making is identical to physical place making because it has been built on communication networks between humans, their environments, social traditions, and other personal experiences. The most defining characteristic of virtual environments would be the addition of digital features to a space. In virtual environments, audio, graphics and digital user interface contributes to the place-making efforts of it. Piercy (2019) suggests that virtual places within games are constituted by three key components: social interactions, audio-visual elements and developer-driven communication. While physical built environments are known for their unchanging nature, virtual realms are commonly regarded as timeless, existing without a linear sense of chronology allowing for the buildings to constantly undergo change to adapt to latest trends or needs of the clients. Moneta (2020), suggested that in the virtual realm, buildings have the flexibility to undergo rapid changes and reconfigurations, functioning as interactive architectural structures that are responsive to user input.

Another key distinction between virtual environments and the physical world is the absence of a singular world. In the Metaverse, numerous worlds coexist, each with its unique context and characteristics. This multiplicity of worlds provides diverse and independent settings for virtual experiences and interactions Benedikt (1991b) had predicted that with the evolution of cyberspace, there will likely be another world of its own within the virtual realm that encompasses its own places, regions and regulations. Moreover, he also discussed the possibility of various types of cyber spaces to which each has its own culture, appearance and regulations. His predictions are now justified with the emergence of the Metaverse as we can now witness various types of virtual worlds like Decentraland, The sanbox and much more. These worlds have their own environments, uses their own currency and has their own unique set of rules and appearance.

Architecture in the Metaverse

Digital Architectonics was established by Professor Ludger Hovestadt in 2009. He proposed this to redefine the concept of architecture within the virtual world as an ever changing and dynamic subject instead of a static one in the physical reality. Architecture in the Metaverse is unorthodox with its own unique qualities that goes beyond what is usually achievable in the physical world. While the idea of virtual built environments seem to be unfamiliar to architects, it is actually not too far from what we have been doing since the start of one's architectural journey. Gerber and Götz (2019) have discussed that architects have long been exploring and crafting virtual spaces without physical limitations, which is the start of every project — the design process. The design process is where usually all the crazy and wild ideas take place. To define architecture in the Metaverse as architecture during the design process, does it suggest that structure, costs and functions can be overlooked? This allows for architects to create architecture that are more utopic in terms of form and spatial creation within the Metaverse.

The Metaverse has introduced new possibilities for spatial design and interaction especially for when one is not able to do such activities in the physical world. With the introduction of virtual built environments within the Metaverse, it allowed people to navigate virtual spaces and connect with one another despite being confined to their own homes. One example seen is Michael Beneville's studio in New York City. During the covid-19 pandemic, the physical studio became underutilized due to lockdowns. However, the studio still managed to operate without a problem in the virtual world of Vatom Spatial Web. The digital studio is a replica to its physical studio but with enhanced accessibility and adaptability (ArchDaily, 2022). With the virtual studio employees can work and interact with each other wherever and whenever.

The emergence of the Metaverse has displayed a growing trend of how daily life activities are being integrated into virtual environments. Spatial.io and Horizon worlds both allow for immersive experiences and virtual interactions (Bertelsen & Vines, 2023) in which you can witness how daily activities like group meetings, shopping and more are taking place in these virtual worlds. The introduction of virtual worlds emphasizes on how the Metaverse creates an endless amount of architectural possibilities. Architects should leverage the abilities of these virtual worlds and use them as supporting tools to create environments that are previously unimaginable.

Examples of Architecture in the Metaverse

To provide a deeper understanding of architecture within the Metaverse in Malaysian contect, a great example is the Sri Perpatih Palace by QWA architects (2022). The Sri Perpatih Palace is a digital nft Gallery designed by QWA for Sanuri Zulkefi to exhibit his NFT works. It is a virtual exhibition spaces and it demonstrates how traditional design thinking can be adapted to a digital context.

Figure 2. Screenshot of Sri Perpatih Palace from Qhawarizmi Architect.

(Qhawarizmi Architect, 2021, 1:05)

In terms of user-centric design, the Sri Perpatih palace has an interactive virtual built environment where users are able to interact with the digital artwork displays. This interactivity of the architecture enhances user engagement with the building and guides the navigation of users through spaces. The virtual space is also designed with the intention of integrating the artist's content into the space. The overall spatial layout complements Sanur's exhibition content. The Sri Perpatih palace also demonstrates a difference in the use of space as compared to physical built environments. The Sri Perpatih Palace's layout it dynamic and flexible allowing for the display of various contents paired with different types of user interactions. Lastly, it was clearly shown in the overall outlook of the building how the architect incorporations symbolism into the design. Traditional design elements are used in the design language to create a cohesive narrative for both the exhibition and the design.

Figure 3. Screenshot of Sri Perpatih Palace's interior art gallery display from Qhawarizmi Architect.

(Qhawarizmi Architect, 2021, 1:32)

However, this study also demonstrates the how architects have not leveraged the unique qualities of the Metaverse to its optimum. Most of the virtual built environment being designed within the Metaverse are still mostly being constrained to exhibition spaces. The possibilities of the spaces that can be designed within the Metaverse remains a mystery.

Figure 4. Masterplan of Metrotopia from Metrotopia.

(Metrotopia, 2024)

One of the more groundbreaking examples of architecture in the Metaverse in a global context is Metrotopia. Metrotopia is a joint Metaverse venture by Zaha Hadid Architects and ArchAgenda. Metrotopia aims to become a future virtual communication hub for the design community and a virtual city to create a wider design ecosystem. Zaha Hadid Architects (2023) explains that Metrotopia aims to bring together all design disciplines including urban design, architecture, interior, magazines, galleries and many more.

The first building within Metrotopia was designed by Zaha Hadid Architects. The virtual city will later be populated by more works by other architects and designers which creates various opportunities for the design industry to showcase their ability in the virtual universe. 'Knowledge Transfer' is a virtual exhibition introduced by Metrotopia during its debut at the 2023 Venice Architecture Biennale. The virtual exhibition allows visitors to immerse themselves in cutting-edge architectural explorations by renowned firms such as OMA, MAD, Sou Fujimoto and many more (Designboom, 2023).

Figure 5. Main Building of Metrotopia by ZHA from Metrotopia.

(Metrotopia, 2024)

Figure 6. Auditorium of Main Building of Metrotopia by ZHA from Metrotopia.

(Metrotopia, 2024)

The launch of Metrotopia plays a significant role for the architecture world. It demonstrates the ability of us architects to actively shape built environments whether physically or virtually. Metrotopia provides architects and designers a platform to expand and explore their design abilities unlocking potential.

Gaming in the Metaverse

Digital Architectonics had long made its presence known in virtual gaming environments. Several existing games like The Sims, Minecraft and Roblox are examples of how users are able to freely create virtual built environments within digital worlds. Analyzing these games reveals stark distinctions between constructing in the physical and digital realms. In the virtual world, the absence of gravity allows users to design freely, yielding more utopian and imaginative creations. Alvarez and Duarte (2021) agree that there is much architects can glean from video games, where serendipitous and non-linear storytelling techniques are employed to shape environments. City-like video games, in particular, often mimic the real world to leverage our mental associations. Architecture has perpetually been intertwined with the creation of imaginative visuals, translating conventional forms of virtuality such as drawings, books, and movies into tangible spaces in reality. Gerber and Götz (2019) explain that the virtual realms within games necessitate careful design, construction, narration, and infusion of action—a process akin to architects predicting how inhabitants will interact with architecture. If reality is a construct, with architecture as an integral part, then games' virtual spaces offer insights into architecture and its societal role.

Concept of Space and Place

While exploring the concept of space, it is understood that space defies simple categorization. Space is not merely an abstract idea or a physical entity; it encompasses conceptual depth, reality, and the diverse elements that unite life. Holgate and Baur (2011) discussed Kant's theory of space in which he stated that we experience space as defined by one's experience instead of the meaning that is already imposed. Kant (1781), in his metaphysical perspective, highlighted the profound epistemological insight that our experience of space transcends the confines of our limited encounters and that space is far more expansive than what we can grasp through our limited experience. The concept of space is still somewhat unclear and has no boundaries to this day. However, Bennedikt (1996b) stated that we can understand space in two ways: by considering how it feels and looks to us and by examining its objective functions and the operations it allows or restricts. Notably, various space theorists have highlighted the flexibility of the space concept, revealing its multifaceted functions and diverse interpretations for individuals. It revolves around the idea of an appearance (form) and its function, shaping human interaction and activities. The concept of space makes the study of both the physical and virtual environment relevant to this study as the different concepts of space result in different human experiences.

Universal Power of Design

Design plays a big role in our daily life. It wields a universal influence, transcending boundaries and languages. As humans, we rely strongly on our five senses to navigate through our daily life, and hence appearances play a big role in guiding our actions. Schumacher (2022) discussed that appearances play a crucial role in social function as they serve as essential guiding forms of communication. Other than that, the author mentioned that everything that surrounds us, meets our senses and, as interface, mediates our communications with the social world.

Design Thinking

Design thinking is a human-centered approach to problem-solving that emphasizes creativity, iteration, and user engagement. Bryan Lawson's "How Designers Think" provides a comprehensive overview of the cognitive processes involved in design, highlighting the importance of understanding users' needs, generating multiple solutions, and refining ideas through iterative prototyping and testing (Lawson, 2006). This methodology involves both divergent and convergent thinking, allowing designers to explore a wide range of possibilities before converging on the most viable solutions. Design thinking's human-centric approach has proven effective in addressing complex and ill-defined problems across various fields, including architecture, product design, and service design. Lawson (2006) also discussed that design thinking is guided by 7 main design thinking principles; content, client, users, practical, radical, formal and symbolic.

Table 1. Design Thinking Guiding Principles. (Lawson, 2006)

Design Thinking Principle	**Explanation**
Content	Refers to the specific subject matter or information that the design addresses. It involves the knowledge, data, and material relevant to the project.
Client	Focuses on the needs and requirements of the client or stakeholder. This principle ensures that designers are able to deliver designs that satisfy the client's brief.
Users	Focuses on the user's needs and how they interact with the space. It emphasizes on how design needs to address user needs and behaviors into its thought process in order to create user friendly spaces.
Practical	Refers to the practical aspects of design such as its efficiency, functionality and usability.
Radical	This principle discusses innovation and creative thinking. It encourages designers to create new ideas by pushing their boundaries in order to create ground-breaking design solutions.
Formal	This principle reinforced the need to incorporate formal elements into the design such as structural and visual elements.
Symbolic	This principle discusses how designers incorporate symbolic elements to evoke memory and emotions.

Recent studies have added on Lawson's design thinking theories, exploring how design thinking can be adapted to technological advancements and such. Smithe and White (2021) proposed that by adding sustainability into design thinking, it provides consideration to the lifespan of buildings and how the built environment affects the natural environment. Moreover, Thompson et al. (2020) discussed that advance and ai technologies can enhance design processes to create more environmentally friendly and responsive built environments.

Furthermore, Johansson-Sköldberg, Woodilla, and Çetinkaya (2013) argues that even though the existing design thinking principles has been proven to be effective in various design fields, design thinking in architecture specially for virtual built environments remains underexplored. Brown (2008) mentioned that design thinking drives innovation which suggests that with a clear design thinking guideline, it can guide architects and designers to discover groundbreaking design solutions for designing virtual built environments within the Metaverse.

Design Thinking in Physical Built Environments

Lawson (2006) discussed the importance of using empathy in design thinking to understand the needs and behaviors of users which plays a crucial role in designing effective user-centric physical built environments. Although visual aesthetics (content) plays a crucial role for design thinking, architects need to address both functional and aesthetic considerations and create spaces that are not only visually appealing but also respond to users' behaviors and needs.

Moreover, in terms of practical design thinking for physical built environments, integration of sustainability, materiality and technology into the design thinking is needed. Smith and White (2012) argued that buildings need to stay resilient which requires sustainable design. They also highlighted how smart technologies are used as data analytical tools to enhance the design process by running real-time simulations. Smart technologies enable architects to gain possible insights on how their designs could perform which allows for architects to design more responsive architecture that are able to adapt to changing user needs and environmental conditions. The role of technology in design thinking in this time reflects the ever-changing nature of the built environment and emphasizes on the importance for architects to be in line with technological advancements.

Brown and Dixon (2017) discussed how design thinking can be used to address social inequalities in a society. This suggests that architects can use design thinking to create more inclusive and accessible spaces that caters to a diverse base of users. Moreover, Oke, Aigbavboa, and Raphiri (2020) mentioned the importance of culture in design thinking which translates into symbolism in Lawson's words. They emphasized how architects should consider local traditions and cultural values in order to create spaces that fit into the context of the existing community on site. By understanding how the existing design principles are being applied to physical built environments, this review allows for the study to identify design thinking strategies that can be applied to design effective and engaging virtual built environments within the Metaverse.

Design Thinking in Digital Gaming Environments

Design thinking for digital gaming environments are similar to physical built environments but has its own set of unique challenges and qualities. Game designers are required to design immersive virtual worlds that are highly interactive for its users. In order to achieve this, they need to incorporate interactiveness, immersiveness and also story-telling elements into the overall gaming environment design to keep its users interested. Brown (2021) mentioned that design thinking is particularly relevant to gaming design due to its iterative nature. Game designers are required to continuously refine the gameplay elements based on the user feedback and tests to ensure the relevance of the game. Some of the key considerations for designing gaming environments are user engagement, the narrative behind, interactiveness of the environment and such which greatly differs from the static nature of physical built environments.

Recently, studies have explored the relationship of designing for gaming and architectural design, suggesting the potential for innovation in design thinking through this crossover. Kim and Lee (2022) suggested that design thinking principles from game design can potentially enhance architectural design in terms of creating immersive and interactive spaces. They discussed that game design thinking principles such as user engagement, interactivity and storytelling can enhance the design of physical and virtual built environments by making them more user-centered and engaging.

What's different from designing a physical built environment compared to digital built environments is their iterative nature. While iterative problem solving exists in designing a physical built environment, it only exists up until the building is constructed. Once the building is constructed, it is difficult for architects to modify the design due to practical and client constraints. Whereas in gaming environments design, it is an ongoing process. The iterative approach of gaming environment designs allows designers to experiment with different design solutions and even change them after they are constructed after obtaining feedback from user analytics. This constant iterative nature in designing virtual environments allows for architects to design environments that are highly tailored to suit its users. Schell (2020) highlights that gaming mechanics plays a role in shaping user behavior within the virtual space and enhances engagement. This suggests that gaming design principles can be applied to design virtual built environments within the Metaverse.

The intersection of design thinking principles being applied to gaming virtual environments and physical built environments is significant to design thinking for virtual built environments within the Metaverse. Virtual built environments within the Metaverse consists of a combination of qualities from both physical built environments and Digital Gaming Environments. By leveraging existing knowledge from these two environments, architects can create fully immersive and interactive virtual built environments within the Metaverse that are innovative and not constrained by physical limitations in terms of its visual content.

Table 2. Literature review findings on traditional design thinking, design thinking in physical built environments and virtual gaming environments. (Author, 2024)

Topic	Findings	Authors	Themes (Lawson,2006)
Traditional Design Thinking	Emphasizes creativity, iteration, and user engagement; involves both divergent and convergent thinking; effective in various fields.	Lawson (2006)	User, Content, Practical
	Expands to contemporary challenges such as sustainability and technological advancements.	Smith and White (2021), Thompson et al. (2020)	Practical, Radical, Client
	Critical review reveals impact on innovation and user-centered design; potential in architecture is still under-explored.	Johansson-Sköldberg, Woodilla, and Çetinkaya (2013)	User, Practical, Client
	Importance of design thinking in driving innovation; application can lead to groundbreaking solutions in physical and virtual environments.	Brown (2008)	Practical, Radical
Design Thinking in Physical Built Environments	Empathy is crucial for understanding user needs; involves functional and aesthetic considerations.	Lawson (2006)	User, Content
	Integration of sustainable design principles enhances long-term viability and resilience of buildings.	Smith and White (2021)	Practical, Client
	Smart technologies enhance design processes, creating more responsive and adaptive environments.	Thompson et al. (2020)	Practical, Radical
	Design thinking addresses social inequalities by creating inclusive and accessible spaces.	Brown and Dixon (2017)	User, Client
	Importance of cultural context in design thinking to create spaces that resonate with the community.	Oke, Aigbavboa, and Raphiri (2020)	User, Client, Symbolic

continued on following page

Table 2. Continued

Topic	Findings	Authors	Themes (Lawson,2006)
Design Thinking in Virtual Gaming Environments	Involves creating immersive virtual worlds with interactive and narrative elements; iterative nature aligns with design thinking principles.	Brown (2021)	User, Content, Practical
	Convergence of gaming and architectural design highlights potential for cross-disciplinary innovation.	Kim and Lee (2022)	Practical, Radical
	Game design techniques can enhance the design of physical and virtual environments, making them more engaging and user-centered.	Kim and Lee (2022)	User, Practical, Radical
	Iterative approach in game design allows continuous refinement and testing, improving user experience.	Flanagan (2009); Schell (2020)	User, Content, Practical

AIM AND OBJECTIVES

Ultimately, this research aims to conduct new knowledge to the built environment field by proposing a new design thinking guideline that integrates existing design thinking knowledge with new design thinking knowledge suited towards the unique qualities of the Metaverse to enrich the role of architects in designing spaces for the future. The subsequent research objectives have been formulated.

1. To explore on how traditional design thinking principles can be altered to suit the unique qualities of the Metaverse or whether new principles are required.
2. To identify key differences in design thinking principles across physical built environments, digital gaming environments, and the Metaverse.

To propose an integrated design thinking guideline that combines insights from physical and digital environments for application in the Metaverse.

RESEARCH METHODOLOGY

Data Collection Method

This research is a qualitative research as it focuses on using qualitative research tools to explore the new frontiers of design thinking for virtual built environments within the Metaverse. Due to the subjective nature of this research, qualitative methods are chosen specifically to tackle the emotional aspects of this research. The objective of this research is to explore the unique considerations and thought process one has to go through when designing the virtual built environment within the Metaverse and use the new knowledge captured to propose a design thinking guideline for built environments within the Metaverse.

The main methodology for this research is a combination of literature review and interviews. The literature review conducted allows the research to have a pre-established understanding of design thinking in order to ensure that the topics being explore in the interviews stays relevant to the study. Alamri (2019) discussed that the flexible nature of interviews as a qualitative tool for research allows the interviewees to express and discuss more of themselves and their experiences, which is the main requirement for this research. Design Thinking is a rather subjective topic as it requires one's personal interpretation of design to solve problems. Hence, due to the subjective nature of design thinking, semi-structured interviews are chosen as the main data collection method as it is the most suitable research tool to capture emotions and thoughts. The semi-structured interviews are conducted in a one-one-one manner and the interview questions are pre-defined from the theoretical framework to ensure that the questions are relevant to the research topic and to ensure that the interview is conducted comprehensively. This approach ensures that the interview sessions are conducted with a guided discussion yet allows for informal expressions and discussions that delves deeper into the topic, providing a deeper understanding of how they design built environments within the Metaverse.

Theoretical Framework

Following a comprehensive review of existing design thinking principles for both physical and digital built environments, a theoretical framework is created to guide the research. The theoretical framework illustrates the relationship between the existing theories to the research aim. By using this framework as a guideline, it allows this research to discover new knowledge and themes of design thinking for virtual built environments within the Metaverse with a structured approach. This creates a base for the development of a new set of design thinking guidelines for virtual built environments within the Metaverse.

Figure 7. Position of research among existing theories of design thinking.

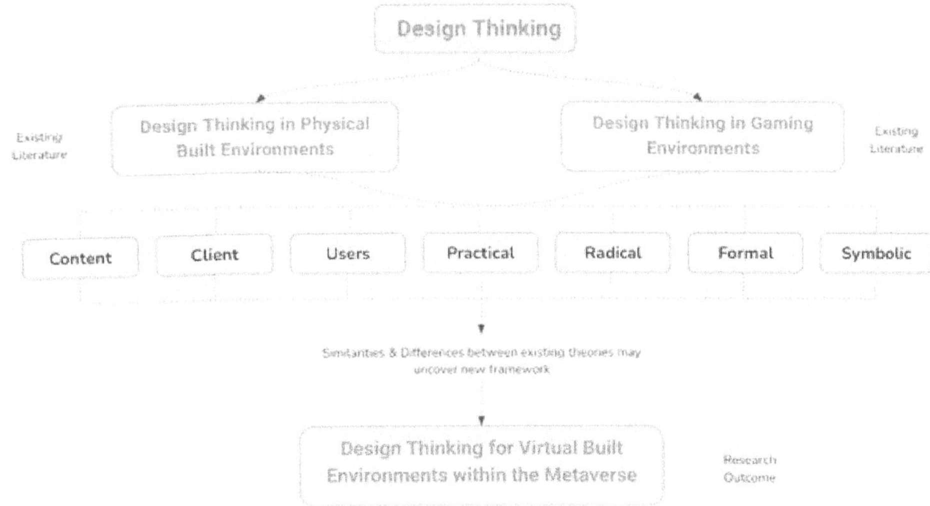

(Author, 2024)

Figure 8. Framework for uncovering design thinking themes for virtual environments within the metaverse.

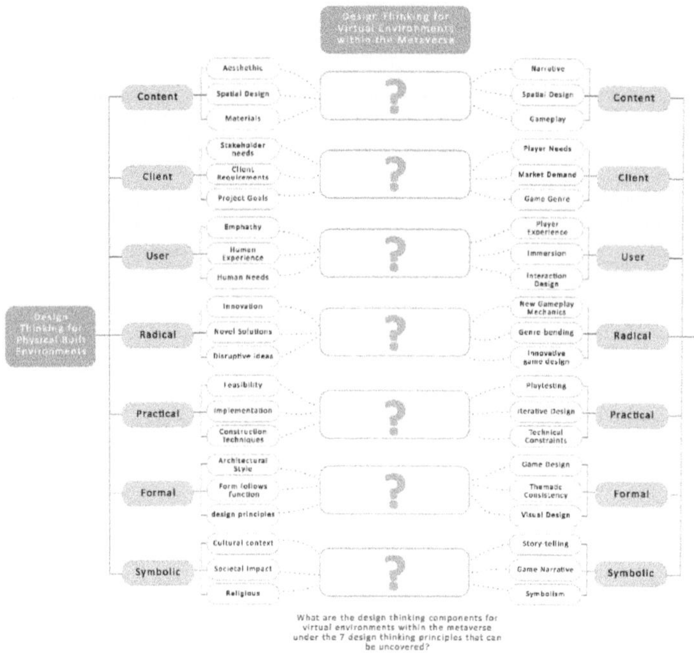

(Author, 2024)

Interview Questions

The predefined interview questions were designed to explore how architects design within the Metaverse with the primary focus of the 7 design thinking principles developed from the theoretical framework. The overall interview is semi-structured so it allows for the interview sessions to be flexible and allows for the participants to elaborate more on their personal experiences and expressions which could significantly benefit the data analysis process due to the subjective nature of this research. The interview questions have been refined after a pilot interview to provide a more direct approach for the research.

Table 3. Tabulation of interview questions based on objectives and research questions. (Author, 2024)

Research Questions	Design Thinking Principles	Interview Questions	Probing Questions
RQ1. How can traditional design thinking principles be adapted and applied to the unique characteristics of the Metaverse, considering its immersive nature, dynamic user interaction, and absence of physical constraints?	User, Content, Practical	**Q1:** How do you understand the needs and behaviors of users to design interactive and engaging spaces in the Metaverse?	• Can you share an experience where user feedback significantly altered the design?
	User, Content, Formal	**Q2:** How does the users' needs influence the content and formal design structure of designing environments in the Metaverse?	• How do you balance user-driven content with aesthetic coherence?
	Client, Practical	**Q3:** How do client expectations in the Metaverse differ from those in physical built environments?	• What are the practical considerations you need to prioritize to meet these expectations? • Can you provide a case where client requirements evolved during the design process? • How do you ensure the technical feasibility of these requests?
RQ2. What are the key differences in design thinking principles between traditional physical built environments, digital gaming environments, and the Metaverse?	User, Radical, Practical	**Q4:** What are the differences of your design process when designing in the Metaverse compared to designing physical built environments?	• How does the unique characteristics of the Metaverse influence your design approach as to the usual one (eg: site analysis, site response) ?
	User, Content, Client	**Q5:** What are the key differences of how users engage and interacts with built environments within the Metaverse compared to physical or gaming environments? How do these differences affect how you design?	• How do you address user needs and encourage user engagement with your design in a digital environment versus a physical environment? • How do you design spaces that provides dynamic user interactions in the Metaverse compared to static experiences in physical environments?

continued on following page

Table 3. Continued

Research Questions	Design Thinking Principles	Interview Questions	Probing Questions
	Radical, Formal, Practical	**Q6:** How does the flexibility of space and time in the Metaverse allow for flexibility in designing? (i.e., the absence of gravity, real-world physics, and physical limits)	• How does the lack of physical constraints and flexibility of time influence how you design in the Metaverse? • Do you think that your design are more interesting compared to physical built environments?
	Content, User, Symbolic	**Q7:** Do you think that narratives are important to designing built environments within the Metaverse? If yes, how do you translate your narratives into your Metaverse design?	• What role does a narrative play in a virtual world? How does having a narration influence how you design spaces, their functions and how users interact with it?
	Symbolic, Formal, Client	**Q8:** How do aesthetic choices in virtual environments like the Metaverse differ from those in physical architecture and gaming, especially when it comes to representing symbolic or cultural elements?	• How do you incorporate symbolism into designing digital built environments? Also, how do you ensure that since these elements maintain their symbolic meaning since they're no longer physical? Or is there now another opportunity to manipulate symbolism (eg: through sound effects) ? •
	Practical, Radical, Formal	**Q9:** How do you balance functional requirements (e.g., usability, navigation) with more fantastical or imaginative elements in the Metaverse, and how does this compare to physical and gaming environments?	• Does the Metaverse give you more freedom to emphasize creativity over functionality? How do you address practical needs (e.g., wayfinding) in the absence of physical limitations?

Data Analysis Method

The qualitative data analysis tool employed for this research is NVivo. Nvivo was used to aid in organizing the collected data by storing the transcripts and also for ease of coding. Bazeley and Jackson (2013) agreed that NVivo can enhance the dept and accuracy of a qualitative research as the tool is effective for coding data, illustrating the relationships within the data collected and also generating detailed reports.

RESULTS AND DISCUSSION

In the results and analysis section, information obtained derived from qualitative interviews are transcribed and meticulously coded using NVivo software. This analysis aims to explore how design thinking principles are applied within the Metaverse, focusing on the integration of content, client needs, user experiences, and innovative solutions. By systematically categorizing and interpreting the data, we reveal key themes and patterns that underscore the unique challenges and opportunities in Metaverse design.

> **Research Question 1:** How can traditional design thinking principles be adapted and applied to the unique characteristics of the Metaverse, considering its immersive nature, dynamic user interaction, and absence of physical constraints?
> Q1: How do you understand the needs and behaviors of users to design interactive and engaging spaces in the Metaverse?

The respondents all emphasized on the importance of understand the needs and behaviors of users during their design process. 3/5 of the respondents discussed how the best way to understand how users interact with their design was through empathy which is to use participatory methods and observation to gather feedback which is highly similar of how empathy works in designing physical built environments. Other than that, Respondent B and E mentioned they have used iterative feedback loops to refine their design until it resonated with its users. This creates a new opportunity for the design process as now designers are allowed to obtain real time and iterative feedback which allows for them to continuously improve or change their design as compared to a physical one that is unchangeable once built. All of the respondents have also agreed that user feedback plays a major role in contributing to their highly immersive and engaging virtual built environments within the Metaverse.

Themes uncovered:

- Empathy in design. (User)
- Iterative feedback loops. (Practical)
- Ever-changing design. (Practical, Content)
- Immersive and engaging design. (User, Content)

Q2: How does the users' needs influence the content and formal design structure of designing environments in the Metaverse?

A majority of the responses discussed how intuitive navigation is one of the key considerations when designing virtual built environments. They have used immersive and engaging elements to create an environment that allows for the users to navigate through seamlessly, balancing form and function. Respondent A mentioned that he incorporated traditional music into the background as well as traditional Malaysian patterns to create emotional connection between the users and the environment. Respondent C articulated how he used narrative elements to enhance the interactivity and engagement of virtual built environments, using content to drive a design. Respondent D talked about how the balance between user preferences and narrative elements are crucial to create content that remains relevant to time and engaging to its users. The designers all recognized that in the Metaverse, design needs to be flexible in content adaptation to ensure that it continually resonates with its users.

Themes uncovered:

- Narrative elements. (Content)
- Intuitive navigation. (Formal, Content)
- Immersive. (User, Content)
- Engaging. (User, Content)
- Emotional Connection. (Symbolic)
- Adaptive Content. (Content)
- Content driven design. (Content, Formal)

Q3: How do client expectations in the Metaverse differ from those in physical built environments?

The respondents have all agreed that there is a shift in client expectations for designing virtual built environments compared to physical built environments. For the Metaverse, clients have constantly emphasized the need for virtual built environments that are highly interactive as well as designs that are innovate that surpasses traditional architectural solutions. Respondent B stressed that educating the client on the abilities of the Metaverse plays a big role in managing client's expectations. It is important for the designers to educate the clients on the possibilities and limitations of the virtual built environment in order to create feasible designs. Educating the clients also manages their expectations in terms of investments as different degree of innovative designs might require for higher investments on technological software and hardware to produce them.

Themes uncovered:

- Interactivity. (Content, Client)
- Innovation. (Content, Client, Radical)
- Awareness. (Client)

- Feasibility. (Client)
- Investment. (Client)

Research Question 2: What are the key differences in design thinking principles between traditional physical built environments, digital gaming environments, and the Metaverse?

Q4: What are the differences of your design process when designing in the Metaverse compared to designing physical built environments?

The respondents have identified several differences in their design process for designing in the Metaverse compared to physical environments. Respondent C highlighted that they have took advantage of the iterative nature of the Metaverse to constant real-time testing which results in the opportunity to explore experimental designs. Respondent D responded that the design process in the Metaverse differs in which the overall process is more non-linear as compared to the usual design process for physical built environments that requires a fixed set of steps before proceeding to the actual design itself. Most of the respondents highlighted that for physical built environments, the creation of the space mostly depends on the input of the consultants and the client himself, whereas now, in the Metaverse, users have the ability to provide input to the design due to the ability of the Metaverse to conduct real-time feedbacks. This has resulted in a collaborative design process between the designer, client and users.

Themes uncovered:

- Iterative. (Practical)
- Collaborative. (Practical, User)
- Experimental design. (Radical)
- Non-linear design process. (Practical)

Q5: What are the key differences of how users engage and interacts with built environments within the Metaverse compared to physical or gaming environments? How do these differences affect how you design?

All respondents have agreed that the way users interact with physical built environment and digital built environment differs greatly in which now users play an actual role in shaping the built environment within the Metaverse through user feedback. Multiple respondents highlighted that immersion is a major element to consider when designing built environments within the Metaverse in order for the users to fully experience the virtual space itself. Interactive elements is also a recurring theme among the conversations as it plays a crucial role in obtaining user feedback.

Themes uncovered:

- Interactive elements. (User, Content)
- Immersion. (User)
- User Feedback. (Practical)

Q6: How does the flexibility of space and time in the Metaverse allow for flexibility in designing? (i.e., the absence of gravity, real-world physics, and physical limits)

Flexibility is another recurring theme in the discussions with the respondents. Respondent D mentioned that the absence of physical constraints has allowed her to create fluid and transformative designs that challenges traditional formal design thinking of static spaces. Respondent E stated that with the flexibility that the Metaverse has provided, environments can now be created to respond to user's interaction, resulting in captivating environments that engages with the users on multiple sensory levels.

Themes uncovered:

- Responsive environments. (Formal, Radical)
- Fluid and transformative (Formal)
- Flexibility. (Formal)

Q7: Do you think that narratives are important to designing built environments within the Metaverse? If yes, how do you translate your narratives into your Metaverse design?

All respondents agreed that narratives are important when designing virtual built environments within the Metaverse as they form the context of the design itself. They discussed that a narrative ties the whole thing together, creating a cohesive design and a narrative experience. Respondent D highlighted how she used visual metaphors to create a narrative through her design, translating abstract concepts into interactive and symbolic designs within the virtual space.

Themes uncovered:

- Cohesive design. (Content, Symbolic)
- Narrative experience. (Content, Symbolic)
- Visual metaphors. (Symbolic)

Q8: How do aesthetic choices in virtual environments like the Metaverse differ from those in physical architecture and gaming, especially when it comes to representing symbolic or cultural elements?

The inputs from several respondents pointed to a similar theme of multi-sensory design elements. The flexibility and physical constraint free nature of the Metaverse allows for the designers to incorporate multi-sensory design elements into their spatial design. This heightens the immersive user-experience and overall narrative of the design. Respondent A added that visual effects and animations adds layers of meaning to the aesthetics of a space, creating a cohesive experience throughout. Respondent D discussed how abstract interpretation creates opportunities for more fluid and dynamic designs.

Themes uncovered:

- Multi-sensory design elements. (Content, Symbolic)
- Abstract interpretation. (Symbolic)
- Visual effects & animations. (Content, Symbolic)

Q9: How do you balance functional requirements (e.g., usability, navigation) with more fantastical or imaginative elements in the Metaverse, and how does this compare to physical and gaming environments?

Respondent E stated that when designing for the functional part of the space, practical concerns such as spatial navigation must be integrated creatively into the design to create a seamless journey for the users. Functional creativity plays a huge role in connecting the users to the intangible space of the Metaverse. Respondent C mentioned that usability of the space is still one of the key design considerations needed to make the space accessible and engaging for users. The balance of usability with imaginative design features creates an environment that is both functional and imaginative.

Themes uncovered:

- Functional Creativity. (Formal, Practical)
- Usability. (Practical)

FINDINGS AND DISCUSSION

The aim of this research is to propose a design thinking guideline for designing built environments within the Metaverse to address the unique qualities of virtual built environments within the Meataverse. Through the comprehensive literature review conducted in Chapter 2.0, seven traditional thinking principles had been identified as follows:

Table 4. Design thinking principles and their explanation. (Author, 2024).

Design Thinking Principles	Explanation
Content	To curate contextually rich virtual built environments in the Metaverse to create a seamless and immersive experience for the users. This can be created through the integration of an overall theme, storytelling, cultural elements, and other visual effects into the design of the virtual built environment.
Client	To address the outcome that the client wants to achieve through the design. In the relatively new context of the Metaverse, it also means to educate the client on the possibilities of virtual spaces and the limits and constraints.
User	To design user-centric virtual built environments. Virtual built environments should be immersive, interactive and intuitive in order for users to fully enjoy themselves and continue to use it.
Radical	To leverage on the limitless ability of the Metaverse to explore architecture designs. The radical principle discusses how the Metaverse allows for architects and designers to create bold and experimental design, redefining the usual architecture norms.
Practical	Practical design in the Metaverse is defined by creating virtual environments that allows for users to navigate themselves with ease. The spatial layout and design should be designed to support intuitive navigation throughout the space with clear interfaces to enhance the usability and function of the virtual built environment.
Formal	To create designs that focuses on creating intuitive navigation through a seamless spatial design. To create flexible designs that are free from the physical constraints of traditional architecture through the limitless capabilities of the Metaverse.
Symbolic	To create designs that fosters emotional connection from users to the virtual built environment through incorporating cultural narratives or elements that are relevant to the brief, interactive elements and dynamic storytelling through spatial design.

Throughout the data collection, the study discovered that although the design thinking principles remains the same, emerging themes from each design thinking principle have been identified. This chapter discusses the relationship between newly identified themes with the traditional design thinking principles. The findings will then establish an integrated design thinking guideline the will provide architects and designers guidance on how to navigate the ever-changing nature of the built environment.

Relationship of Emerging Themes to Theoretical Framework

Upon uncovering the emerging themes from the interviews, the relationship between these themes and the existing design thinking principles has been identified. This analysis explores how the newly identified themes align with and enhance the seven design thinking principles established by Lawson. By examining these relationships, we gain a deeper understanding of how contemporary design practices in the Metaverse can be guided by and expand upon traditional design thinking frameworks. This integration of new insights with established principles provides a comprehensive approach to Metaverse design, ensuring that it remains user-centric, innovative, and practically feasible.

Role of Content in Design Thinking of Built Environments within the Metaverse

Figure 9. Themes uncovered under content design thinking principle.

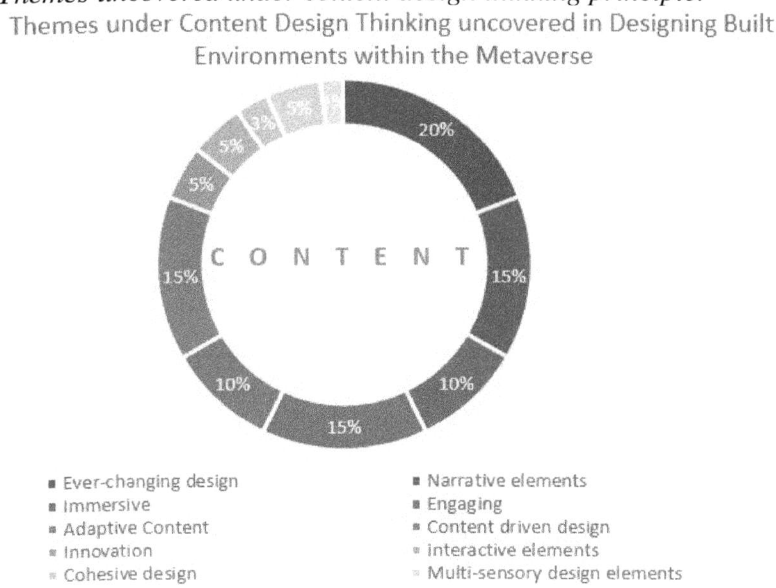

(Author, 2024)

Several new themes have been identified under the Content design thinking principle in the context of the Metaverse. One of the newly discovered theme for designing built environments within the Metaverse is to have an ever-changing design. The said theme is relatively new for designing built environments as now the Metaverse allows architects and designers to have the ability to incorporate adaptive content in the design of the spaces. The design elements can be used to enhance the dynamism of the spaces within the Metaverse and respond to the specific needs and preferences of the users. The adaptive design of spaces will provide users with fresh and exciting experiences upon every interaction.

Without the physicality of space in the Metaverse, architects and designers are required to use narrative elements in their design to create a cohesive journey for the virtual built environment. These design narratives are required in order to guide user through the space, creating a more immersive environment and to ensure that each design element aligns with the overall design and aesthetic of the space. Content-driven design also plays a huge part in creating an immersive environment. Through a well curated or themed design, architects and designers are able to curate a unified thematic experience for its users.

Multi-sensory design elements also enhances immersion. The ability of the Metaverse to have visual effects and animations allows for architects and designers to design spaces in a way that connects the users to the virtual built environment through various different sensory levels. Visuals, sounds, and animations, add an additional dimension to the spatial experience of the built environment within the Metaverse, enhancing the user's experience within the space. These elements are required in the design to serve as a guiding tool for users to navigate through the space as well.

Lastly, it is important for architects and designers to prioritize innovation when designing for the Metaverse. With the ability of the Metaverse to have interactive elements, built environments within the Metaverse can now be designed to allow users to navigate their own surroundings with the use of design elements. Through allowing users to interact with the virtual space, it creates an overall more immersive experience for its users despite the lack of physicality of space.

The key conclusion for the emerging themes under Content shows that architects and designers should design with adaptability, engagement and sensory richness in mind while designing built environments within the Metaverse.

Role of Client in Design Thinking of Built Environments within the Metaverse

Figure 10. Themes uncovered under client design thinking principle.

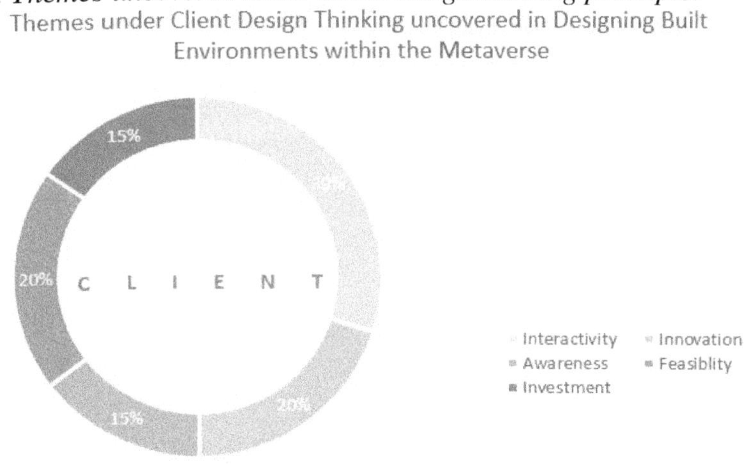

(Author, 2024)

In terms of the Client design thinking principle, the themes that falls under differs significantly. Client requirements for traditional built environments have been constantly focused around aesthetics, cost, and function. In the Metaverse, due to it being relatively new, it is important for architects and designers to actually educate and inform client on a project's feasibility and the investment required. Cost considerations on manpower and materials in traditional built environments have evolved to the client's willingness in invest on technological software and hardware to achieve the outcome they want. Due to the Metaverse's unlimited capabilities, clients might expect for architects and designers to design extravagant and high quality spaces without the consideration of the limitation in terms of software and hardware. It is the architect and designer's responsibility to manage the client's expectation through educating them and creating awareness on the requirements needed to create such designs.

Moreover, another new theme that emerged under Client is interactivity. Clients who engage architects and designers to design built environments within the Metaverse often prioritize interactivity of the space. Architects and designers are expected to design spaces that are highly innovative with design elements that surpass traditional architectural solutions to enhance the interactivity of the space. Through a highly interactive space, only then users will be attracted to visit the space. These findings demonstrates the evolving client requirements for designing the Metaverse which requires a new design approach to adapt to the evolving needs and preferences of users in the Metaverse as well as the capability of emerging technologies.

Role of User in Design Thinking of Built Environments within the Metaverse

Figure 11. Themes uncovered under user design thinking principle.

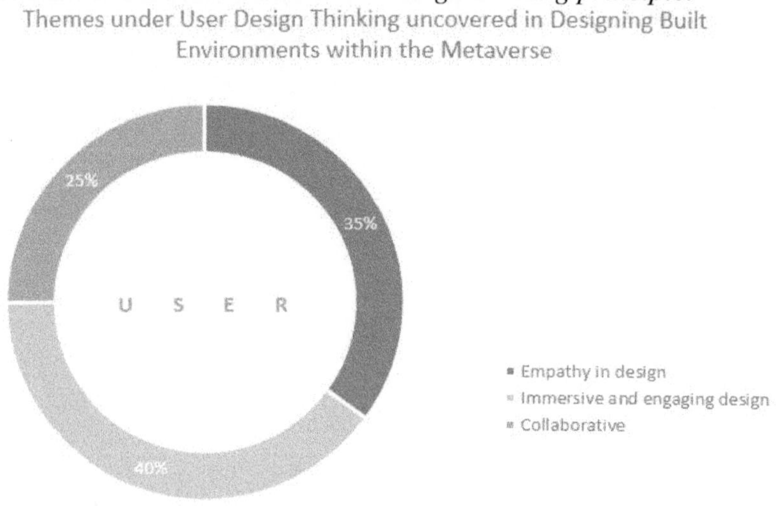

(Author, 2024)

The User design principle in the context of the Metaverse is similar to the traditional themes as it discusses the importance of empathy in design. Architects and designers are required to understand the needs and behaviors of users when designing for the Metaverse. In order to achieve empathy in designing for the Metaverse, architects and designs can take advantage of the technology by using real-time user observation as well as real-time feedback data from the built environments within the Metaverse to craft their design.

In the Metaverse, instead of needing a tangible space like physical built environments, users now require for highly immersive and engaging design. The spaces within the Metaverse should be designed to allow the users to immerse themselves as if they're part of the digital space to make up for the absence of tangible cues. By designing spaces that are highly immersive and engaging, architects and designers will be able to keep the users engaged with the digital space.

Lastly, collaborative design has emerged under the User principle for designing the Metaverse. Usually, when designing a physical building, collaboration happens usually in the form of input from clients during the design proposal or schematic design phase. In the context of the Metaverse, collaborative means that users actually get to play a part in modifying the digital built environment even after it is built. The co-creation process between the architect and the user goes beyond passive interaction. Architects or designers can now provide interactive features and thematic elements that allows users to customize the digital space according to their own preferences. For example, the digital space allows for users to customize wall colors, lighting tones, background music, and more. Through this personalization, a sense of ownership is fostered among the users which also enhances their experience of being in the digital space.

Role of Radical in Design Thinking of Built Environments within the Metaverse

Figure 12. Themes uncovered under radical design thinking principle.

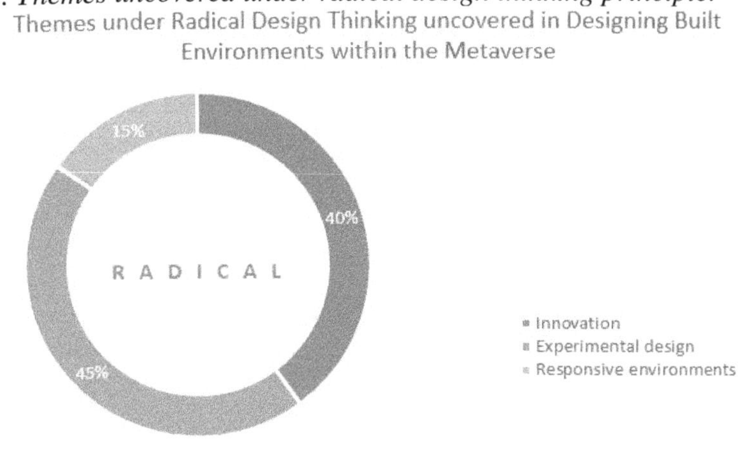

(Author, 2024)

The Radical design principle in the context of the Metaverse emphasizes on Innovation, experimental design and responsive environments. The architects discussed that the Metaverse allows for experimental design that goes beyond the physical constraints of a physical built environment. Architects and designers are now able to design unprecedented forms and structures without the physical limitations of real life. The lack of physicality also allows for architects and designers to constantly experiment with innovative form making without worrying about having to rebuilt or demolish the space.

In terms of responsive environments, the Metaverse provides real-time adjustments that responds to user actions. This feature enriches the design thinking process of architects and designers by constantly providing opportunities for them to obtain feedback to brainstorm and improve their design. It also encourages for the architects and designers to continuously innovate within the digital landscape.

Role of Practical in Design Thinking of Built Environments within the Metaverse

Figure 13. Themes uncovered under practical design thinking principle.

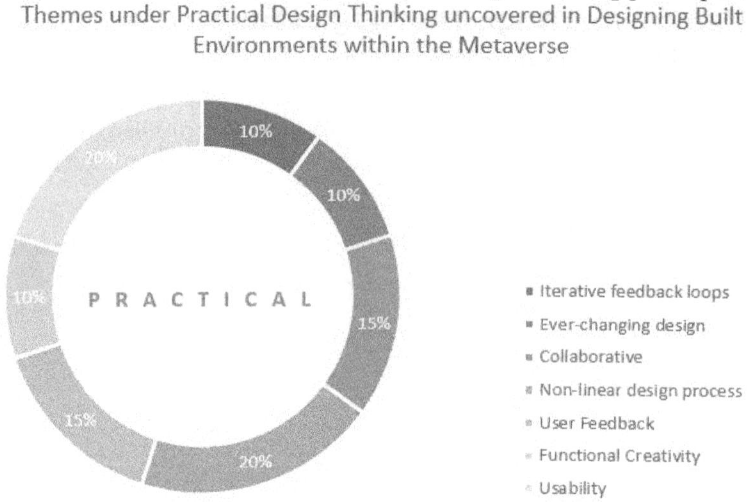

(Author, 2024)

The role of the Practical design principle in the Metaverse is multifaceted. The findings discovered that most of the themes uncovered under practicality are new themes that have not been used when designing for physical built environments. The allowance of iterative feedback loops with technology allows for the design to constantly go through constant updates and refinements according to the information obtained from real-time user feedback. This allows for the architects and designers to create ever-changing designs. This continuous cycle of designing and refining also suggests that the design process when designing for built environments within the Metaverse is a non-linear design process as compared to when designing for physical built environments which follows a somewhat fixed procedure.

When designing for the Metaverse, it is important for achitects and designers to balance functional creativity and usability. Due to its intangible nature, built environments within the Metaverse requires to be designed with practical solutions that creates intuitive environments for users to engage with. While architects and designers are free to explore imaginative and fanciful designs, they should always keep in mind that the spatial design of the digital space should be intuitive to its users.

Role of Formal in Design Thinking of Built Environments within the Metaverse

Figure 14. Themes uncovered under formal design thinking principle.

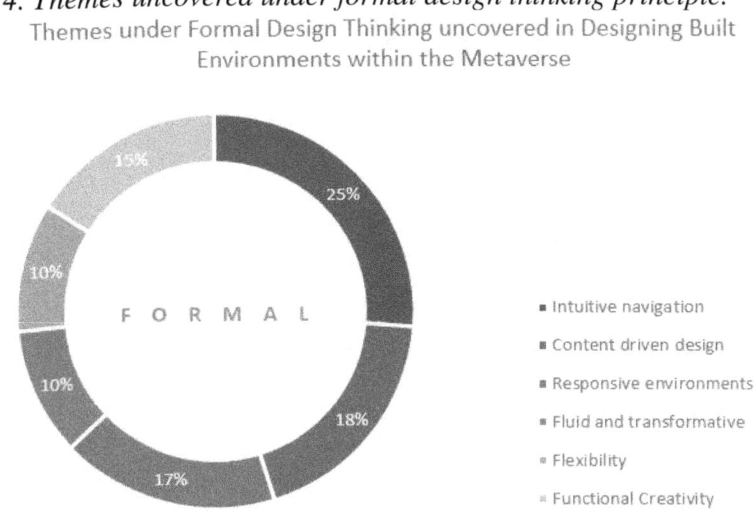

(Author, 2024)

The Formal design thinking principle in the Metaverse is defined by designing spaces that provides a sense of intuitive nagivation for its users as well as responsive environments. Architects and designers are required to design spaces with elements that provides intuitive guidance to its users and spaces that feel natural and accessible.

Moreover, in the Metaverse, it is important to design spaces that are fluid and flexible. The fluidity and flexibility themes highly contrasts with traditional architecture where form is usually controlled by structural requirements, building laws, material costs and other physical constraints. Due to the absence of such constraints in the Metaverse, architects and designers are now able to design with greater creative freedom transforming the Formal design thinking principle. Instead of rigid structural forms that creates order and scale, formal design in the Metaverse can be understood as designing spaces that flows seamlessly and cohesively, incorporating functional creativity into their design similar to the Practical Design thinking. Users should be able to navigate themselves through the space naturally.

Role of Symbolic in Design Thinking of Built Environments within the Metaverse

Figure 15. Themes uncovered under symbolic design thinking principle.

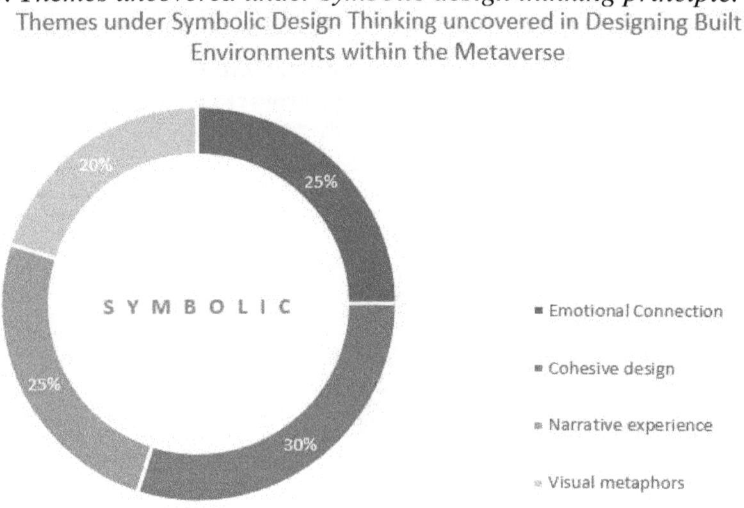

(Author, 2024)

The Symbolic design thinking principle in the Metaverse emphasizes the importance for architects and designers to create emotional connections through narrative experiences. The findings has suggested that much like traditional architecture, architects and designers can include visual metaphors that resonates with its target user into the spatial design. Culturally relevant symbols and patterns can be incorporated with the spatial design to deepen user engagement, creating a cohesive design that aligns with the overall narrative of the design.

In traditional architecture, symbolism is often represented in the form of materiality or contextual references. While the Metaverse allows for the same, it also allows for higher dynamism. Architects and designers can now create interactive functions within the space like added sound effects and animations to strengthen symbolism. The ability of the Metaverse to create multidimensional digital built environments redefines the idea of symbolism in architecture. Spaces can now be designed by blending architectural elements with virtual expressiveness which creates a layered storytelling for the users as they navigate through the space, increasing emotional engagement.

CONCLUSION

The study has led to a proposed design thinking guideline for designing built environments within the Metaverse. Through a comprehensive literature review, traditional themes of design thinking have been uncovered to act as a base guidance for this research. Through the qualitative data gathered from this study, it shows that the main design thinking principles remains the same whereas the sub-themes within have transformed to adapt to the new qualities of the Metaverse. The proposed guideline highlights the possible aspects for architects and designers to consider under each design thinking principle.

The proposed guideline shows a clear relationship of between the 3 different types of design thinking among 3 different types of built environment. With the basis of design thinking for physical built environments and virtual environments, the guideline for design thinking for the Metaverse becomes more informed with its past precedents combined with the new data collected through this study. When designing built environments within the Metaverse, architects and designers should combine aspects of physical and digital design. This is to craft spaces that are engaging and immersive to its users which in return responds to the diverse needs of digital users in a digital and virtual space.

This study lays the groundwork for a new architectural paradigm in the hopes to bridge the gap between traditional design thinking and the boundless capabilities of the Metaverse. As the Metaverse continues to grow, the knowledge from this study acts as a tool for architects and designers to use and develop in this new and growing virtual era. This ensures that the architecture profession remains relevant in both the physical and virtual landscape, creating more opportunities for architects in the future. This guideline also hopes to aid architects and designers in the creation of groundbreaking multidimensional designs of built environments within the Metaverse.

Figure 16. Proposed design thinking guideline for built environments within the metaverse.

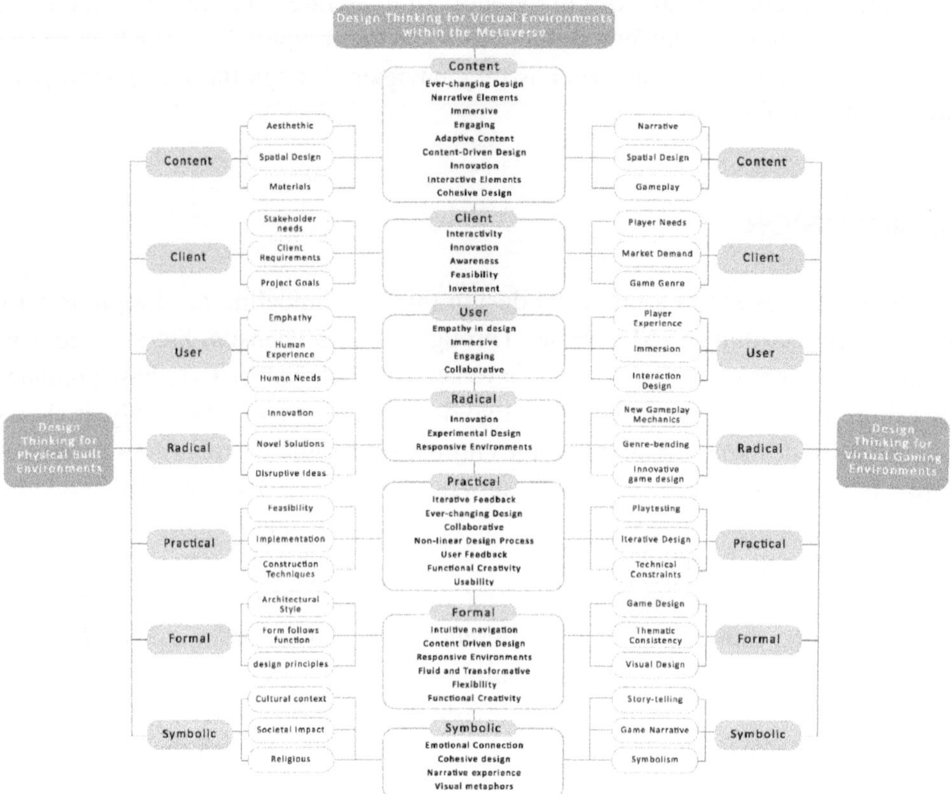

(Author, 2024)

ACKNOWLEDGMENT

The researchers of this article would like to thank the journal's editors and the anonymous reviewers for their positive feedback and suggestions for the improvement of this journal.

This research received no specific grant from any funding agency in the public, commercial, or not-for-profit sectors.

REFERENCES

Al-Qawasmi, J., & Hadjri, K. (2007). Architecture in the digital age: The effect of digital media on design, production and evaluation of the built environment. *Open House International.*

Alamri, A. W. (2019). *Effectiveness of qualitative research methods: Interviews and diaries.* University of Jeddah.

Alexenberg, M. (2011). *The future of art in a post-digital age.* Intellect Books. DOI: 10.2307/j.ctv36xvpkc

Alvarez, R., & Duarte, F. (2021). *Urban play: Make believe, technology, and urban space.* The MIT Press.

Architects, Q. W. A. (2022). Sri Peraptih Palace: An exhibition space in the metaverse. Retrieved from www.qwaarchitects.com

Bazeley, P., & Jackson, K. (2013). *Qualitative data analysis with NVivo* (2nd ed.). SAGE Publications.

Benedikt, M. (1991). *Cyberspace: First steps* (Benedikt, M., Ed.). MIT Press.

Bertelsen, O. W., & Vines, J. (2023). Virtual architecture: Designing for the metaverse. *International Journal of Architectural Computing, 21*(1), 45–60.

Bielefeld, B. (2013). *Basics: Architectural design.* Birkhäuser.

Brown, A., & Dixon, M. (2017). Design thinking for social equity. *Journal of Architectural Education.*

Brown, T. (2008). Design thinking. *Harvard Business Review.* PMID: 18605031

Browne, J., & Russell, S. (2003). The significance of sample size in qualitative research. *Qualitative Inquiry, 9*(2), 129–148.

Canter, D. (1977). *The psychology of place*. The Architectural Press.

Charmaz, K. (2014). *Constructing grounded theory* (2nd ed.). SAGE Publications.

Corbin, J., & Strauss, A. (2008). *Basics of qualitative research: Techniques and procedures for developing grounded theory* (3rd ed.). SAGE Publications. DOI: 10.4135/9781452230153

Designboom. (2023). *Zaha Hadid explores the future of architecture in the metaverse with Metrotopia venture*. Retrieved from https://www.designboom.com/architecture/zaha-hadid-future-architecture-metaverse-metrotopia-venture-06-01-2023/

Ewing, R., & Thomas, R. (2021). Architectural innovation in augmented and virtual realities. *Journal of Digital Architecture and Design*, *12*(3), 78–92.

Falahat, M. (2006). Sense of place and the factors shaping it. *HONAR-HA-YE-ZIBA*, *26*, 57–66.

Flanagan, M. (2009). *Critical play: Radical game design*. MIT Press. DOI: 10.7551/mitpress/7678.001.0001

Flick, U. (2014). *An introduction to qualitative research* (5th ed.). SAGE Publications.

Friese, S. (2019). *Qualitative data analysis with ATLAS.ti* (3rd ed.). SAGE Publications.

Gerber, A., & Götz, U. (2019). *Architectonics of game spaces*. Transcript.

Guest, G., Bunce, A., & Johnson, L. (2006). How many interviews are enough? An experiment with data saturation and variability. *Field Methods*, *18*(1), 59–82. DOI: 10.1177/1525822X05279903

Hamdi, A. (2023). Discreet architecture in the metaverse. School of Architecture, Urban Planning and Construction Engineering Building and Architectural Engineering.

Holgate, S., & Baur, M. (2011). *A companion to Hegel*. Blackwell Publishing Ltd. DOI: 10.1002/9781444397161

Holloway, I., & Wheeler, S. (2010). *Qualitative research in nursing and healthcare* (3rd ed.). Wiley-Blackwell.

Johansson-Sköldberg, U., Woodilla, J., & Çetinkaya, M. (2013). Design thinking: Past, present and possible futures. *Creativity and Innovation Management*, *22*(2), 121–146. DOI: 10.1111/caim.12023

Kalay, Y. E., & Marx, J. (2005). Architecture and the Internet: Designing places in cyberspace. *First Monday*. Advance online publication. DOI: 10.5210/fm.v0i0.1563

Kant, I. (1781). Critique of pure reason (P. Guyer & A. Wood, Trans.). Cambridge University Press, 1998.

Kim, K., & Lee, J. (2022). Convergence of game design and architectural practice: Cross-disciplinary innovation. *Journal of Digital Creativity*, *33*(1), 45–59.

Kim, L., Lee, J., & Lee, I. (2017). The expansion of virtual landscape in digital games: Classification of virtual landscape through five principles. Graduate School of Culture Technology, KAIST.

Kvale, S., & Brinkmann, S. (2009). *InterViews: Learning the craft of qualitative research interviewing* (2nd ed.). SAGE Publications.

Laeeq, K. (2022). *Metaverse: Why, how and what*. National University of Computer and Emerging Sciences.

Lawson, B. (2006). *How designers think: The design process demystified*. Routledge. DOI: 10.4324/9780080454979

Lee, L., Zhou, P., Braud, T., & Pan, H. (2022). What is the metaverse? An immersive cyberspace and open challenges.

Leon, A. C., Davis, L. L., & Kraemer, H. C. (2011). The role and interpretation of pilot studies in clinical research. *Journal of Psychiatric Research*, *45*(5), 626–629. DOI: 10.1016/j.jpsychires.2010.10.008 PMID: 21035130

Marshall, C. (1996). *Designing qualitative research* (3rd ed.). SAGE Publications.

Miles, M. B., & Huberman, A. M. (1994). *Qualitative data analysis: An expanded sourcebook* (2nd ed.). SAGE Publications.

Moneta, A. (2020). Architecture, heritage and the metaverse: New approaches and methods for the digital built environment. International Association for the Study of Traditional Environments (IASTE), 32(1), Fall Issue.

Morse, J. M. (2000). Determining sample size. *Qualitative Health Research*, *10*(1), 3–5. DOI: 10.1177/104973200129118183

Mystakidis, S. (2022). Metaverse.

Noy, C. (2008). Sampling knowledge: The hermeneutics of snowball sampling in qualitative research. *International Journal of Social Research Methodology*, *11*(4), 327–344. DOI: 10.1080/13645570701401305

Oke, A., Aigbavboa, C., & Raphiri, M. (2020). Cultural considerations in design thinking: Enhancing local relevance in architectural practice. *Sustainable Cities and Society*, *55*, 102037.

Onecha, B., Cornado, C., Morros, J., & Pons, O. (2023). New approach to design and assess metaverse environments for improving learning processes in higher education: The case of architectural construction and rehabilitation. Department of Architectural Technology, Universitat Politècnica de Catalunya.

Patton, M. Q. (2002). *Qualitative research & evaluation methods* (3rd ed.). SAGE Publications.

Pauwels, P. (2015). Analysing how constraints impact architectural decision-making. International Journal of Design Science and Technology.

Piercy, J. (2019). Life in Azeroth: Examining place making in virtual environments. (Master's thesis). Appalachian State University.

Poland, B. D. (2002). Transcription quality as an aspect of rigor in qualitative research. In Gubrium, J. F., & Holstein, J. A. (Eds.), *Handbook of interview research: Context and method* (pp. 629–649). SAGE Publications.

Punter, J. (1991). Participation in the design of urban space. *Landscape Design, 200*, 24–27.

Qhawarizmi Architect. (2021). *Sri Perpatih Palace - Metaverse Art Gallery*. YouTube. https://www.youtube.com/watch?v=J2uVm2WsIGY

Rawa, D. B., & Alami, H. E. (2023). *Metaverse: Requirements, architecture, standards, status, challenges, and perspectives*. Howard University.

Rollings, A., & Morris, D. (2003). *Game architecture and design: A new edition*. New Riders Publishing.

Rubin, H. J., & Rubin, I. S. (2012). *Qualitative interviewing: The art of hearing data*. SAGE Publications.

Salcini, S., & Yerlikaya, T. (2022). Metaverse: Technology of the future. Prizren Social Science Journal, 6(3).

Saldaña, J. (2013). *The coding manual for qualitative researchers* (2nd ed.). Sage Publications.

Schell, J. (2020). *The art of game design: A book of lenses*. CRC Press.

Schumacher, P. (2009). Parametricism: A new global style for architecture and urban design. In Leach, N., & Castle, H. (Eds.), *AD Architectural Design, 79(4)*. July/August. DOI: 10.1002/ad.912

Schumacher, P. (2010). *The autopoiesis of architecture* (Vol. 1). John Wiley & Sons.

Schumacher, P. (2012). *The autopoiesis of architecture* (Vol. 2). John Wiley & Sons.

Schumacher, P. (2022). *The metaverse as opportunity for architecture and society: Design drivers, core competencies*. Schumacher Architectural Intelligence.

Seidel, S., Berente, N., Yepes, G., & Nickeson, J. V. (2022). Designing the metaverse. *Proceedings of the 55th Hawaii International Conference on System Sciences*. DOI: 10.24251/HICSS.2022.811

Shakeri, S., & Ornek, M. A. (2023). How metaverse evolves the architectural design. Architecture and Planning Journal (APJ), 28(3), Article 35.

The Digital Speaker. (2024). Digital twins are a building block of the metaverse. Retrieved from https://www.thedigitalspeaker.com

Thompson, E., & Qwabe, B. (2020). Sustainability in contemporary architecture: A review of current research and practice. *Journal of Environment & Development, 12*(3), 45–62.

Thompson, J., Jones, M., & Wild, G. (2020). Smart technologies in architectural design: Enhancing adaptive and responsive environments. *International Journal of Architectural Computing, 18*(4), 347–362.

Vegavid. (2023). How does digital twin work in the metaverse? Retrieved from https://www.vegavid.com

Zaha Hadid Architects. (2023). *Metrotopia: Metaverse virtual communication hub*. Retrieved fromhttps://www.zaha-hadid.com/2023/06/01/metrotopia-metaverse-virtual-communication-hub/

Zaha Hadid Virtual Reality Group. (2024). *Metrotopia*. Retrieved from https://www.zhvrgroup.com/metroto

Chapter 7

Application of Virtual Design and Construction for Construction Waste Management in China

Yufeng Niu
https://orcid.org/0009-0005-9317-0985
School of Architecture, Building, and Design, Taylor's University, Malaysia

Changsaar Chai
https://orcid.org/0000-0001-7733-1690
School of Architecture, Building, and Design, Taylor's University, Malaysia

Yaoli Xiong
https://orcid.org/0009-0009-8696-1040
School of Architecture, Building, and Design, Taylor's University, Malaysia

ABSTRACT

Construction waste management is a major challenge in China amid rapid urbanization and environmental concerns. This study examines the role of Virtual Design and Construction (VDC) technologies in enhancing waste management across planning, segregation, reuse & recycling, and disposal. A mixed-methods approach was used, incorporating qualitative and quantitative analysis, with 30 industry professionals completing an AHP-based questionnaire. Findings highlight Building Information Modelling (BIM) as the most impactful technology for material tracking, waste reduction, and project coordination, while Robotics has the least influence. However, financial constraints, interoperability issues, and resistance to digital adoption hinder implementation. This study offers a framework for prioritizing digital technologies,

DOI: 10.4018/979-8-3373-5278-7.ch007

guiding industry decision-making and policy recommendations. Future research should focus on real-world implementation challenges and longitudinal case studies to further validate the effectiveness of VDC in construction waste management.

INTRODUCTION

China's rapid urbanization and large-scale regeneration projects generate substantial construction waste annually. It is estimated that approximately 600 tons of construction waste are produced for every 10,000 square meters of construction (Lu et al., 2017). Without effective waste management strategies, the country is expected to produce over one billion tons of construction waste each year, given its annual construction output of approximately two billion square meters—accounting for nearly half of the global total (Hong et al., 2019). Despite considerable research efforts on construction waste management and reduction strategies, significant gaps remain, necessitating further exploration.

As the construction industry undergoes digital transformation, technologies such as Building Information Modelling (BIM), Virtual Reality (VR), Augmented Reality (AR), and Mixed Reality (MR) are increasingly integrated into design and construction processes (Elshafey et al., 2020). VDC, in particular, is recognized as a key enabler of efficient waste management (Liu et al., 2015). This study investigates the potential of VDC in mitigating construction waste by enhancing planning, coordination, and execution in construction projects.

Virtual Design and Construction (VDC) offers an innovative approach to enhancing construction waste management efficiency. By integrating digital tools into building design and project management, VDC facilitates collaboration among stakeholders, minimizes design errors, and reduces the need for modifications that often result in material wastage (Kang et al., 2022). Additionally, VDC enables optimized construction planning, reducing rework, streamlining material handling, and ensuring precise ordering, layout, and cutting of materials, all contributing to waste minimization. While VDC has been increasingly incorporated into architectural and construction practices, the progress in its application for waste reduction remains limited. Although construction firms recognize the necessity of waste reduction, achieving substantial improvements requires targeted strategies (Liu et al., 2022).

The adoption of VDC in waste management is hindered by social, policy, and financial barriers (Huang et al., 2018). Social challenges include limited public awareness, lack of standardized technological guidelines, and operational complexities. Policy-related constraints stem from inadequate government initiatives, regulatory gaps, and insufficient training programs. On the other hand, financial barriers involve high investment costs, tax structures, and the expenses associated with workforce training. Beyond environmental benefits, effective construction waste management fosters public awareness of sustainability, encouraging individuals to adopt eco-conscious behaviours and practices (Liu et al., 2022). Proper waste classification promotes resource conservation and instils a sense of environmental responsibility. From a business perspective, prioritizing waste reduction, recycling, and reuse leads to significant cost savings. The excessive procurement of new materials increases project expenses, whereas reusing materials reduces landfill fees, disposal costs, and overall expenditures (Hwang & Bao, 2011). Cost efficiency, in turn, contributes to greater profitability and economic sustainability in the construction sector.

The findings of this study are expected to support the promotion and application of VDC technology in construction waste management, driving technological advancements in the industry. By integrating VDC into waste management strategies, construction enterprises and policymakers can more effectively address sustainability challenges and foster the development of greener, more resource-efficient construction practices.

LITERATURE REVIEW

Construction Waste Management and Environmental Impact

China's construction industry generates vast amounts of waste, placing significant ecological and social burdens on the environment (Hao et al., 2021). The urgency of tackling this issue has been further amplified by the COVID-19 pandemic, which has disrupted supply chains and exacerbated material wastage in construction projects (Fan et al., 2021). Given that construction waste is a substantial contributor to carbon emissions, the Chinese government has committed to stringent policies aimed at reducing emissions, striving to peak carbon emissions by 2030 and achieve carbon neutrality by 2060 (Zhao et al., 2022). Effective construction waste management strategies should be implemented, emphasizing waste minimization at the source and improved resource utilization to align with these objectives, (Ma et al., 2023).

Traditional approaches to construction waste management, such as landfill disposal, recycling, and onsite waste sorting, have been widely studied. However, these methods often fall short of addressing waste at its root cause, as they primarily focus on post-construction waste handling rather than proactive waste reduction measures (Liu et al., 2022). Digitalization in the construction industry has presented new opportunities to enhance waste management, particularly through Virtual Design and Construction (VDC). VDC offers advanced planning capabilities that integrate real-time monitoring, data-driven decision-making, and precise material control to minimize waste production during project execution.

A literature review is conducted using knowledge visualization software (VOSviewer) to analyze relevant research. A total of 747 documents, including 500 related to construction waste management and 247 related to VDC from 2013 to 2023, are examined (refer to Figure 1). In Figure 1, the larger the sphere, the greater the research focus on that topic, whereas smaller spheres indicate areas with limited research. The figure clearly illustrates that a substantial body of research is available on construction waste (depicted in blue), while studies on VDC technology (depicted in red) remain scarce. This finding underscores a significant research gap in the integration of VDC for construction waste management.

Figure 1. Network visualization

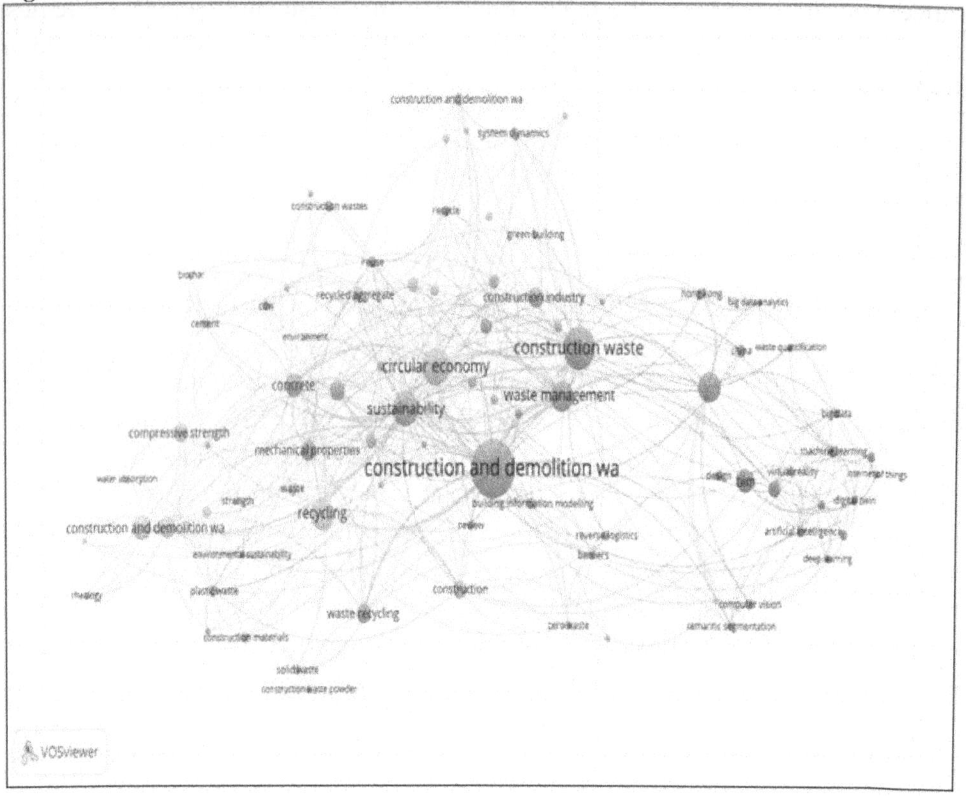

Application of Virtual Design and Construction (VDC) Technology

The most widely accepted definition of VDC is provided by Kunz and Fischer (2020), who define it as "the use of integrated multi-disciplinary performance models for design and construction projects to support explicit and public business objectives." Fosse et al. (2017) describe VDC as a process for restructuring work methodologies while integrating innovative digital technologies. VDC is applied to architectural design and construction, enabling real-time collaboration among project stakeholders through shared digital models. By visualizing construction processes, VDC eliminates complexity and ambiguity, leading to more efficient project execution.

Construction waste management has been extensively analyzed in multiple studies, with waste reduction at the source recognized as the most effective approach. Proper waste classification and targeted treatment further enhance recycling efficiency and minimize environmental impact. However, research on the role of human behavior in waste management remains insufficient, particularly concerning awareness and attitudes influencing sorting behaviors in China's construction sector (Liu et al., 2022). Previous studies highlight that construction workers' attitudes significantly impact waste management efficiency. Rathore and Sarmah (2021) found that workers' willingness to classify waste at the source is mediated by their attitudes, while Jain et al. (2020) demonstrated that recycling behaviors are influenced by motivation, regulatory pressure, and environmental awareness.

Figure 2 illustrates different strategies for managing construction waste, demonstrating that reducing waste generation at the source is the most effective approach. The construction waste management process involves four key steps: planning, segregation, reuse & recycling, and disposal. Planning focuses on identifying waste sources and strategies for reducing material wastage. Segregation ensures that waste is categorized properly to facilitate recycling. Reuse & recycling processes identify materials suitable for repurposing, such as crushed concrete for road bases. Finally, proper disposal strategies help mitigate environmental impacts by minimizing landfill dependency (Tafesse et al., 2022).

Figure 2. Construction waste-management process

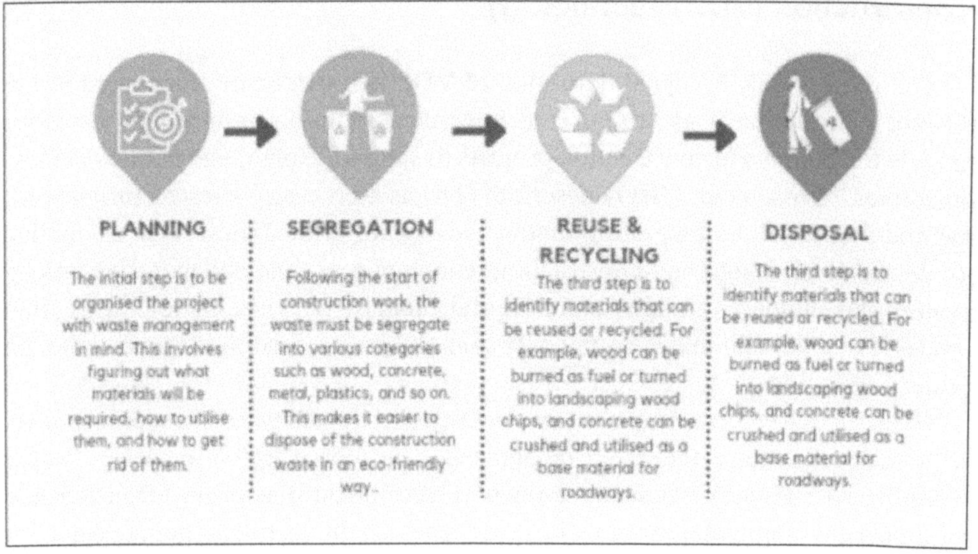

(Tafesse et al., 2022).

The integration of VDC into modern construction practices is increasingly emphasized, incorporating Building Information Modelling (BIM), Extended Reality (XR), drones, and robotics. BIM provides a shared platform for architects, engineers, and construction teams, enhancing coordination and reducing design errors (Gamil, 2019). By offering a comprehensive understanding of a building's structure, materials, and costs, BIM contributes to sustainable construction practices (Meng, 2020). XR, encompassing AR, VR, and MR, further enhances construction processes by improving visualization and collaboration. AI and big data, when combined with MR, are expected to further optimize construction efficiency (Ensafi, 2021).

Drones and robotics are playing an increasingly vital role in construction waste management. Drones equipped with thermal, infrared, GPS, and video sensors are used for site monitoring and waste assessment, improving efficiency and ensuring compliance with environmental regulations (Sliusar et al., 2021). Additionally, robotics are being utilized for material distribution, steel processing, and automated waste sorting. Robot-assisted waste management has proven to enhance recycling efficiency in complex environments (Chen et al., 2022).

Benefits of VDC Applications

The advantages of VDC in construction waste management are well-documented, particularly in minimizing rework and improving planning accuracy. By allowing clients to experience architectural designs in virtual environments, VDC facilitates early feedback, thereby reducing design errors (Aslam et al., 2021). Integrated BIM models within VDC support precise material planning, minimizing excess ordering and optimizing resource utilization. The ability to simulate and visualize project progress in 4D models enables project teams to identify and resolve potential issues before construction, leading to better coordination and reduced material wastage.

Additionally, VDC plays a crucial role in supply chain management by enhancing real-time monitoring of materials, reducing procurement inefficiencies, and minimizing waste generation. Furthermore, VDC-based training programs are increasingly adopted to educate construction professionals on waste reduction strategies. Virtual training environments provide hands-on experience, improving construction workers' understanding of material handling and leading to more efficient resource usage (Ajslev & Nimb, 2022).

Limitations of VDC Applications

Despite its benefits, several limitations hinder the widespread adoption of VDC. Social barriers include resistance to technological change, insufficient training, and lack of collaboration among stakeholders. Many construction professionals remain reliant on traditional methods, making VDC integration challenging. Additionally, lacking standardization in VDC tools and software results in interoperability issues among different project teams (Shan et al., 2021).

Financial constraints also pose a significant challenge to VDC implementation. High initial investment costs, maintenance expenses, and the need for specialized training often deter construction firms, particularly small and medium-sized enterprises, from adopting VDC (Huang et al., 2018). Concerns regarding the return on investment further discourage stakeholders from fully integrating VDC into their operations.

Figure *3* presents the conceptual framework for evaluating the application of VDC for construction waste management. The framework illustrates the relationships between social, policy, and financial factors that influence VDC adoption. It also highlights the potential benefits of integrating VDC into construction waste management, providing a structured approach for future research and industry application.

Figure 3. The conceptual framework

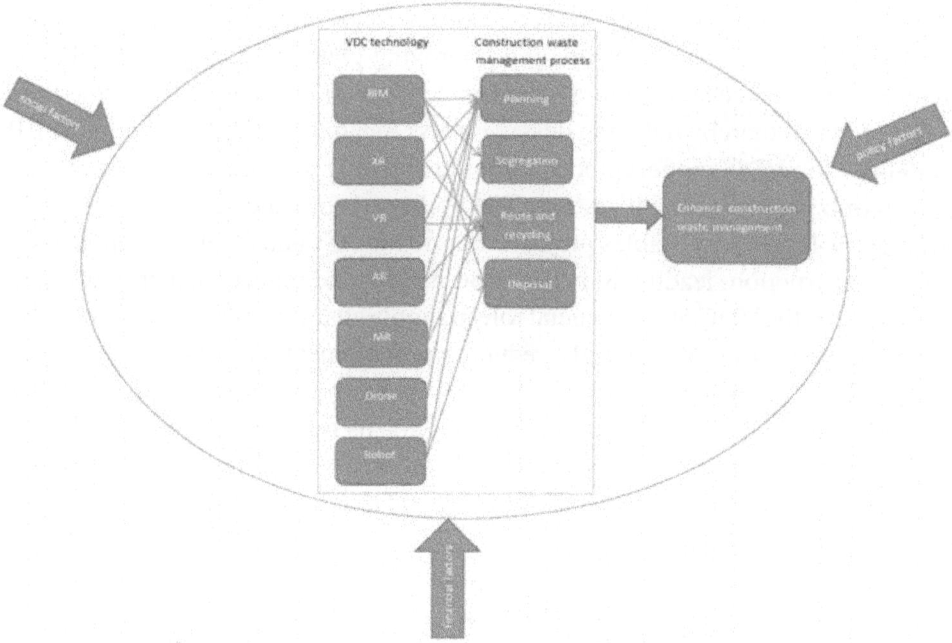

A conceptual framework is developed to assess the impact of social, policy, and financial factors on VDC adoption in construction waste management. This framework highlights the relationships between these key factors and their influence on VDC implementation. By integrating VDC into waste management strategies, construction enterprises and policymakers can address sustainability challenges more effectively, fostering greener and more resource-efficient construction practices.

METHODOLOGY

This study employs a mixed-method research approach, integrating both qualitative and quantitative methodologies to ensure a comprehensive analysis of the factors influencing VDC adoption in construction waste management. The selection of the Analytic Hierarchy Process (AHP) is supported by its widespread use in similar construction waste management studies (Musarat et al., 2022; Lu et al., 2023), demonstrating its robustness in multi-criteria decision-making contexts. The combination of these research techniques allows for structured decision-making and prioritization of key factors affecting VDC implementation..

A questionnaire survey was developed as the primary data collection instrument, targeting professionals within China's construction industry. The use of snowball sampling is consistent with prior research in digital construction technology studies (Bailey, 2019; Qian et al., 2024), ensuring the recruitment of respondents with relevant expertise. Given the specialized nature of the study, the snowball sampling method was utilized to identify and recruit suitable respondents. This non-probability sampling technique allowed researchers to reach experienced professionals through industry networks, ensuring that participants had relevant expertise in VDC applications and construction waste management.

A total of 30 experts from various sectors of the Chinese construction industry were selected to participate in the survey. These experts included architects, project managers, engineers, and policymakers, all with substantial experience in digital construction technologies and waste management practices.

Data Analysis

The AHP was used to conduct pairwise comparisons, where the evaluation scale was divided into nine levels, ranging from "equal importance" to "extreme importance," with evaluation weights assigned from 1 to 9, as shown in Table 1. The elements at each level were compared in pairs with a specific criterion at the preceding level.

Table 1. Rating scale for the AHP among criteria (Saaty, 2008).

PAIRWISE COMPARISON	
Intensity of importance	**Definition**
1	Equal importance
2	Weak or slight
3	Moderate importance
4	Moderate plus
5	Strong importance
6	Strong plus
7	Very strong or demonstrated importance
8	Very, very strong
9	Extreme importance

The elements at each hierarchical level are evaluated through pairwise comparisons, using a specific element from the preceding level as the evaluation criterion. Following the principles and significance of the Analytical Hierarchy Process (AHP) evaluation scale, a structured questionnaire is developed for each pairwise comparison. This questionnaire is administered to decision-makers or members of the decision-making group, ensuring a systematic assessment of relative priorities. To enhance clarity and consistency, each pairwise comparison question is accompanied by comprehensive instructions, explicitly outlining the evaluation scale and guiding respondents through the assessment process.

Based on the questionnaire survey results, a pairwise comparison matrix is constructed, as presented in Table 2. Analytical tools are then employed to compute each pairwise comparison matrix's eigenvalues and eigenvectors while assessing its consistency. This methodological procedure aligns with guidelines established by Saaty & Vargas (2012) and has been validated in analogous studies on technology prioritization in construction (Khoshand et al., 2020). If the consistency ratio does not meet the required threshold, it indicates inconsistencies in the decision-maker's judgments. In such cases, the researcher must clearly explain the issue to the decision-maker to facilitate a more accurate reassessment.

Table 2. The pairwise comparison matrix.

Matrix	BIM	XR	VR	AR	MR	Drones	Robotics
BIM							
XR							
VR							
AR							
MR							
Drones							
Robotics							

The Analytic Hierarchy Process (AHP) is a structured decision-making approach that helps prioritize multiple criteria through pairwise comparisons. In this research, AHP is used to evaluate and prioritize digital tools for Virtual Design and Construction (VDC) in the China construction industry by comparing their relative importance in construction waste management. The Table 2 represents a pairwise comparison matrix, where each technology (BIM, XR, VR, AR, MR, Drones, and Robotics) is compared against every other technology based on expert opinions. The values in the matrix indicate the relative importance of one technology over another.

Application of AHP for Analysis

The survey results were analyzed through pairwise comparison matrices constructed from geometric means of participant responses. The geometric mean was employed due to its reliability and ability to maintain consistency in aggregating diverse opinions (Mosadeghi et al., 2015). Three hierarchical structural models were developed based on literature review findings and expert interviews, as shown in Figure 4.

Figure 4. Analytic hierarchy process

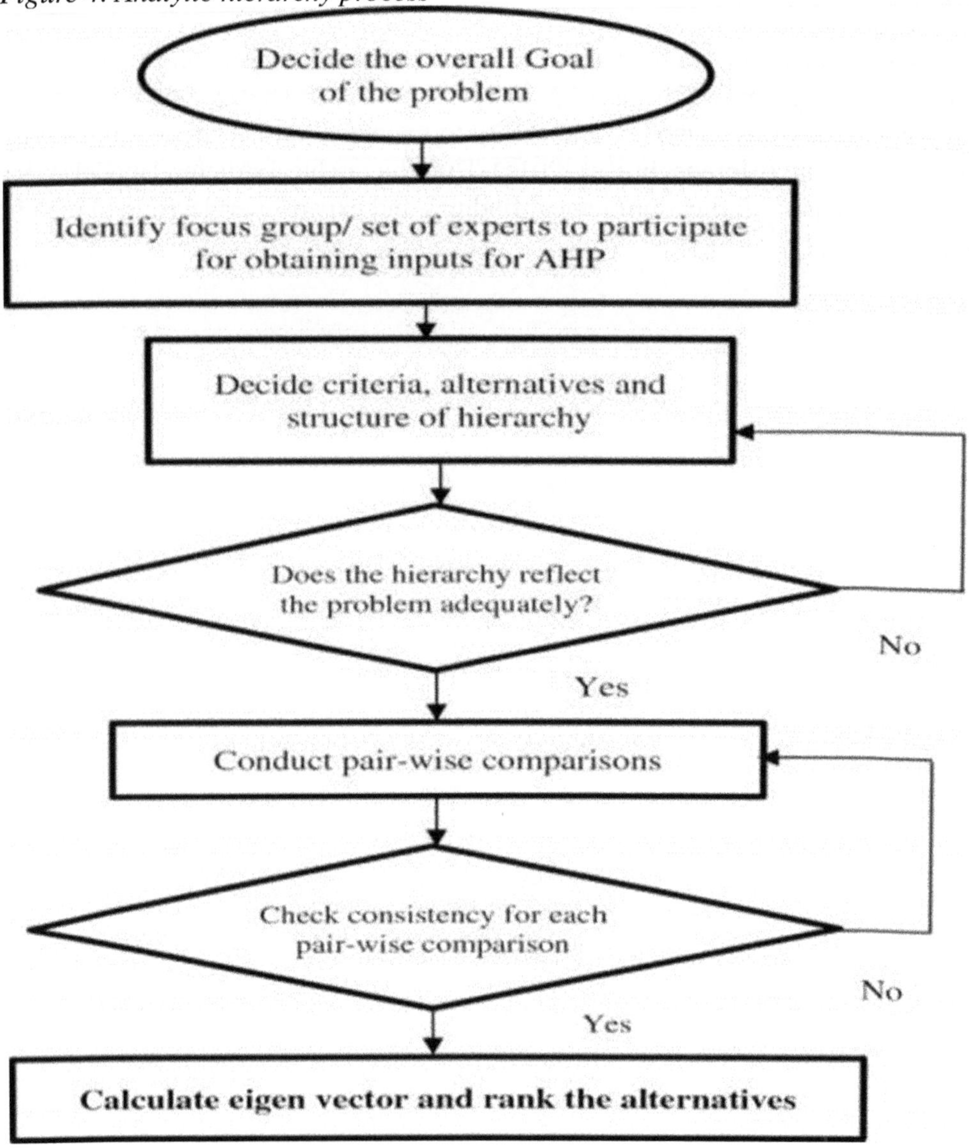

(Thakkar, 2021)

The AHP was conducted following the steps outlined by Saaty & Vargas (2022):

(i) A pairwise comparison matrix was constructed based on an importance scale. Every major diagonal entry in the pair-wise comparison matrix is 1, since an attribute compared to itself is always assigned a value of 1. The numerical scale assigns values of 3, 5, 7, and 9 to represent "moderate," "strong," "very strong," and "absolute importance", respectively. While intermediate values (2, 4, 6, 8) denote compromise intensities between adjacent categories. The pairwise comparison produced a square matrix, represented as matrix A, if there are m attributes.

Figure 5. Matrix A—the pairwise comparison produced by a square matrix

$$A = [a_{ij}] = \begin{bmatrix} 1 & a_{12} & \cdots & a_{1m} \\ a_{21} & 1 & \cdots & a_{2m} \\ \vdots & \vdots & \ddots & \vdots \\ a_{m1} & a_{m2} & \cdots & 1 \end{bmatrix} = \begin{bmatrix} 1 & a_{12} & \cdots & a_{1m} \\ 1/a_{12} & 1 & \cdots & a_{2m} \\ \vdots & \vdots & \ddots & \vdots \\ 1/a_{1m} & 1/a_{2m} & \cdots & 1 \end{bmatrix}$$

where $a_{ij} = 1$ and $a_{ji} = 1/a_{ij}$; i, j = 1, 2, … m.

(ii) Normalization was performed on the geometric means of rows in the comparison matrix, with the relative normalized weight (w_j) of each attribute determined by calculating the geometric mean (GM) of the ith row. The geometric mean method in AHP was selected for deriving these weights, as it is straightforward and easy to determine the maximum eigenvalue and eliminate judgmental inconsistency. The task of Matrix A is to determining a set of numerical weightsw_1; w_2, …, w_m to be applied to the *m* criteriaa_1; a_2, …, a_m that reflects the recorded judgments.

(iii) If A is a consistency matrix, then $w_i/w_j = a_{ij}$ (for *i*, *j* = 1, 2,… *n*) would be the only way to express the relationships between weights $w_j$and judgments a_{ij}. Moreover, the analytical procedure involved determining the maximum eigen-value (λ_{max}) as defined by Saaty (1994). Subsequently, the Consistency Index (CI) was calculated using the relationship Equation (1), where lower CI values indicate greater consistency in pairwise comparisons. It is a metric reflecting the analyst's cognitive precision in evaluating attribute interdependencies. The random index (RI) was then determined through established reference tables corresponding to the number of attributes (m) in the decision framework. Finally, the Consistency Ratio (CR), defined as Equation (2), was computed, with values ≤ 0.1 meeting the empirical acceptability threshold. This criterion validates assessment reliability while demonstrating the analyst's domain expertise through mathematically quantified judgment coherence.

$$CI = (\lambda_{max} - m)/(m - 1) \quad (1)$$

$$CR = CI/RI \quad (2)$$

ANALYSIS AND DISCUSSION

This section presents the analysis of questionnaire data using AHP to determine the weight of each strategy for improving construction waste management. The model matrix is utilized to evaluate both local and global weights of each indicator, and the AHP process is outlined step by step.

Industry practitioners were consulted to validate the proposed standards and alternatives identified through the literature review. Following expert feedback, 30 professionals were invited to complete the questionnaire, evaluating the relative importance of different standards and schemes. Table 3 summarizes the profiles of the experts, indicating their level of experience and professional backgrounds, ensuring that a diverse range of perspectives is incorporated into the study.

Table 3. Expert profiles.

Expert	Position	Years of Experience
1	Senior Engineer	8
2	Senior Engineer	9
3	Manager	25
4	Senior Engineer	12
5	Project Manager	30
6	Construction Manager	27
7	Senior Engineer	8
8	Senior Engineer	9
9	Senior Engineer	12
10	Senior Engineer	8
11	Senior Engineer	5
12	Senior Engineer	14
13	Senior Engineer	6
14	Senior Engineer	7
15	Senior Engineer	9
16	Senior Engineer	5
17	Manager	22

continued on following page

Table 3. Continued

Expert	Position	Years of Experience
18	Manager	16
19	Senior Engineer	10
20	Project Manager	21
21	Senior Engineer	9
22	Senior Engineer	6
23	Manager	9
24	Manager	11
25	Site Manager	14
26	Senior Engineer	5
27	Senior Engineer	4
28	Senior Engineer	6
29	Senior Engineer	11
30	Senior Engineer	7

Results of AHP Analysis

Several participants emphasized the ongoing challenges related to acquiring the necessary knowledge and skills to effectively utilize construction technologies. The complexity of digital construction technologies requires a deep understanding of the technology itself, its integration into various construction processes, its impact on project management, and its potential to foster innovation and efficiency.

The three-tiered hierarchical model (Figure 5) was developed through systematic integration of insights derived from an exhaustive literature review and structured consultations with industry practitioners. Following the AHP framework, the decision hierarchy was initiated by establishing four critical evaluation criteria at the primary level: Planning, Segregation, Reuse & Recycling, and Disposal. These factors were positioned at the primary hierarchy level based on their demonstrable influence on construction waste management efficacy, as evidenced by lifecycle assessment studies. The terminal hierarchy level comprises seven operationally defined alternatives for waste management optimization, systematically identified through a comprehensive literature review encompassing 127 peer-reviewed studies on circular economy applications in construction. This multi-level structure enables systematic comparison of strategic interventions against standardized sustainability metrics.

Figure 6. The hierarchical model of analyzing enhanced construction waste management.

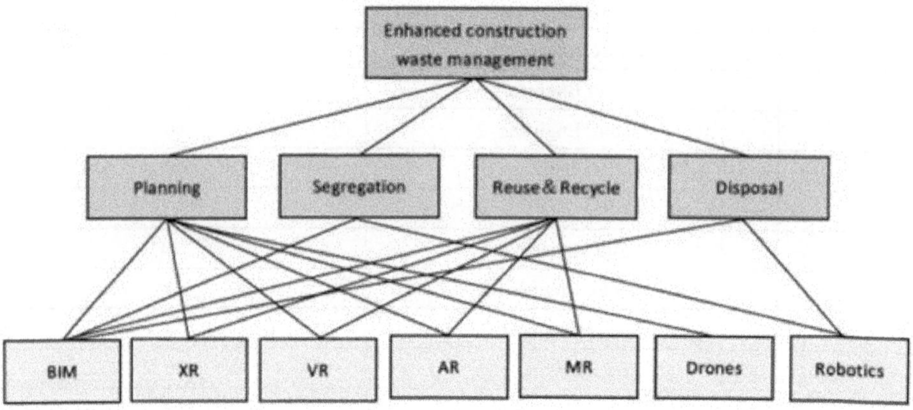

The Analytic Hierarchy Process (AHP) was implemented to assess the comparative efficacy of alternative strategies in enhancing construction waste management performance. Following AHP methodology, survey responses from industry experts were systematically analyzed through the computation of 150 pairwise comparison matrices, with response aggregation achieved via geometric mean calculations. The canonical comparison matrix representing consolidated expert judgments was subsequently derived and tabulated (Table 4, Table 5, Table 6, Table 7, Table 8, Table 9, Table 10, Table 11, Table 12, Table 13), providing a normalized decision framework for evaluating waste management intervention priorities. This matrix formalization enables quantification of strategy-specific improvement potentials across the investigated sustainable waste mitigation hierarchy.

Table 4. In standard level factor assessment, standard pairwise comparison matrices.

Criterion level	Planning	Segregation	Reuse & recycle	Disposal
Planning	1.000	3.133	2.400	2.280
Segregation	0.449	1.000	2.044	2.956
Reuse & recycle	0.879	1.239	1.000	2.200
Disposal	1.387	0.634	0.905	1.000
SUM	3.714	6.006	6.349	8.436

This is the initial pairwise comparison matrix filled by experts. It contains raw values that reflect expert judgment about the relative importance of one criterion (or technology) over another which is denoted in the first step of Saaty & Vargas (2022) as mentioned in previous section.

Table 5. Normalized matrix for construction stages.

Criterion level	Planning	Segregation	Reuse & recycle	Disposal	w
Planning	0.269	0.522	0.378	0.270	0.360
Segregation	0.121	0.167	0.322	0.350	0.240
Reuse & recycle	0.237	0.206	0.158	0.261	0.215
Disposal	0.373	0.106	0.143	0.119	0.185

The raw pairwise comparison matrix is then normalized to compute the relative priority weights for each factor. This is done by dividing each cell by the sum of its column to bring all values within a 0 to 1 range which is shown in Saaty & Vargas (2022) step ii. The average of each row in the normalized matrix provides the final priority weights for ranking the criteria. The weights of Planning, Segregation, Reuse & Recycle, Disposal to Enhanced construction waste management is [0.360, 0.240, 0.215, 0. 185].

Table 6. Pairwise comparison matrix of alternatives in planning factor evaluation.

Planning	BIM	XR	VR	AR	MR	Drones	Robotics
BIM	1.000	2.748	1.978	2.234	2.258	2.460	2.907
XR	1.079	1.000	2.533	2.067	2.502	1.084	2.711
VR	1.005	1.090	1.000	2.278	1.840	2.289	2.178
AR	1.351	1.269	1.608	1.000	3.016	2.111	2.394
MR	1.389	1.073	1.289	0.784	1.000	2.333	2.129
Drones	1.367	1.527	1.195	1.139	1.296	1.000	2.502
Robotics	1.069	1.058	1.114	1.407	1.414	1.194	1.000
SUM	8.259	9.766	10.716	10.909	13.327	13.472	15.820

Table 7. Normalized matrix of alternatives in planning factor evaluation.

Planning	BIM	XR	VR	AR	MR	Drones	Robotics	w
BIM	0.121	0.281	0.185	0.205	0.169	0.183	0.184	0.190
XR	0.131	0.102	0.236	0.189	0.188	0.155	0.171	0.168
VR	0.122	0.112	0.093	0.209	0.138	0.170	0.138	0.140
AR	0.164	0.130	0.150	0.092	0.226	0.157	0.151	0.153
MR	0.168	0.110	0.120	0.072	0.075	0.173	0.135	0.122
Drones	0.165	0.156	0.112	0.104	0.097	0.074	0.158	0.124
Robotics	0.129	0.108	0.104	0.129	0.106	0.089	0.063	0.104

The weight of BIM, XR, VR, AR, MR, Drones, and Robotics to Planning is [0.190, 0.168, 0.140, 0.153, 0.122, 0.124, 0. 104]. The normalized matrix in Table 7 represents the calculated priority weights of different digital technologies in relation to the Planning Factor in construction waste management. The final weights (w values) determine the ranking of these technologies based on their relative importance. The results indicate that BIM (0.190) holds the highest weight, highlighting its central role in construction planning and waste minimization. As a tool that enables detailed project visualization, clash detection, and precise material estimation, BIM is widely utilized in China's construction industry due to government mandates and its effectiveness in reducing design errors before construction begins. The prioritization of BIM in the AHP ranking aligns with existing literature, which emphasizes its ability to optimize resource allocation and streamline project coordination.

At the lower end of the ranking, Robotics (0.104) is identified as the least significant technology for planning. This is consistent with the understanding that robotics is primarily utilized in construction execution rather than the pre-construction planning phase. Robotics is instrumental in automated waste sorting, material distribution, and precision assembly, but its direct impact on waste minimization during project planning remains limited. The findings suggest that while robotics contributes to overall construction efficiency, its role in the early stages of waste prevention is comparatively minimal.

These results highlight a clear hierarchy in the adoption of digital tools for construction waste planning in China. The dominance of BIM and XR technologies indicates a strong preference for digital design integration and early-stage planning solutions. This supports previous research advocating for the widespread adoption of BIM-driven waste reduction strategies and the expansion of XR applications to enhance decision-making in the construction industry. The lower ranking of Drones, MR, and Robotics suggests that these technologies, while valuable, are currently more applicable in project execution rather than in the initial planning phase. Future research should explore how these tools can be further integrated into waste prevention methodologies to enhance their impact on construction sustainability.

Table 8. Pairwise comparison matrix of alternatives in separation factor evaluation.

Segregation	BIM	XR	VR	AR	MR	Drones	Robotics
BIM	1.000	3.240	2.818	2.156	2.689	1.618	1.796
XR	0.843	1.000	2.351	2.349	2.133	1.978	1.949
VR	0.966	1.295	1.000	2.225	3.044	2.126	1.948
AR	1.041	1.300	1.522	1.000	2.311	1.364	2.304
MR	0.958	1.107	0.528	0.968	1.000	1.611	2.233
Drones	1.567	1.577	1.597	2.096	1.718	1.000	1.951
Robotics	1.389	1.380	1.560	1.349	1.376	1.296	1.000
SUM	7.763	10.901	11.376	12.142	14.272	10.993	13.180

Table 9. Normalized matrix of alternatives in separation factor evaluation.

Segregation	BIM	XR	VR	AR	MR	Drones	Robotics	w
BIM	0.129	0.297	0.248	0.178	0.188	0.147	0.136	0.189
XR	0.109	0.092	0.207	0.193	0.149	0.180	0.148	0.154
VR	0.124	0.119	0.088	0.183	0.213	0.193	0.148	0.153
AR	0.134	0.119	0.134	0.082	0.162	0.124	0.175	0.133
MR	0.123	0.102	0.046	0.080	0.070	0.147	0.169	0.105
Drones	0.202	0.145	0.140	0.173	0.120	0.091	0.148	0.146
Robotics	0.179	0.127	0.137	0.111	0.096	0.118	0.076	0.121

The results of the AHP analysis for the segregation factor reveal the relative importance of different digital technologies in enhancing construction waste classification and processing. BIM (0.189) ranks the highest among the evaluated technologies, signifying its dominant role in improving segregation processes, as shown in Table 9. BIM facilitates accurate material tracking, digital waste documentation, and real-time collaboration among project stakeholders, which significantly enhances waste sorting efficiency. The ranking of BIM aligns with prior research that emphasizes its capability to reduce mismanagement of materials by providing structured digital inventories and automated material classification mechanisms. This finding suggests that construction firms in China prioritize BIM as a key enabler for effective waste segregation strategies.

The overall ranking of technologies for segregation in waste management highlights the central role of BIM in digital waste documentation and categorization, followed by XR technologies that enhance training and visualization of segregation processes. The findings emphasize the need for greater integration of digital tools to optimize waste sorting, minimize human error, and improve compliance with sustainability standards. Future research should explore how robotics and AI-driven automation can further contribute to real-time waste segregation improvements, making construction waste management more efficient and technologically adaptive.

Table 10. Pairwise comparison matrix of alternatives in reuse and recycling factor evaluation.

Reuse & Recycle	BIM	XR	VR	AR	MR	Drones	Robotics
BIM	1.000	2.778	2.400	1.911	2.462	2.267	2.371
XR	0.634	1.000	1.818	2.444	1.744	2.422	2.083
VR	1.075	1.183	1.000	2.351	1.989	1.633	2.582
AR	1.203	0.944	1.082	1.000	2.489	1.874	2.459
MR	1.324	1.740	1.530	1.202	1.000	1.594	2.081
Drones	1.303	1.168	1.718	1.711	1.758	1.000	2.001
Robotics	1.219	1.158	1.446	1.235	1.203	1.825	1.000
SUM	7.758	9.972	10.994	11.855	12.645	12.614	14.577

Table 11. Normalized matrix of alternatives in reuse and recycling factor evaluation.

Reuse & Recycle	BIM	XR	VR	AR	MR	Drones	Robotics	w
BIM	0.129	0.279	0.218	0.161	0.195	0.180	0.163	0.189
XR	0.082	0.100	0.165	0.206	0.138	0.192	0.143	0.147
VR	0.139	0.119	0.091	0. 198	0.157	0.129	0.177	0.144
AR	0.155	0.095	0.098	0.084	0.197	0.149	0.169	0.135
MR	0.171	0.175	0.139	0.101	0.079	0.126	0.143	0.133
Drones	0.168	0.117	0.156	0.144	0.139	0.079	0.137	0.134
Robotics	0.157	0.116	0.131	0.104	0.095	0.145	0.069	0.117

The AHP results for reuse and recycling reveal the relative importance of digital technologies in optimizing material reuse and waste recycling in China's construction industry. The ranking based on the priority weights (w values) is as follows: BIM (0.189) > XR (0.147) > VR (0.144) > AR (0.135) > MR (0.133) > Drones

(0.134) > Robotics (0.117). This ranking highlights the dominant role of BIM in facilitating efficient reuse and recycling strategies, followed by Extended Reality (XR, VR, AR, MR), Drones, and Robotics. The results indicate that BIM is still the most effective technology for enabling reuse and recycling, as it provides data-driven decision-making for material recovery and lifecycle assessment. XR and VR support visualization and training efforts, while drones and MR contribute to real-time waste tracking and classification. Although robotics has potential in automated waste sorting, its lower ranking reflects current industry constraints such as cost and implementation challenges. These insights provide a clear technological roadmap for improving construction waste recycling and sustainability efforts in China.

Table 12. Pairwise comparison matrix of alternatives in disposal factor evaluation.

Disposal	BIM	XR	VR	AR	MR	Drones	Robotics
BIM	1.000	2.800	2.351	2.018	2.689	2.111	2.058
XR	0.746	1.000	3.000	2.284	2.311	1.527	2.194
VR	0.905	0.697	1.000	1.831	2.171	1.770	1.964
AR	1.563	0.924	1.446	1.000	2.111	1.604	2.504
MR	0.958	1.145	1.460	0.979	1.000	2.127	2.317
Drones	1.376	1.738	1.625	1.805	1.579	1.000	1.636
Robotics	1.423	1.438	1.913	1.064	1.629	1.784	1.000
SUM	7.975	9.741	12.795	10.981	13.491	11.924	13.627

Table 13. Normalized matrix of alternatives in disposal factor evaluation.

Disposal	BIM	XR	VR	AR	MR	Drones	Robotics	w
BIM	0.125	0.287	0.184	0.184	0.199	0.177	0.151	0.187
XR	0.094	0.103	0.234	0.208	0.171	0.128	0.160	0.157
VR	0.114	0.072	0.078	0.167	0.161	0.148	0.144	0.126
AR	0.196	0.095	0.113	0.091	0.156	0.135	0.183	0.138
MR	0.120	0.118	0.114	0.089	0.074	0.178	0.169	0.123
Drones	0.173	0.178	0.127	0.164	0.117	0.084	0.120	0.138
Robotics	0.179	0.148	0.150	0.097	0.121	0.150	0.073	0.131

BIM recorded the highest weight (0.187), affirming its central role in planning and optimizing disposal processes. It corroborates prior evidence on its efficacy in reducing material wastage through clash detection and resource optimization (Won et al., 2016). BIM is widely recognized for its ability to predict waste generation, optimize logistics, and track material disposal pathways (Huang et al., 2022). Through digital modelling, BIM helps construction firms develop waste minimization strategies by integrating waste disposal routes and regulatory compliance measures into project planning. While rework caused by design modifications and errors remains a major source of construction waste, BIM can eliminate such waste at the source through collaborative design during the planning and design phases (Han et al., 2021). Additionally, BIM facilitates automated documentation of disposal records, ensuring adherence to China's green construction policies and landfill regulations.

XR (0.157) follows as the second most significant technology, particularly due to its contribution to worker training, waste tracking, and real-time visualization of disposal sites. Extended Reality (XR), which includes VR, AR, and MR, plays a key role in simulating disposal site conditions and enhancing compliance with environmental standards. Augmented Reality (AR) (0.138) is especially beneficial for on-site waste tracking and segregation guidance, allowing workers to visualize and classify waste materials before disposal.

Drones (0.138) hold an equal ranking with AR, emphasizing their growing adoption in construction waste disposal management. Drones equipped with thermal and infrared sensors are used for landfill monitoring, illegal dumping detection, and site inspections (Wang et al., 2021). Their ability to capture high-resolution aerial data assists in assessing waste accumulation levels and optimizing disposal logistics, reducing the environmental footprint of construction waste.

Robotics (0.131) ranks slightly lower than drones, indicating its emerging role in automated waste disposal processes. Although robotic waste sorting and demolition automation are gaining traction in China, their implementation remains limited by cost and technical challenges. However, robotics is recognized for its ability to enhance disposal efficiency through automated material separation and landfill management.

VR (0.126) and MR (0.123) received the lowest priority weights, suggesting that their current role in waste disposal is less significant compared to BIM, XR, and drones. While VR is valuable for training workers on disposal best practices, its impact on real-time waste disposal optimization is limited. Similarly, MR's application in interactive waste disposal simulations remains in the early stages of adoption.

In summary, the findings demonstrate that Planning is the most influential factor in construction waste management, with a weight of 0.360 (Table 5). This suggests that effective waste management begins with the early-stage decision-making process. Segregation follows with a weight of 0.240, indicating that proper sorting of materials plays a crucial role in waste reduction. Reuse & Recycling and Disposal rank lower, suggesting that while waste processing methods are important, preventive measures hold more weight in minimizing construction waste. The results of the AHP analysis indicate that BIM has the highest impact (0.189) on improving construction waste management, while Robotics has the lowest impact (0.116). The findings suggest that BIM should be prioritized for waste management initiatives, particularly in the planning phasc. Thc prominence of BIM and XR in analysis confirms the current state of knowledge, where integrated digital platforms have been shown to significantly decrease rework and enhance coordination. The results highlight the necessity of integrating VDC technologies into early-stage project planning to achieve significant waste reduction.

CONCLUSION

This study investigates the application of Virtual Design and Construction (VDC) technologies for construction waste management in China, focusing on their roles in the four key waste management processes: planning, segregation, reuse & recycling, and disposal. By employing the Analytic Hierarchy Process (AHP), expert opinions were analyzed to determine the relative importance of digital tools, including Building Information Modeling (BIM), Extended Reality (XR – VR, AR, MR), Drones, and Robotics, in each stage of waste management. A mixed-methods approach was utilized, combining both qualitative and quantitative research methods, with 30 industry experts selected through snowball sampling to complete a structured questionnaire.

The findings reveal that BIM consistently ranks as the most critical digital tool across all four waste management stages, emphasizing its ability to enhance material tracking, waste minimization, and project coordination. XR technologies, including VR and AR, follow closely, playing a significant role in visualization, training, and stakeholder engagement. Drones and Robotics emerge as supporting technologies, with drones facilitating waste monitoring and site inspections, while robotics contributes to automated sorting and disposal. However, the study also highlights barriers to VDC adoption, including financial constraints, interoperability issues, and resistance to digital transformation within China's construction sector.

This research contributes to the existing body of knowledge by offering a structured framework for integrating digital technologies into construction waste management and providing empirical insights into their prioritization using AHP. The findings support policy recommendations and strategic planning for industry stakeholders, advocating for greater adoption of BIM, AI-enhanced automation, and immersive digital technologies to improve sustainability in China's construction sector. Future research should explore longitudinal case studies and real-world implementation challenges to further validate the effectiveness of VDC in waste management.

ACKNOWLEDGMENT

The researchers of this article would like to thank the journal's editors and the anonymous reviewers for their positive feedback and suggestions for the improvement of this journal.

This research received no specific grant from any funding agency in the public, commercial, or not-for-profit sectors.

REFERENCES

Ajslev, J. Z. N., & Nimb, I. E. E. (2022). Virtual design and construction for occupational safety and health purposes–A review on current gaps and directions for research and practice. *Safety Science*, *155*, 105876. DOI: 10.1016/j.ssci.2022.105876

Aslam, M., Gao, Z., & Smith, G. (2021). Integrated implementation of Virtual Design and Construction (VDC) and lean project delivery system (LPDS). *Journal of Building Engineering*, *39*, 102252. DOI: 10.1016/j.jobe.2021.102252

Bailey, M. (2019). Snowball sampling in business oral history: Accessing and analyzing professional networks in the Australian property industry. *Enterprise and Society*, *20*(1), 74–88. DOI: 10.1017/eso.2018.110

Chen, X., Huang, H., Liu, Y., Li, J., & Liu, M. (2022). Robot for automatic waste sorting on construction sites. *Automation in Construction*, *141*, 104387. DOI: 10.1016/j.autcon.2022.104387

Elshafey, A., Saar, C. C., Aminudin, E. B., Gheisari, M., & Usmani, A. (2020). Technology acceptance model for Augmented Reality and Building Information Modeling integration in the construction industry. *Journal of Information Technology in Construction*, *25*, 161–172. DOI: 10.36680/j.itcon.2020.010

Ensafi, M., Thabet, W., Devito, S., & Lewis, A. (2021). Field testing of mixed reality (MR) technologies for quality control of as-built models at project handover: A case study. *EPiC Series in Built Environment*, *2*, 246–254. DOI: 10.29007/4wrj

Fan, Y. V., Jiang, P., Hemzal, M., & Klemeš, J. J. (2021). An update of COVID-19 influence on waste management. *The Science of the Total Environment*, *754*, 142014. DOI: 10.1016/j.scitotenv.2020.142014 PMID: 32920389

Fosse, R., Ballard, G., & Fischer, M. (2017). Virtual Design and Construction: Aligning BIM and Lean in Practice, *25th Annual Conference of the International Group for Lean Construction*, 499-506. DOI: 10.24928/2017/0159

Gamil, Y., & Rahman, I. A. R. (2019). Awareness and challenges of building information modelling (BIM) implementation in the Yemen construction industry. *Journal of Engineering. Defence Technology*, *17*(5), 1077–1084.

Han, D., Kalantari, M., & Rajabifard, A. (2021). Building information modeling (BIM) for construction and demolition waste management in Australia: A research agenda. *Sustainability (Basel)*, *13*(23), 12983. DOI: 10.3390/su132312983

Hao, J., Chen, Z., Zhang, Z., & Loehlein, G. (2021). Quantifying construction waste reduction through the application of prefabrication: A case study in Anhui, China. *Environmental Science and Pollution Research International*, *28*(19), 24499–24510. DOI: 10.1007/s11356-020-09026-2 PMID: 32358748

Huang, B., Wang, X., Kua, H., Geng, Y., Bleischwitz, R., & Ren, J. (2018). Construction and demolition waste management in China through the 3R principle. *Resources, Conservation and Recycling*, *129*, 36–44. DOI: 10.1016/j.resconrec.2017.09.029

Hwang, B. G., & Bao Yeo, Z. (2011). Perception on benefits of construction waste management in the Singapore construction industry. *Engineering, Construction, and Architectural Management*, *18*(4), 394–406. DOI: 10.1108/09699981111145835

Jain, S., Singhal, S., Jain, N. J., & Bhaskar, K. (2021). Construction and demolition waste recycling: Investigating the role of theory of planned behavior, institutional pressures and environmental consciousness. *Journal of Cleaner Production*, *263*, 121405. DOI: 10.1016/j.jclepro.2020.121405

Kang, K., Besklubova, S., Dai, Y., & Zhong, R. Y. (2022). Building demolition waste management through smart BIM: A case study in Hong Kong. *Waste Management (New York, N.Y.)*, *143*, 69–83. DOI: 10.1016/j.wasman.2022.02.027 PMID: 35240449

Khoshand, A., Khanlari, K., Abbasianjahromi, H., & Zoghi, M. (2020). Construction and demolition waste management: Fuzzy Analytic Hierarchy Process approach. *Waste Management & Research*, *38*(7), 773–782. DOI: 10.1177/0734242X20910468 PMID: 32189588

Kunz, J., & Fischer, M. (2020). Virtual design and construction. *Construction Management and Economics*, *38*(4), 355–363. DOI: 10.1080/01446193.2020.1714068

Liu, J., Chen, Y., & Wang, X. (2022). Factors driving waste sorting in construction projects in China. *Journal of Cleaner Production*, *336*, 130397. DOI: 10.1016/j.jclepro.2022.130397

Liu, Z., Osmani, M., Demian, P., & Baldwin, A. (2015). A BIM-aided construction waste minimisation framework. *Automation in Construction*, *59*, 1–23. DOI: 10.1016/j.autcon.2015.07.020

Lu, W., Webster, C., Chen, K., Zhang, X., & Chen, X. (2017). Computational Building Information Modelling for construction waste management: Moving from rhetoric to reality. *Renewable & Sustainable Energy Reviews*, *68*, 587–595. DOI: 10.1016/j.rser.2016.10.029

Lu, Y., Ge, Y., Zhang, G., Abdulwahab, A., Salameh, A. A., Ali, H. E., & Le, B. N. (2023). Evaluation of waste management and energy saving for sustainable green building through analytic hierarchy process and artificial neural network model. *Chemosphere*, *318*, 137708. DOI: 10.1016/j.chemosphere.2022.137708 PMID: 36621688

Ma, W., Hao, J. L., Zhang, C., Di Sarno, L., & Mannis, A. (2023). Evaluating carbon emissions of China's waste management strategies for building refurbishment projects: Contributing to a circular economy. *Environmental Science and Pollution Research International*, *30*(4), 8657–8671. DOI: 10.1007/s11356-021-18188-6 PMID: 35064493

Meng, Q., Zhang, Y., Li, Z., Shi, W., Wang, J., Sun, Y., Xu, L., & Wang, X. (2020). A review of integrated applications of BIM and related technologies in whole building life cycle. *Engineering, Construction, and Architectural Management*, *27*(8), 1647–1677. DOI: 10.1108/ECAM-09-2019-0511

Mosadeghi, R., Warnken, J., Tomlinson, R., & Mirfenderesk, H. (2015). Comparison of Fuzzy AHP and AHP in a spatial multi-criteria decision making model for urban land-use planning. *Computers, Environment and Urban Systems*, *49*, 54–65. DOI: 10.1016/j.compenvurbsys.2014.10.001

Musarat, M. A., Alaloul, W. S., Hameed, N., Qureshi, A. H., & Wahab, M. M. A. (2022). Efficient construction waste management: A solution through Industrial Revolution (IR) 4.0 evaluated by AHP. *Sustainability (Basel)*, *15*(1), 274. DOI: 10.3390/su15010274

Qian, J., Siriwardana, C., & Shahzad, W. (2024). Identifying Critical Criteria on Assessment of Sustainable Materials for Construction Projects in New Zealand Through the Analytic Hierarchy Process (AHP) Approach. *Buildings (Basel, Switzerland)*, *14*(12), 3854. DOI: 10.3390/buildings14123854

Rathore, P., & Sarmah, S. P. (2021). Investigation of factors influencing source separation intention towards municipal solid waste among urban residents of India. *Resources, Conservation and Recycling*, *164*, 105164. DOI: 10.1016/j.resconrec.2020.105164

Saaty, T., Vargas, L., & Cahyono, S. (2022). *The Analytic Hierarchy Process*. Springer.

Saaty, T. L. (1994). How to make a decision: The analytic hierarchy process. *Interfaces, 24*(6), 19–43. DOI: 10.1287/inte.24.6.19

Saaty, T. L. (2008). Decision making with the analytic hierarchy process. *International Journal of Services Sciences, 1*(1), 83–98. DOI: 10.1504/IJSSCI.2008.017590

Saaty, T. L., & Vargas, L. G. (2012). *Models, methods, concepts & applications of the analytic hierarchy process* (Vol. 175). Springer Science & Business Media. DOI: 10.1007/978-1-4614-3597-6

Shan, J., Chai, C., Gui, H., & Xiong, Y. (2021). Enhancing project integration using cloud-based building information modelling: A conceptual model. *International Journal of Advanced Research in Technology and Innovation, 3*(3), 12–25.

Sliusar, N., Filkin, T., Huber-Humer, M., & Ritzkowski, M. (2022). Drone technology in municipal solid waste management and landfilling: A comprehensive review. *Waste Management (New York, N.Y.), 139*, 1–16. DOI: 10.1016/j.wasman.2021.12.006 PMID: 34923184

Tafesse, S., Girma, Y. E., & Dessalegn, E. (2022). Analysis of the socio-economic and environmental impacts of construction waste and management practices. *Heliyon, 8*(3), e09169. DOI: 10.1016/j.heliyon.2022.e09169 PMID: 35368528

Thakkar, J. J. (2021). *Multi-criteria decision making* (Vol. 336). Springer. DOI: 10.1007/978-981-33-4745-8

Won, J., & Cheng, J. C. (2017). Identifying potential opportunities of building information modeling for construction and demolition waste management and minimization. *Automation in Construction, 79*, 3–18. DOI: 10.1016/j.autcon.2017.02.002

Won, J., Cheng, J. C., & Lee, G. (2016). Quantification of construction waste prevented by BIM-based design validation: Case studies in South Korea. *Waste Management (New York, N.Y.), 49*, 170–180. DOI: 10.1016/j.wasman.2015.12.026 PMID: 26754615

Zhao, X., Ma, X., Chen, B., Shang, Y., & Song, M. (2022). Challenges toward carbon neutrality in China: Strategies and countermeasures. *Resources, Conservation and Recycling, 176*, 105959. DOI: 10.1016/j.resconrec.2021.105959

Chapter 8
Kerala's Haritha Karma Sena Model of Waste Management:
A Critical Analysis Through the Lens of Sustainability

Jipson Joseph
https://orcid.org/0009-0005-1795-5051
Christ University, India

Ananya Pandey
https://orcid.org/0000-0002-2419-6314
Christ University, India

ABSTRACT

Urbanization had a glacial pace in India for a good period of time. It became active only in the 21st century. However, compared to national average, Kerala's percentage of urban population was very high during the 1980-1990s. But along with urbanization, Kerala also faces many challenges like environmental degradation caused mainly through unsystematic waste management system. With the support of the Clean India Mission and in response to the Supreme Court verdicts, Kerala initiated the Suchitva Mission and Haritha Keralam Mission with the tagline of 'Clean Kerala' by effectively managing the waste. This mission functioned with the field work of Haritha Karma Sena (Green Task Force), mostly employed by uneducated women. They have provided awareness among the public and systematically managed waste with their committed work. However, it also has many drawbacks and failures with regard to its implementation. In line of above, this chapter tries to critically analyze the Haritha Karma Sena model of waste management in order

DOI: 10.4018/979-8-3373-5278-7.ch008

to examine whether it is a suitable model of waste management.

1 INTRODUCTION

The globe has experienced a significant evolution and growth in the post-Industrial Revolution period. One of the prime changes is in the nature of society, from agrarian to industrial. This also shifted the society from rural to urban (Cochran & O'Kane, 1977). Realizing the potential of urban society for educational, economic, and social benefits, many people from rural areas moved to urban areas. This became intense during the post-Second World War period. The second half of the 20th century further witnessed a huge shift of rural population into the urban areas. It started in the Global North but towards the end of the 20th century, with the support of globalization, liberalization, and open economy this move became global in nature. Besides, technically improved transport and communication facilities facilitated domestic and external migration for economic advantages and employment opportunities. Slowly, the number of cities and urban areas grew enormously in size and demography. That is, the world population became more urbanized than ever (UNDESA, 2018).

Along with industrialization and urbanization, many vital concerns like unemployment, housing facilities, education, healthcare, sanitation, water supply, and waste management became the center of the discussions on policy-making and socioeconomic development. All these required more coordinated action from the State agencies. Among all these complexities, the issue of waste management became the most prioritized concern of both urban local bodies (ULBs) and the people living in urban areas (Benny, 2022). It is more complex in urban areas compared to rural sector. Inappropriate waste management systems cause severe health hazards and environmental issues. It even affects the satisfaction of basic rights of people. The sustainable development paradigm demands that the development should be oriented to build up a sustainable society, sustainable economy, and sustainable environment (WCED, 1987). But in the absence of proper waste management system, this objective becomes meaningless.

Urbanization and waste generation are serious concerns of the most populated country, India (Dutta & Jinsart, 2020). Throughout the world, many waste management systems can be seen. India also has a variety of waste management systems. The city of Indore, the cleanest city in India, has its own very effective and efficient waste management system (Pradhan et al., 2025). But this chapter focuses on Haritha Karma Sena (HKS), a waste management system designed by the government of Kerala, India. The State of Kerala is acclaimed for its 100% literacy and effective measures on healthcare. Together with these achievements, it also has a strong plan of waste management which is claimed to properly function in many of the urban areas. This chapter thus makes a critical analysis of this model in order to examine how far this model fulfills its objective and advances as a suitable policy model for other Indian states and developing countries.

The chapter revolves around a problem statement that unscientific and unsystematic urbanization causes enormous complexities including waste management, which become an obstacle for the realization of sustainable development paradigm. The Indian state of Kerala is no exception for this basic understanding. The HKS model is acclaimed as a perfect model of waste management, but in reality, it functions with many drawbacks and complexities. Accordingly, the chapter primarily explores the evolution of urbanization around the globe explaining its benefits and challenges. It then examines urbanization and waste management in the specific context of India. It further analyzes urbanization and waste management more specifically within the context of Kerala and explains the details of the HKS model of waste management through the lens of sustainable development. Finally, the chapter critically evaluates the feasibility of this model to be accepted as a suitable model of waste management.

2 URBANIZATION AND GOVERNANCE

The globe, from the first half of the 20th century to the first quarter of the 21st century, witnessed massive growth in urbanization. Economic growth, population explosion, and migration caused such an increase in urbanization at a geometric rate (Nhamo et al., 2021). It is estimated that compared to the 746 million in 1950, 4.2 billion people in 2018 lived in urban areas. That is, in 1950 just 30% of the global population were urban but in 2018 almost 55% belonged to urban population (Hong et al., 2021). Besides, it is projected that almost 60% of global population will be urban in 2050. Urbanization is more rapid in Asia compared to other parts of the globe. In 1950, only 17.5% of population lived in the urban areas, whereas in 2018 it became 49.9% and it is projected to become 66.2% by 2050 (UNDESA, 2019). Urbanization was strengthened by economic liberalization and its structural reforms boosting rural-urban migration (Kundu, 2003). It causes socioeconomic and cultural changes in the society, which no one in the beginning of the 20th century could ever imagine.

Systematic governance is a vital requisite of urbanization. Governance with the qualifying adjective 'good' is essential for the development of cities. Governance is a broader term than government. Governance is an analytical framework consisting of a set of rules and processes (Obeng-Odoom, 2012). However, urban governance is very vague and most of the definitions either directly or indirectly calls for 'good planning' (Nuissl & Heinrichs, 2011). Urban governance needs to be understood in the context of decentralization, entrepreneurialism, and democratization (Obeng-Odoom, 2012). Urban governance was equated with urban management for a good period. This became a worldwide agenda with the launch of World Bank's Urban Management Program (UMP) in 1986, the United Nations Development Program (UNDP), and the United Nations Human Settlements Program (UN-HABITAT). Urban management primarily focused on economic development, social interaction, and poverty reduction (Davey, 1993). But many criticized this limited perspective of urban management as insignificant, vague, and narrow (McGill, 2001). Realizing the inefficiency of the urban management schemes, the UN started to popularize urban governance at the beginning of the third millennium with the launch of Global Campaign for Good Governance in 2001 (Tibaijuka, 2009). With this movement, the UN prioritized inclusive and collaborative governance to improve urban population's quality of life and the country's overall growth.

2.1 Transition from Agrarian to Industrial Society

The most significant transition in the history of humanity after the stone age was the Industrial Revolution, happened towards the end of the 18th century. This was mainly due to the technological advancements in the United Kingdom, Germany, and France (Bourdelais, 1987). With the Industrial Revolution, urbanization became a widely accepted phenomenon as more and more people migrated from the rural to the industrial areas. Such a migration and the increased employment opportunities and income of the area called for more infrastructure development and other facilities leading to urbanization. Urban development played a key role in the transition of the society from agrarian to industrial (Fields, 1999). Whether city is a product of modern industrialization is a complex point of discussion that requires a detailed examination (Hohenberg, 1990). According to Adam Smith (1776), increased production along with trade is the main agent of urbanization. Division of labor was central to his theory of production on the basis of which he differentiated agriculture and industry. Whereas Henri Pienne, one of the eminent historians of medieval times, credited trade and maritime commerce as the vital elements of urbanization (Fields, 1999). Accordingly, cities were mainly considered as the places in which manufacturing units are situated that give comparatively higher income than the agricultural sector. Hence, both urbanization and industrialization grew simultaneously enlarging the horizon of each other and delimiting the horizon of rural and agrarian sector.

2.2 Urbanization in the 20th Century

Various factors contributed to the increasing urbanization of the 20th century. Urban areas refer to the large concentration of people, political mechanisms, industries, production units, suppliers, and a variety of possibilities (Song et al., 2023). Scientific and technological advancements, progress in healthcare, increase in agricultural production, and advancement in basic education levels contributed to population growth, demanding more refined possibilities for survival. Urbanization became more intense along with this growing population (Hong et al., 2021). Two factors are mainly responsible for increased urbanization: rural-to-urban outflow of people and the reclassification of rural land into urban land (World Bank, 2013). Besides, the difference in wages and basic utilities at the rural and urban areas also cause urbanization. Rural workers are less paid compared to their urban counterparts and the basic utilities provided for their standard life are enormously insignificant than the urban population. A decrease in rural income is also a vital reason for the growing urbanization (Barrios et al., 2006; da Mata et al., 2007). Additionally, as technological advancements contributed to increased agricultural productivity with less human labor, more people migrated to the cities in search of other opportunities (Glaeser, 2014). Cities slowly became an important space for economic growth, social and political interaction, and global connectivity (Song et al., 2023). Due to the slowing down of agricultural growth and its lack of economic benefits, many people from the agrarian sector of rural areas were forced to migrate to the urban areas. This led to rapid growth in urbanization. However, most of these migrants were absorbed only in the informal economy (Kundu, 2003).

Urbanization marks a significant transition from rural to urban lifestyle together with a cultural transition (David & Henderson, 2003). This induces billions of people to move towards the urban areas resulting in a remarkable escalation in the number of urban populations that almost reaches more than half of the world population (UN World Urbanization Prospects, 2018). This is mainly because of the attractive facilities of urban areas like quality education possibilities, good healthcare facilities, and enormous employment opportunities (V & C A, 2024).

Urbanization from the last decades of the 20th century was closely linked with the idea of sustainable development. The concept of 'sustainable cities' originated from the paradigm of sustainable development. Previously, urbanization was associated with increased consumption and ecological degradation. But with the idea of sustainable urbanization, it has become an inclusive concept of sustainability, development, equitable distribution, and human well-being. Sustainable urbanization focuses on people-centered development with the support of decentralized government arguing that the developmental aspirations should be broad, bottom-up, just, redistributive, equitable, empowering, eco-friendly, and seeking to satisfy the needs of both present and future generations (Mahadevia, 2001).

The idea of sustainable cities is popularized within sustainable development paradigm and ecofriendly movements. It underlined that urbanization should consider its economic and environmental costs, self-reliance in production, consumption, and waste-management, self-reliance in energy usage, and balanced protection of the rights of all its people (Haughton, 1997). Proponents of this Northern ideology of sustainable urbanization forced the Southern countries to embrace the concept of sustainable cities through the Northern lens demanding to reduce resource consumption, use of renewable resources, and local waste absorption. But they ignored the vital element of meeting the basic needs of people (Satterthwaite, 1998). Urbanization should not be imposed through the Northern or Southern lenses. It should evolve and develop within the context of space and time while respecting the needs and aspirations of people. Urbanization in any part of the world should revolve around social equity, environmental sustainability, political decentralization, and equitable redistribution of the benefits of economic growth (Mahadevia, 2001).

2.3 Benefits and Challenges of Urbanization

Urbanization promotes economic growth and increase in Gross Domestic Product (GDP). There are many factors responsible for this financial advantage. Firstly, majority of the migrants from the rural to urban areas are educated people and their qualifications and skills contribute to the growth of the economic sector. Qualified human capital is an essential concern of urbanization. In countries like China, the government thus took initiative to drive rural-urban migration to energize the economy (Hong et al., 2021). Secondly, denser cities provide apparently higher wages to the workers (Henderson et al., 2021). Thirdly, urbanization brings firms, suppliers, and consumers in a close vicinity reducing the costs of transportation. Fourthly, specialization is an essential component of urbanization, which strengthens productivity. Specialization leads to more complex goods and services and quick consumption (Hong et al., 2021).

Urbanization also causes various challenges to human well-being and societal welfare. With the growing urbanization, the demand for water, energy, food, and other resources also increases. At certain times, it becomes a huge challenge to the society (David et al., 2022). Water scarcity is one of the major issues with urbanization. During the summer season, many cities face water scarcity. In 2024 summer, the city of Bangalore, India, faced huge water scarcity leading to restrictions on water supply (just 1-2 hours for 1-2 days in a week). He et al. (2021) argued that by 2050, half of the world's largest cities and more than 2 billion urban population will face water scarcity. Food consumption in urban areas is two times higher than the rural areas (Zhou & Staatz, 2016). Urban population consumes 75% of global energy supply (Chowdhury et al., 2019). Air pollution is another serious drawback of increased urbanization. It is said that India's urban areas are becoming "lethal gas chambers" and Delhi is considered as the most polluted capital city in the world. There are also severe disparities in the living situations of the urban population due to differences in employment and income. Exorbitant increase in land and house plot prices cause issues with regard to housing and many are forced to stay in slums with subhuman conditions (Chakrabarty, 2001). It also causes anthropogenic environmental degradation affecting the availability of resources, reduction of agricultural land, biodiversity, and sanitation and waste management issues (Nhamo et al., 2021). According to the World Bank (2023), overcrowding, insufficient infrastructure, and environmental problems cause more than 50% of forcibly displaced people to live in urban areas without proper identity and means of survival. Urbanization also changes the consumption patterns of people causing more waste generation (Zhang et al., 2019). All these expose the unplanned and uncontrollable growth of large cities and waste accumulation in most parts of the world, including India.

2.4 Urbanization and Governance

Urbanization poses severe concerns on the efficiency of governance as cities are identified with complexity and pluralism. Urban governance can be defined as the sum total of collective work arrangements, mostly formal and informal government actions, that focus on the development and proper functioning of the cities (Song et al., 2023). Urban governance is a vital element for the effective functioning of any country as the majority of population lives in the cities. Accordingly, some authors argue that 'mayors rule the world' (Barber, 2013). The present globe cannot think of governance, development, and well-being without proper urban planning, development, and governance systems. Society's shift from rural to urban nature insisted on good urban governance. However, as stated above, urban management was the modus operandi for a significant period of time. Urban management was a unilateral approach, whereas urban governance stands for multilateral, plural, and diverse ways of administration consisting of stakeholders at different levels (Obeng-Odoom, 2012).

Urban governance is comparatively a new area of research. It has not been established as a systematic discipline. Until the 1950s, urban governance was done through the traditional governance system where the government focused on the construction and maintenance of urban areas for increasing its efficiency and productivity. During the 1950-1980s, urban governance emphasized the interaction between government, market, and society. It devised appropriate planning in view of practical methods that can strengthen the development of urban areas. From the 1980s to late 1990s, urban governance was transformed from traditional model to urban network model, effectively adopting private capital and market mechanisms. From the 21st century onwards, urban governance is more characterized by the participation of multi-subjects as multiple actors actively contribute to the planning and development of urban areas (Song et al., 2023). It devises various plans and policies to improve the society, economy, and environment. For effective functioning, urban governance requires the collaboration of both government and non-government actors. Urban governance requires the active interaction between urban local governments, businessmen, landowners, NGOs, and civic society. That is, urban governance is a wider term compared to urban government (Obeng-Odoom, 2012). Urban governance is a vital point of attraction in the context of migration, climate change, security, and a fragile global economy.

Urban governance becomes meaningful within the context of appropriate support from the higher forms of government like central, provincial, or state governments. The urban administrative bodies should be empowered with political, administrative, and fiscal powers necessary for an effective governance. The higher forms of government should coordinate strategies and policies and provide appropriate human and other resources for a better governance (da Cruz et al., 2019). Decentralization helps to strengthen urban governance. It provides increased attention to the basic requirements of the cities and its development. Since governance calls for the collaboration of both government and non-government actors, it can promote sustainable development of the cities with a progressive vision. Compared to the broader program of action of the higher-level government, urban government systems can focus more systematically on the well-being of urban population. It helps them to initiate many welfare programs and policies. In this context, publicity is an important area of urban governance as it has to reach to maximum number of people through its social benefits and social welfare schemes and policies (Song et al., 2023). However, urban governance becomes effective only when there is sufficient allocation of funds. Besides, it also has to consider issues like corruption and exploitation (da Cruz et al., 2019). A concentrated approach towards human well-being and urban welfare, with the support of proper allocation of funds and other mechanisms, can make wonders in urban governance. However, effective steps for empowering urban governance are very rare as most countries do not significantly appreciate this move.

3 INDIA'S URBANIZATION AND DEVELOPMENT

India is one of the fastest growing economies in the world. With its liberalization policies, India's economic growth became explicit from the last decade of the 20th century. India slow pace of urbanization was acknowledged by the 11th five-year plan (Planning Commission, 2008). Besides, India's political leadership largely focused on developing rural areas inspired by the idea of Gandhian Gram Swaraj. Gandhi's vision that India lives in its rural regions heavily influenced the political leadership and it resulted in prioritizing the demands and concerns of the rural India (Garg & Raut, 2015; Kazmin, 2016). Ani-urban bias was very much visible for long period. But urbanization became the center of attraction from the shift towards open market and liberalized economic policies and with the growing influence of multinational and transnational companies in the urban areas of the country. However, India's urbanization lacks proper planning and policies resulting in unsustainable urbanization (Gore & Gopakumar, 2015).

Inspired by the 74th Constitutional Amendment of 1992, more autonomy and authority are provided to ULBs. By providing constitutional status, it expanded the organization, powers, functions, and jurisdiction of ULBs and freed them from the control of central and state governments. However, in practice it all depended on the vision of centralization. Urbanization became a major concern of Indian polity only in the 21st century. India's urbanization experienced a significant surge during 2001-2011, when the urban-rural ratio doubled around 45% (V & C A, 2024). As of now, India holds the second position after China with regard to urban population. However, compared to many other parts of the world India is still one of the least urbanized countries (Chakrabarti, 2001). The 11th five-year plan affirmed that urbanization is a natural outcome of development. Realizing the importance of urbanization, the government of India initiated the Jawaharlal Nehru National Urban Renewal Mission (JNNURM) in 2005 to create economically productive, sustainable, efficient and responsible cities. It calls for providing more central assistance to the development and maintenance of urban areas (Planning Commission, 2008). Although the JNNURM had a significant beginning and functioning, it was closed in 2014 with the change of government.

3.1 Sustainable Development and Urbanization in India

Indian administrative attention and political decentralization concentrated on the growth of its rural areas. The rapidly growing urban areas and urban population demand more systematic reforms and it would be possible only through institutional reforms and changes in mindset and policies. Indian policy-makers concentrated on improving the quality of people's urban life and the society's well-being with better living conditions and balanced urban-rural growth only in the beginning of the 21st century (Ahluwalia, 2017). This balanced approach became crucial in the context of UN's sustainable development paradigm. The interest in urbanization should focus on the three pillars of sustainable development such as, environment, society, and economy (Akinsete et al., 2022).

A sustainable city is an inclusive concept that stands for societal (human), economic, and environmental standards of development and maintenance with the required capacities, resources, and capabilities. Building up of sustainable city calls for independent functioning of urban local governments. India's urbanization and sustainable city development had insignificant growth due to the unnecessary control and regulations from the central and state governments. Sustainable city development and systematic urbanization require the involvement, initiatives, commitment, and efforts of urban population, progressive vision and enhanced understanding of their elected representatives, capacity and capability of local administrators, and active participation of the urban businessmen, industrialists, and other influential people (Chakrabarty, 2001).

With the 74th Constitutional Amendment, the sustainable development of the urban areas became a prerogative of the elected ULBs. In previous times, the state government had the power to suspend and dissolve urban governments even for a prolonged period. But with the amendment, it has become a responsibility to constitute the new urban government within six months of the dissolution. It shows the importance and constitutional protection given to ULBs for advancing sustainable urban development. The amendment also included the 12th schedule in the Constitution that enlists the powers and functions of ULBs. It consists of 18 powers and functions that are included as per the provisions of Article 243-W of the Constitution. These powers cover a wide range of activities like urban planning including town planning, regulation of land-use and construction of buildings, planning for economic and social development, roads and bridges, water supply for domestic, industrial, and commercial purposes, public health, sanitation, and solid waste management, fire services, protection of urban forestry, environment, and promotion of ecological aspects, safeguarding the interests of the weaker sections of the society including the handicapped and mentally retarded, improvement and upgradation of slums, urban poverty alleviation, provision for urban facilities like parks, gardens, and play-grounds, promotion of urban cultural, educational and aesthetic aspects, provisions with regard to burial, burial grounds, and crematorium, prevention of cruelty to animals, preservation of statistics like the registration of births and deaths, public facilities like street lights, parking lots, bus stops and public conveniences, and regulation of slaughter houses and tanneries. All the 18 functions of ULBs point to their assigned role for sustainable urban development. Besides, the Constitution also stated that the ULBs should be assigned the power to levy, collect, and appropriate certain taxes, duties, and fees within certain conditions and limits for its discharge of duties. It also noted that the state government should provide appropriate grants-in-aid from its consolidated fund (Article 243-X). Furthermore, the Constitutional amendment provided that the State Finance Commission should be constituted with the mandate to address the financial position of ULBs and give suggestions about the allocation of funds (Article 243-Y). This provision was in view that without enough funds, ULBs will not be able to achieve its objective of sustainable development. However, still now, no proper financial backing and capacity for planning and management is given to ULBs (Ahluwalia, 2017).

3.2 Urbanization and Waste Management

One of the worst issues with Indian urbanization is waste generation and unsystematic waste management. Only in the context of systematic town planning, waste management becomes meaningful and significant (Ahluwalia, 2017). Waste includes all those materials that have no further use regarding production, transformation, or consumption. It can only be disposed of permanently (Govt of Kerala, 2022). But, among the so-called waste, there will be certain materials that can be used further, if there is a systematic segregation. Waste management is an umbrella term that includes various activities like collection, transport, processing, recycling or disposal, and monitoring of waste (G., 2014). Millions of tons of solid and other waste are generated in the urban areas. It is estimated that almost 44% of the waste generated in the country is produced by cities with a population of more than 100000 (Singh, 2021). Among this, almost 60% are collected daily. But the rest is accumulated and decomposed in public spaces causing many health issues and environmental crises (Chakrabarty, 2001). Waste management is a part of environment management dynamics. Proper and systematic waste processing is essential for ensuring environmental and human securities, which are highly lacking in many urban areas. Right to clean environment, derivative of right to life as per Article 21 of the Constitution, is considered as a fundamental right in India.

Towards the realization of proper and systematic waste management objective, many local urban governments tried to acquire the support of local communities, NGOs, and private agencies (Mahadevia, 2001). Each city in India has its own unique way of waste management. One of the earliest significant urban waste management systems in India was in Chennai, where Exnora, an NGO initiative, advanced an action plan for garbage collection together with awareness campaign. They renamed the rag-pickers as city-beautifiers and provided them loans to buy tricycles for door-to-door garbage collection and disposal. The residents paid them monthly salaries from which they repaid the loans. This initiative also took care of eco-friendly activities and awareness programs, promoting a culture of sustainable waste management (Anand, 1999). Inspired by this movement, Vadodara city started a similar waste management scheme in 1992 engaging the unemployed youth and rag-pickers with a monthly salary from the local residents. The waste collectors were allowed to sell the recyclable waste (plastic, paper, and metals). They also converted the degradable waste to manure with the financial and technical support of the local government (Cherail, 1994). Thereafter, various ULBs started initiatives of waste management. In the present times, specific focus is given to waste management with the launch of Clean India (Swachh Bharat) program of 2014 that has an urban component as well. The urban perspective of this mission calls for Open Defecation Free (ODF) towns and cities by developing various awareness programs in relation to toilet uses and waste disposal, and solid waste management networks (Singh, 2021). However, the success of this program highly depends on the behavioral changes of people (Ahluwalia, 2017).

Indore is acclaimed as the best waste management city in India thanks to the wonderful coordinative efforts of its local government. In 2016, Indore was in the 25th position with regard to waste management and cleanliness in Indian cities. But in 2017, it achieved the number one position. Thereafter, it retained this position with the collaborative efforts of local government, NGOs, businessmen, and the public. Indore has a huge collection of user fees, high punishment charges, and also makes revenues by selling fertilizers and recyclable waste. It promotes decentralized waste processing, which helps to function more effectively and efficiently. It devised a community-oriented initiative by giving proper awareness to people for making the city clean by collecting door-to-door assorted waste without dumping into city's open territories. When the initiative began in 2016, it was not an easy process. But with the growing awareness of people, it became easy and efficient. The city disposes waste in various ways like gasification, incineration, composting, landfill, and recycling. Systematic planning and implementation help the city to benefit from proper waste management for better governance of the city as well as to make the city clean (Singh, 2021).

4 HARITHA KARMA SENA MODEL OF WASTE MANAGEMENT

The term "Haritha Karma Sena" means "green task force" and they are entrusted with the responsibility of greening the society. HKS was launched as part of Haritha Keralam (Green Kerala) mission, which was one of the four significant missions undertaken by the government of Kerala to build up a new Kerala. This mission was launched in 2016 with the objective of advancing cleanliness, water security, water abundance, and safe food production. It functions with the motto of water, hygiene, and harvest, and with three sub-missions related to water conservation, agriculture, sanitation and waste management. This mission is mainly carried out through the initiatives of local self-government bodies (Govt. of Kerala, n.d.). For a good period, Kerala's waste management system was centralized, but with the latest developments and the Supreme Court verdicts, it prioritizes decentralization. However, the government of Kerala still holds the coordinating role. The Supreme Court of India in 2003 affirmed that the local bodies cannot claim lack of funds as a valid reason for the failures in waste management. They have the responsibility to implement an effective waste management system as a response to decentralization and improved governance possibilities.

4.1 Urbanization in Kerala

Kerala is the second most urbanized state in India, after Goa. Three of Kerala's cities—Malappuram, Kozhikode, and Kollam—are among the world's top 10 fast-growing urban areas. Usually, urbanization is equated with city concentration. Still, the case of Kerala shows that its urbanization pattern depends on the expansion of urban areas. The peripheries of cities are swiftly becoming urban. Since Kerala is small in size with a vast population, its free and open space is comparatively very less. Throughout the State, a semi-urban pattern is very much visible. Besides, Kerala has many quality educational institutions and healthcare facilities throughout the State, and there is no much distinction between rural and urban areas. Both transport and communication facilities are widely available in almost all parts of the State. Kerala's urbanization is primarily resulted from the shift in occupational structure and not from migration. Compared to many other parts of the world, Kerala has no metropolitan city as such but many significant urban areas (V & C A, 2024). Accordingly, Kerala's urban population constitutes almost 50% of its total population (Govt of Kerala, 2022).

Lal and Nair (2017) recorded that since 1901, there has been a steady increase in the percentage of Kerala's urban population. "Kerala's urbanization is notable not just for its pace but also for the quality of its urban development, with improvements in infrastructure, healthcare, and education in urban areas" (V & C A, 2024, p. 6). It was during the 1980s that Kerala recorded a high rate of urbanization and it was mainly due to the transition of the workforce from the agricultural to the tertiary sector (Lal & Nair, 2017). Among the 14 districts in Kerala, Ernakulam is the most urbanized district. More than 60% of population of the four districts—Ernakulam, Thrissur, Kozhikode, and Kannur—are urban. Wayanad and Idukki are the less urbanized districts in Kerala mainly due to their geographical complexities (V & C A, 2024). Kerala's history of urbanization explains that in the post-Independence period it witnessed a rapid increase compared to India's slow pace of urbanization. This has resulted in a situation where almost all parts of Kerala are urbanized. Kerala has, at present, six municipal corporations and 87 town municipalities covering a large urban area.

4.2 Waste Management Problems in Kerala

Both unsystematic urbanization and increase in urban population cause huge waste accumulation. Economic development and the rising living standards also increase waste generation (G., 2014). Among the total waste generated in the State, almost 69% is biodegradable and 31% is non-biodegradable (Govt of Kerala, 2022). Population density in Kerala is also very high compared to the national average. As per the 2011 census, Kerala's population density is 860 persons/km^2. This causes lack of free space in the State, especially in urban areas, affecting the waste management programs of the corporations and ULBs. The rapid population growth is challenging ULBs' capacity to significantly manage even the solid waste. Although Kerala is highly acclaimed for its educational level and social sensitivity, it shows poor performance with regard to waste management. Both urban local authorities and public are responsible for this irresponsible behavior. ULBs lack professionalism and commitment and the public lack proper awareness and responsible action in relation to waste management (Ajith, 2016). In 2002, the State launched clean Kerala mission with the objective to make a free waste Kerala (Tom & Kamat, 2020). Although there are rules and regulations with regard to single use plastic materials, Kerala records high-use of single use plastic materials in the absence of proper implementation mechanisms (Benny, 2022).

Kerala is having very strong monsoon and summer seasons. Waste management services cannot be equally applied to both these seasons. The monsoon season requires a more concentrated and effective method of waste disposal mechanism compared to the summer season. Waste collection and disposal frequency should be high during the monsoon season. But such an effort is not well-established. People's waste disposal methods are characterized by three common means: open dumping to a common place, depositing in community bins, and door-to-door collection. Among these methods, both depositing in community bins and door-to-door collection systems are good. But they need to be properly implemented. Open dumping to a common place was the main method followed in the State for a long period of time. Such dumped waste caused many health issues and environmental concerns. Although many signboards and regulations are placed against open dumping, many people still practice it as they are not provided with other effective ways of waste disposal systems. Depositing waste in community bins requires proper awareness among people, necessary bins for depositing waste, and effective mechanisms for the timely transfer and cleaning of bins. Although many urban bodies are rather well-equipped with waste collection, they lack systems for transferring, recycling, and disposing waste (Ajith, 2016).

As we have seen above in the context of Indore, door-to-door collection of waste is the most effective system. It requires high awareness and sensitivity from the public and responsible commitment from the ULBs (Benny, 2022). Unsystematic segregation of waste at source and secondary levels cause a crisis with its management causing flow of mixed waste to the processing units (Govt of Kerala, 2022). Kerala also lacks proper mechanisms to convert waste into energy and other materials. Accordingly, ULBs try to dump the collected waste, without much initiative for its separation and conversion, into the peripheries of urban or rural areas. But the severe protest from the local people in such areas causes more obstacles to waste management (G., 2014). Local people protest against such dumping of waste because of the realization that it causes severe health issues and environmental concerns. Local people's protest against Brahmapuram waste disposal plant is a vital example.

4.3 Evolution of Haritha Karma Sena

Govt of Kerala (2022) estimated that annually 3.70 million tons of solid waste is generated in Kerala, of which almost 59% comes from urban areas. Solid waste management is one of the 18 subjects allotted to the responsibilities of ULBs. They have the responsibility to facilitate scientific processing and infrastructure development with regard to collection, segregation, storage, transportation, processing, conversion, and disposal of waste. The government of Kerala formed HKS for fulfilling this assignment. However, solid waste management is done not merely by the individual efforts of HKS but by the collective efforts of local self-government department, environment department, Kerala State Pollution Control Board (KSPCB), Suchitwa Mission, Haritha Keralam Mission, and Clean Kerala Company Limited (CKCL). Although all these departments, programs, and companies do the coordination, HKS does the ground activities.

HKS evolved as a micro-entrepreneurial initiative under Haritha Keralam Mission in 2017. It is responsible for greening the Kerala society by collecting solid waste with the support of local self-governments. HKS is one of the pioneers of decentralized waste management practices. HKS is the field level waste management system. Most of its workers are women, recruited from Kudumbasree, an NGO for the empowerment and welfare of women (Nishad, 2024). By giving appropriate training, it empowers women to manage waste, earn income, and promote civic sense (Praveen, 2024). At present, more than 36,000 people are employed in HKS. Local self-governments provide them basic infrastructure facilities, vehicles, uniforms, and identity cards.

4.4 Working of Haritha Karma Sena

In order to address the complex issue of solid waste management, the government of Kerala instituted HKS. It functions with the state government order under the supervision of the local self-government bodies. HKS members mainly collect cleaned non-organic waste from households and institutions by door-to-door collection system at regular intervals. The collected non-organic waste is transferred to the material collection center, where it is sorted and send to recycling or reuse. They occasionally collect organic waste, if required by the local bodies. For the collection of waste, the local bodies have fixed a nominal user fee as per the government order in support of HKS. With the support of many NGOs and companies, they try to create new products from waste materials, promote recycling, and advance innovative green initiatives in the State. With the working of HKS, tons of waste are efficiently disposed of (Govt of Kerala, n.d.). In the absence of effective solid waste management system, environmental degradation, pollution, and health hazards will become a common thing (Nishad, 2024).

The normal working of HKS with regard to solid waste management consists of door-to-door collection of waste from households, shops, and institutions during 10-16th of every month. After the six days of waste collection, two days are spent for waste segregation. Altogether its operation consists of eight days. Thereafter, the plastic is given to the CKC company, which uses it for the construction of roads. The company pays HKS for the purchase of plastic. Besides, HKS members make products from recyclable materials for selling to the public. This depends on the choice and initiative of each HKS unit (Nishad, 2024). The coordinating authorities give them required funding, infrastructure, and training. Currently, HKS collects 50 from households and 100 from shops and institutions (for two sacks) as user's fee. This fee is credited in the HKS unit account and at the end of every month, they share the amount among themselves (Shameerdas, 2024).

HKS's waste to wealth project functions with strong awareness program and active commitment of the HKS members (Praveen, 2024). HKS members educate the residents of a locality to segregate waste at sources differentiating them as biodegradable, recyclables (paper, plastic, glass, and metal), and non-recyclables (sanitary waste). Waste collection is mostly done by door-to-door collection and sometimes it is collected through fixed collection points. As of January 2023, there are 114 Material Collection Facilities (MCF) in municipalities and 21 in corporations (Shameerdas, 2024). With regard to recyclable waste materials, HKS either sets up its own small-scale recycling industries or collaborate with such companies. Organic waste is converted into energy or compost by installing such units at community centers. Waste that cannot be recycled or disposed is used for landfill, but HKS tries to reduce its quantity through awareness creation. The most important work that HKS does with regard to waste management is the creation of awareness among the public (Nishad, 2024).

4.5 Advantages and Challenges

HKS's functioning affirms that it is a beacon of hope in systematic waste management in the State of Kerala (Ravikumar, 2024). Since Kerala's free space is much less, it is essential to reduce the amount of solid waste by effectively employing methods of recycling and conversion. The collaborative and coordinated efforts of many agencies in the field and public-private partnerships help to strengthen the project's efficiency. Solid waste segregation helps to channelize recyclable waste for recycling and plastic waste to road tarring. Organic waste is used to make biogas and compost at sources, helping to make the State energy efficient. Since HKS members provide training and support to households and institutions to convert biodegradable waste to gas and compost, people can easily cope up with the new technologies for survival and ecofriendly living (Shameerdas, 2024).

Giving the example of 50-year-old Vasanthakumari, a class X dropout, who earns 25,000/month for her work in HKS, Praveen (2024) says that it empowers women with monetary benefits. Since majority of HKS workers are uneducated women, they get a chance to earn good amount of money with some socially valuable works. They can also become more creative as government provides them financial and technical support to start up some small-scale units for converting recyclable materials into usable products. This continuous support of ULBs and the State government is an energizer for HKS.

Although HKS has many advantages, it also confronts certain challenges. Lack of highly qualified waste management equipment, technologies, and waste treatment centers is a serious drawback of Kerala's waste management system. Lack of experienced and skilled personnel in waste management processing units and treatment plants also causes serious issues (Benny, 2022). The financial capacity of ULBs, who are supposed to support HKS, is very low and it sometimes affects its functioning. Lack of awareness and support from the public is another major drawback that many of the green warriors (HKS) face in their daily activities.

4.6 Haritha Karma Sena through the Lens of Sustainability

Sustainability refers to the ability to maintain at a certain level. This can be applied to society, environment, and economy. Drawing a fixed line for sustainability would be difficult as it varies according to culture and circumstances. Sustainability level depends on each country's level of development (Chakrabarty, 2001). Sustainable society refers to a society whose members are free from fear, want, and indignity.

Societal sustainability prioritizes human security demands. Sustainable economy insists that the benefits of economic growth should be equitably distributed. Economic sustainability stands against the concentration of income and wealth into the hands of a few and the uncontrollable increase in the gap between rich and poor. Sustainable environment stands for a more neutral and stringent standard that aims at establishing ecological balance by controlling the depletion of natural resources (UNCHS, 1998).

HKS is a model of sustainable development (Ravikumar, 2024). HKS is designed to function primarily for the benefit of society. Its objective was to reduce health hazards that are caused by the accumulation of waste in the public and improve the living standards of people. But whether this model was really successful in achieving this objective is a challenging question. Its success depends on the interest of ULBs and the commitment of HKS members. In certain areas, it functions well, in some areas it functions well for a specific period of time, and in some other areas it remains only in policy guidelines. There are diverse opinions with regard to its success (Thomas et al., 2023, Ravikumar, 2024). Some argue that it is beneficial to the society (Ravikumar, 2024), some other argue that it helps to provide economic sustainability to its members (Thodukayil et al., 2024). As stated above, majority of HKS workers are uneducated women, and with the help of this project, they become economically sustainable. Furthermore, the most advantageous area of HKS environmental sustainability. By systematic and scientific solid waste management system, HKS reduces the ecological burden that happens through the accumulation of non-degradable waste (Chathukulam, 2024). If properly implemented, HKS model has the potential to become an effective mechanism for ensuring sustainability in the society, economy, and environment.

5 HARITHA KARMA SENA MODEL: A CRITIQUE

According to the latest policy of the State of Kerala, local bodies are responsible for taking care of the solid waste generated in the State. Scientific processing and recycling of solid waste is an effective means of waste management. Meanwhile, citizens have the responsibility to manage biodegradable waste at its source. It functions under the campaign "my waste, my responsibility" (Govt of Kerala, 2022). As stated above, a major part of the waste is biodegradable and its disposal would not be an issue in rural areas. But in the urban areas, where the majority of people live in apartments, it requires efficient planning and coordination. Besides, the biodegradable waste from the markets, restaurants, and other public places are also a cause of concern. Here, ULBs cannot escape from its responsibility. The HKS system here functions by educating and encouraging the households, community centers, and institutions to plant biogas plants and composting systems that can address the issue of biodegradable waste. It supports them to use such waste for their benefit instead of dumping it into the public. But how far this becomes effective is a convoluted eroteme. Ravikumar's (2024) quantitative study claims that the public believes that HKS is only moderately successful. It shows that the idea behind this inspirational move is laudable, but it cannot be considered as a fully successful project due to lack of implementation mechanisms and policies.

HKS model of solid waste management affirms that door-to-door waste collection or the collection from fixed collection points on fixed dates by giving proper awareness to the public and making use of certain regulations and fines can solve the issues of environmental degradation and unsystematic waste management. But it does not properly function in all urban areas of the state. It all depends on the commitment of HKS members and the availability of infrastructures. Ravikumar (2024) thus claims that about 40% are not satisfied by the functioning of HKS. She further states that HKS is only partially successful in managing e-waste. All these are matters of serious concern. On the one side, there is an efficient policy initiative, but on the other side, there is lack of implementation mechanism. Just like any other policy initiative, HKS can also become a truly successful project only through stringent measures for its implementation.

By giving employment to thousands of uneducated women, HKS shows that together with waste disposal it also provides employment opportunities for the less educated and less qualified sections of the society. However, they are not the formal workers of the State. There should be proper state initiative to shift them from the informal to formal sector, realizing the importance of their service (Thodukayil, 2024). They are not provided with insurance and other benefits. Most of them are not even aware of many governmental schemes and policies. Many times, they are not supported by the local bodies. Sometimes, they also face discrimination from the society. They are not provided with good buckets for waste collection, appropriate transport facilities, and raincoats during the monsoon time (Thomas et al., 2023). Without enhancing the working conditions of HKS members through policy interventions, this model cannot be made into an effective waste management system (Thodukayil et al., 2024). The State and local bodies should respectfully address their concerns and needs in order to make Haritha Keralam mission successful.

Finally, HKS is often acclaimed for its decentralized way of functioning. Decentralization in waste management is very effective for the developing world as it can reduce the burden on the central or provincial governments by actively employing the involvement of the local self-governments. With the decentralized functioning, the waste management system will be more effective as the local authorities can be easily approached and questioned by the public. However, the problem with HKS model is that the local bodies do not have enough financial resources and personnel to support the concerns of HKS members. They require financial support and coordination from the state government. It shows that though decentralization is propagated in theory, in practice it does not happen.

6 CONCLUSION

Urbanization has become very active in the aftermath of the Industrial Revolution that led to the transition from agrarian to industrial society. This transition became more explicit in the aftermath of the Second World War. Although urbanization became a common phenomenon in the North in the second half of the 20th century, it became common in the South only towards the end of the 20th century. Compared to many other parts of the world, India is less urbanized. After independence, India focused on the villages due to its allegiance towards Gandhian idea of Gram Swaraj. India's urbanization is mainly a 21st century phenomenon. However, the State of Kerala has a long history of urbanization. At certain points, its percentage of urban population even surpassed national average. At present, it is the second urbanized State in India after Goa. However, along with urbanization, many challenges also emerged, including waste management.

For a good period, Kerala had a centralized waste management system and it did not become a success. But towards the second decade of the 21st century, Kerala initiated a decentralized waste management system under the Suchitva Mission and Haritha Keralam Mission, called Haritha Karma Sena. It mainly consists of uneducated women. With the support of ULBs, they are given training and they become agents and educators of Kerala's waste management system. It also empowered them with monetary benefits and skills in the form of small-scale industries. This model aimed to provide social, economic, and environmental sustainability to the State. Although this model succeeded in giving some impact on urban waste management issue, it cannot be considered as a fully successful model. The HKS policy initiative is a well-designed model, but the problem lies in its implementation process. It is highly depended on the interest of the local bodies and the commitment of HKS members. The interest of the local bodies depends on the interest of the persons in authority and it changes with government. Since the government does not treat HKS members as formal sector workers and does not provide them enough support, they also fail to show commitment. Besides, the lack of awareness and commitment of people also make this model to partially fail. Although it stands as a good project in theory, it lacks success in practice. But if proper attention and financial support is given, it can become a perfect policy model for waste management that can even influence the whole developing world.

REFERENCES

Ahluwalia, I. J. (2017). Urban Governance in India. *Journal of Urban Affairs*, *41*(1), 83–102. https://www.tandfonline.com/doi/full/10.1080/07352166.2016.1271614. DOI: 10.1080/07352166.2016.1271614

Ajith, P. S. (2016). A Study on the Solid Waste Collection, Treatment and Disposal System of Municipalities in Kerala. *JETIR* 3(12), 702-712. https://www.jetir.org/papers/JETIR1701510.pdf

Akinsete, E., Koundouri, P., Kartala, X., Englezos, N., Lautze, J., Yihdego, Z., Gibson, J., Scholz, G., van Bers, C., & Sodoge, J. (2022). Sustainable WEF Nexus Management: A Conceptual Framework to Integrate models of Social, Economic, Policy, and Institutional Developments. *Frontiers in Water*, *4*, 727772. https://www.frontiersin.org/journals/water/articles/10.3389/frwa.2022.727772/full. DOI: 10.3389/frwa.2022.727772

Anand, P. B. (1999). Waste Management in Madras Revisited. *Environment and Urbanization*, *11*(2), 161–176. https://journals.sagepub.com/doi/abs/10.1177/095624789901100214. DOI: 10.1177/095624789901100214

Arya, V., & Rejuna, C. A. (2024). Unveiling Kerala's Distinct Urbanization: A Comparative Analysis within India. *The Indian Economic Journal*, •••, 1–16. https://journals.sagepub.com/doi/full/10.1177/00194662241278066

Barber, B. (2013). *If Mayors Ruled the World: Dysfunctional Nations and Rising Cities*. Yale University Press.

Barrios, S., Bertinelli, L., & Stribl, E. (2006). Climate Change and Rural-Urban Migration: The Case of Sub-Saharan Africa. *Journal of Urban Economics*, *60*(3), 357–371. https://www.sciencedirect.com/science/article/abs/pii/S0094119006000398. DOI: 10.1016/j.jue.2006.04.005

Benny, V. (2022). Practices and Challenges of Household Plastic Waste Disposal: An Evaluation of Waste Management System in Kerala. *Orissa Journal of Commerce* 43(2), 27-44. https://ojcoca.org/volume43_issue2/OJC_2022_43_2_03.pdf

Bourdelais, P. (1987). Transitions from Agricultural to Industrial Societies: Some Introductory Remarks. In Schröder, W. H., & Jarausch, K. H. (Eds.), *Quantitative History of Society and Economy: Some International Studies* (pp. 113–119). Scripta Mercaturae Verl.

Chakrabarti, P. G. D. (2001). Urban Crisis in India: New Initiatives for Sustainable Cities. *Development in Practice*, *11*(2-3), 260–272. https://ideas.repec.org/a/taf/cdipxx/v11y2001i2-3p260-272.html. DOI: 10.1080/09614520120056397

Chathukulam, J. (2024). Climate-Resilient Governance: Nurturing Eco-friendly Public Services at Grassroots in Kerala, India. *Unisia* 42(2), art.21. https://journal.uii.ac.id/Unisia/article/view/37651

Cherail, K. (1994). Haul Your own Garbage. *Down to Earth*, *3*(8), 10. https://www.downtoearth.org.in/environment/haul-your-own-garbage-32442

Chowdhury, P. K. R., Weaver, J. E., & Bhaduri, B. L. (2019). Electricity Consumption Patterns Within Cities: Application of a Data-Driven Settlement Characterization Method. *International Journal of Digital Earth*, *13*(1), 119–135. https://www.tandfonline.com/doi/full/10.1080/17538947.2018.1556355. DOI: 10.1080/17538947.2018.1556355

Cochran, L. T., & O'Kane, J. M. (1977). Urbanization-Industrialization and the Theory of Demographic Transition. *Pacific Sociological Review* 20(1), 113.134. https://www.jstor.org/stable/1388908

da Cruz, N. F., Rode, P., & McQuarrie, M. (2019). New Urban Governance: A Review of Current Themes and Future Priorities. *Journal of Urban Affairs*, *41*(1), 1–19. https://www.tandfonline.com/doi/full/10.1080/07352166.2018.1499416. DOI: 10.1080/07352166.2018.1499416

da Marta, D., Deichmann, U., & Wang, H. G. (2007). Determinants of City Growth in Brazil. *Journal of Urban Economics*, *62*(2), 252–272. https://ideas.repec.org/a/eee/juecon/v62y2007i2p252-272.html. DOI: 10.1016/j.jue.2006.08.010

Davey, K. (1993). *Elements of Urban Management*. World Bank. DOI: 10.1596/0-8213-2424-1

David, L. O., Nwulu, N. I., Aigbavboa, C. O., & Adepoju, O. O. (2022). Integrating Fourth Industrial Revolution (4IR) Technologies into the Water, Energy, and Food Nexus for Sustainable Resource Security. *Clean Technologies and Environmental Policy*, *26*(3), 803–820. https://pure.uj.ac.za/en/publications/integrating-fourth-industrial-revolution-4ir-technologies-into-th. DOI: 10.1007/s10098-023-02651-8

Davis, J. C., & Henderson, J. V. (2003). Evidence on the Political Economy of the Urbanization Process. *Journal of Urban Economics*, *53*(1), 98–125. https://www.sciencedirect.com/science/article/abs/pii/S0094119002005041. DOI: 10.1016/S0094-1190(02)00504-1

Dutta, A., & Jinsart, W. (2020). Waste Generation and Management Status in the Fast-Expanding Indian Cities: A Review. *Journal of the Air & Waste Management Association, 70*(5), 491–503. https://www.tandfonline.com/doi/pdf/10.1080/10962247.2020.1738285. DOI: 10.1080/10962247.2020.1738285 PMID: 32150518

Fields, G. (1999). City Systems, Urban History, and Economic Modernity: Urbanization and the Transition from Agrarian to Industrial Society. *Berkeley Planning Journal, 13*, 102–128. https://escholarship.org/uc/item/7242m3b1

Garg, B. S., & Raut, A. V. (2015). Adarsh Gram: A Gandhian Dream of Gram Swaraj. *Indian Journal of Community Medicine : Official Publication of Indian Association of Preventive & Social Medicine, 40*(1), 1–4. https://pmc.ncbi.nlm.nih.gov/articles/PMC4317974/. DOI: 10.4103/0970-0218.149260 PMID: 25657504

Glaeser, E. L. (2014). A World of Cities: The Causes and Consequences of Urbanization in Poorer Countries. *Journal of the European Economic Association, 12*(5), 1154–1199. https://academic.oup.com/jeea/article-abstract/12/5/1154/2319678?redirectedFrom=fulltext. DOI: 10.1111/jeea.12100

Gore, C. D., & Gopakumar, G. (2015). Infrastructure and Metropolitan Reorganization: An Exploration of the Relationship in Africa and India. *Journal of Urban Affairs, 37*(5), 548–567. https://www.tandfonline.com/doi/full/10.1111/juaf.12180. DOI: 10.1111/juaf.12180

Govt of Kerala. (2022). *Report of the Comptroller and Auditor General of India on Waste Management in Urban Local Bodies*. Govt of Kerala.

Harikrishnan, G. (2014). Solid Waste Management: A Comparative Study between Kerala and Tamil Nadu. *IJRSSH* 4(4), 29-37. https://www.ijrssh.com/admin/upload/05%20Harikrishnan.pdf

Haughton, G. (1997). Developing Sustainable Urban Development Models. *Cities (London, England), 14*(4), 189–195. https://www.sciencedirect.com/science/article/abs/pii/S0264275197000024. DOI: 10.1016/S0264-2751(97)00002-4

He, C., Liu, Z., Wu, J., Pan, X., Fang, Z., Li, J., & Bryan, B. (2021). Future Global Urban Water Scarcity and Potential Solutions. *Nature Communications, 12*(1), 4667. https://www.nature.com/articles/s41467-021-25026-3. DOI: 10.1038/s41467-021-25026-3 PMID: 34344898

Henderson, J. V., Nigmatulina, D., & Kriticos, S. (2021). Measuring Urban Economic Density. *Journal of Urban Economics, 125*, 103188. https://ideas.repec.org/a/eee/juecon/v125y2021ics0094119019300658.html. DOI: 10.1016/j.jue.2019.103188

Hohenberg, P. M. (1990). *The City: Agent or Product of Urbanization. Urbanization in History*. Clarendon Press.

Hong, T., Yu, N., Mao, Z., & Zhang, S. (2021). Government-Driven Urbanization and its Impact on Regional Economic Growth in China. *Cities (London, England), 117*, 103299. https://www.sciencedirect.com/science/article/abs/pii/S0264275121001992?via%3Dihub. DOI: 10.1016/j.cities.2021.103299

Kazmin, A. (2016). India's Population Shift Spaks a Rethink of Rural Strategy. *Financial Times*. https://www.ft.com/content/950a3f1c-ef47-11e5-9f20-c3a047354386

Kundu, A. (2003). Urbanization and Urban Governance: Search for a Perspective beyond Neo-Liberalism. *EPW* 38(29), 3079-3087. https://www.epw.in/journal/2003/29/special-articles/urbanisation-and-urban-governance.html

Lal, P., & Nair, S. B. (2017). Urbanization in Kerala—What does the Census Data Reveal? *Indian Journal of Human Development, 11*(3), 356–386. https://www.researchgate.net/publication/326106606_Urbanization_in_Kerala-What_Does_the_Census_Data_Reveal. DOI: 10.1177/0973703018763241

Mahadevia, D. (2001). Sustainable Urban Development in India: An Inclusive Perspective. *Development in Practice, 11*(2-3), 242–259. https://www.jstor.org/stable/4029451. DOI: 10.1080/09614520120056388

McGill, R. (2001). Urban Management Checklist. *Cities (London, England), 18*(5), 347–354. https://www.sciencedirect.com/science/article/abs/pii/S0264275101000270. DOI: 10.1016/S0264-2751(01)00027-0

Nhamo, L., Rwizi, L., Mpandeli, S., Botai, J., Magidi, J., Tazvinga, H., Sobratee, N., Liphadzi, S., Naidoo, D., Modi, A. T., Slotow, R., & Mabhaudhi, T. (2021). Urban Nexus and Transformative Pathways towards a Resilient Gauteng City Region, South Africa. *Cities (London, England), 116*, 103266. https://www.sciencedirect.com/science/article/pii/S0264275121001669. DOI: 10.1016/j.cities.2021.103266 PMID: 37674556

Nishad, A. (2024). Haritha Karma Sena: Transforming Waste Management. *IJEBM* 8(3), 19-25. https://aipublications.com/uploads/issue_files/3IJEBM-SEP20241-Haritha.pdf

Nuissl, H., & Heinrichs, D. (2011). Fresh Wind or Hot Air — Does the Governance Discourse have Something to Offer to Spatial Planning? *Journal of Planning Education and Research, 31*(1), 47–59. https://journals.sagepub.com/doi/10.1177/0739456X10392354. DOI: 10.1177/0739456X10392354

Obeng-Odoom, F. (2012). On the Origin, Meaning, and Evaluation of Urban Governance. *Norsk Geografisk Tidsskrift*, *66*(4), 204–212. https://www.tandfonline.com/doi/abs/10.1080/00291951.2012.707989. DOI: 10.1080/00291951.2012.707989

Planning Commission. Govt. of India. (2008). *Eleventh Five Year Plan (2007-2012): Inclusive Growth.* Oxford University Press.

Pradhan, S., Bhatt, M., & Gautam, R. S. (2025). Social Interventions for Liquid Waste Management on Livelihood: A Case of Indore, India. *Journal of Development Policy and Practice*. https://journals.sagepub.com/doi/epub/10.1177/24551333241304538

Praveen, S. R. (2024). Foot Soldiers of Kerala's Silent Revolution against Waste. *The Hindu*. https://www.thehindu.com/news/national/kerala/foot-soldiers-of-keralas-silent-revolution-against-waste/article68106197.ece

Ravikumar, P. (2024). Public Perception Regarding the Role of Harithakarma Sena in Waste Management at Panchayat Level. *IJRASET* 12(4), 574-578. https://www.ijraset.com/best-journal/perception-regarding-haritha-karma-sena-in-the-sustainable-waste-management

Satterthwaite, D. (1998). Environmental Problems in Cities in the South: Sharing my Confusions. In Fernandez, E. (Ed.), *Grassroots Environmental Action—People's Participation in Sustainable Development* (pp. 62–82). Routledge.

Shameerdas, T. (2024). Haritha Karma Sena and Non-Biodegradable Waste Management. *IJIRT* 11(3), 1455-1462. https://ijirt.org/publishedpaper/IJIRT167527_PAPER.pdf

Singh, R. (2021). Municipal Solid Waste Management in the City of Indore — A Case Study. *Journal of Civil Engineering and Environmental Sciences*, *7*(1), 8–17. https://www.engineegroup.us/articles/JCEES-7-139.pdf

Smith, A. (1776). *An Inquiry into the Nature and Causes of the Wealth of the Nations*. Clarendon Press. DOI: 10.1093/oseo/instance.00043218

Song, K., Chen, Y., Duan, Y., & Zheng, Y. (2023). Urban Governance: A Review of Intellectual Structure and Topic Evolution. *Urban Governance* 3, 169-185. https://www.sciencedirect.com/science/article/pii/S2664328623000670

Thodukayil, F. S. O., Raseena, R. S., Palaniswamy, U., & Panneer, S. (2024). Challenges of Women Workers in the Solid Waste Management Sector of Kerala: on the Path to Formalization. In Vinodan, A., Mahalakshmi, S., & Rameshkumar, S. (Eds.), *Informal Economy and Sustainable Development Goals: Ideas, Interventions, and Challenges* (pp. 373–397). Emerald Publishing., https://www.emerald.com/insight/content/doi/10.1108/978-1-83753-980-220241020/full/html DOI: 10.1108/978-1-83753-980-220241020

Thomas, E., Nelson, N. S., Saju, N., Menon, A., & Aleena, P. (2023). A Study on the Socio-Economic Profile of Haritha Karma Sena Workers with Special Reference to Cheranalloor Panchayat. http://117.239.78.102:8080/jspui/bitstream/123456789/2856/1/PROJECT%20COMBINED.pdf

Tibaijuka, K. (2009). *Building Prosperity: Housing and Economic Development.* Earthscan.

Tom, T., & Kamat, R. (2020). A Study on Waste Management of Household with Special Reference to Uzhavoor Panchayat — Kottayam District, Kerala. *Alochana Chakra Journal, 9*(5), 2533–2543.

UNCHS (United Nations Centre for Human Settlements). (1998). *Sustainable Cities Program.* UNCHS.

UNDESA. (2019). *World Urbanization Prospects 2018: Highlights.* UN.

UNDESA (United Nations Department of Economic and Social Affairs). (2018). *World Urbanization Prospects: The 2018 Revision.* UN.

World Bank (2013). *Global Monitoring Report 2013: Rural-Urban Dynamics and the Millennium Development Goals.*

World Bank. (2023). *World Bank Urban Development Review: The 2023 Revision.* World Bank.

World Urbanization Prospects, U. N. (2018). Population Dynamics. https://www.un.org/development/desa/pd/sites/www.un.org.development.desa.pd/files/files/documents/2020/Feb/un_2018_wup_highlights.pdf

Zhang, S., Yi, B., Worrell, E., Wagner, F., Crijns-Graus, W., Purohit, P., Wada, Y., & Varis, O. (2019). Integrated Assessment of the Resource-Energy-Environment Nexus in China's Iron and Steel Industry. *Journal of Cleaner Production, 232*, 235–249. https://www.sciencedirect.com/science/article/abs/pii/S0959652619319353. DOI: 10.1016/j.jclepro.2019.05.392

Zhou, Y., & Staatz, J. (2016). Projected Demand and Supply for Various Foods in West Africa: Implications for Investments and Food Policy. *Food Policy*, *61*, 198–212. https://www.sciencedirect.com/science/article/abs/pii/S0306919216300185?via%3Dihub. DOI: 10.1016/j.foodpol.2016.04.002

Chapter 9
Water and Waste Management:
Innovations for Sustainable Cities

Treesa Sindhu P. Thomas
https://orcid.org/0009-0002-7405-5162
St. Teresa's College, Ernakulam, India

Leena Leon
Rajagiri College of Social Sciences, Kalamasseri, India

ABSTRACT

Urban water and waste management are critical in achieving sustainability, particularly in coastal cities that face unique challenges from rapid urbanization and climate change. These pressures demand innovative approaches to resource management that balance environmental stability, economic resilience, and social well-being. This article explores the intersection of cutting-edge technologies, adaptive urban planning, and participatory governance models, offering insights into how these approaches contribute to the Sustainable Development Goals (SDGs), specifically SDG 6 and SDG 11. The findings emphasize the importance of adopting holistic strategies that integrate zero-waste policies, enhanced recycling programs, extended producer responsibility (EPR), and community engagement. By leveraging innovation, circular economy principles, and active community participation, cities can build adaptable systems that ensure long-term sustainability, improve quality of life, and support global efforts toward environmental and social progress

DOI: 10.4018/979-8-3373-5278-7.ch009

INTRODUCTION

Water and waste management are critical pillars of urban sustainability, particularly for coastal cities grappling with the complex challenges of rapid urbanization and climate change. These interconnected issues demand innovative approaches to resource management to ensure environmental stability, economic resilience, and social well-being. Coastal urban areas, with their high population densities and vulnerability to climate impacts such as rising sea levels and extreme weather events, face unique pressures on water resources and waste systems.

The growing demand for water, coupled with inefficient waste disposal practices, exacerbates environmental degradation and public health concerns. This chapter delves into strategies to address these challenges, focusing on cutting-edge technologies, adaptive urban planning practices, and participatory governance models. The discussion emphasizes how effective water and waste management can contribute to achieving sustainable development goals (SDGs), specifically SDG 6 (Clean Water and Sanitation) and SDG 11 (Sustainable Cities and Communities).

Objectives

1. **To provide a comprehensive overview** of the current state of water and waste management in urban areas, highlighting trends, challenges, and opportunities.
2. **To explore innovative technologies and practices** that contribute to sustainable urban development, including resource recovery, circular economy models, and climate-adaptive infrastructure.
3. **To present real-world case studies** that illustrate the successful application of these innovations in diverse urban settings, with a special focus on coastal communities.
4. **To examine the role of community engagement** and public-private partnerships in driving sustainable water and waste management initiatives.
5. **To offer actionable recommendations** for policymakers, urban planners, and practitioners to integrate sustainable water and waste management solutions into urban strategies.

METHODOLOGY

The development of this chapter employs a multi-faceted approach, combining an extensive review of academic literature, analysis of real-world case studies, and expert interviews. Each method contributes to a holistic understanding of water and waste management innovations in urban settings.

1. Literature Review

A thorough review of recent research, policy documents, and technical reports forms the foundation of the chapter. The literature review synthesizes key findings on:

- Urban water scarcity and distribution challenges.
- Waste generation patterns, including solid and liquid waste streams.
- Technological advancements such as decentralized water treatment, smart waste management systems, and resource recovery methods.
- Best practices for integrating sustainability into urban water and waste management.

2. Case Study Analysis

The chapter incorporates detailed case studies from various regions, offering practical insights into the implementation of innovative practices. Case studies are selected based on the following criteria:

- Geographic diversity to ensure representation from developed and developing regions.
- Focus on coastal urban areas to address specific challenges faced by these communities.
- Demonstrated success in deploying sustainable water and waste management technologies or practices.

For example:

- A case study on Singapore's *NEWater* initiative highlights cutting-edge water recycling technologies and their impact on urban water resilience.
- Analysis of Sweden's circular economy model demonstrates effective strategies for transforming waste into energy and other resources.
- Examples from coastal cities in India and Indonesia showcase participatory waste management practices that engage local communities.

3. Expert Interviews

Interviews with subject matter experts, including urban planners, environmental scientists, and policymakers, provide valuable insights into emerging trends and challenges. These interviews:

- Highlight the practical implications of implementing innovations in water and waste management.
- Offer context-specific solutions tailored to the needs of urban coastal settings.
- Shed light on the role of governance, financing, and stakeholder collaboration in ensuring project sustainability.

4. Comparative Analysis

The chapter integrates a comparative analysis of the findings from the literature, case studies, and expert interviews. This approach identifies commonalities, differences, and lessons learned across diverse urban contexts.

5. Synthesis of Actionable Recommendations

Drawing on the above methodologies, the chapter concludes by offering evidence-based recommendations tailored for key stakeholders, including:

- Policymakers, for enacting supportive legislation and allocating resources.
- Urban planners, for designing adaptable and resilient infrastructure.
- Private sector entities, for investing in innovative technologies and fostering public-private collaborations.
- Community leaders, for mobilizing grassroots participation and fostering behavioral change.

By leveraging a robust methodological framework, this chapter provides a comprehensive exploration of water and waste management innovations for sustainable cities. Through a mix of theoretical insights and practical applications, it underscores the importance of adopting holistic, inclusive, and scalable approaches to address the dual challenges of urbanization and climate change in coastal communities.

RESULTS AND DISCUSSION

Water Management: A Pillar of Urban Sustainability

Water management is fundamental to sustaining urban life by ensuring the availability of clean water for consumption, sanitation, agriculture, industry, and the functioning of urban ecosystems. Urban water resources are increasingly strained due to factors such as population growth, pollution, over-extraction of groundwater, and climate change. As cities expand, the competition for water resources intensifies, demanding innovative and integrated water management practices. These include water conservation, recycling, rainwater harvesting, and wastewater treatment. Collectively, these strategies play a critical role in achieving sustainable urban development, balancing environmental needs with the demands of growing populations.

Waste Management: The Backbone of Sustainable Cities

Waste management is equally essential for urban sustainability, given the vast quantities of municipal solid waste (MSW), industrial waste, and hazardous materials generated by cities. Improper disposal leads to severe environmental pollution and public health risks. Implementing effective waste management strategies, such as reduction, recycling, composting, and safe hazardous waste disposal, can mitigate these impacts.

Sustainable waste management is central to the circular economy, aiming to minimize waste generation through reduction, reuse, and recycling. Proper waste management not only conserves resources and reduces landfill dependency but also fosters economic growth by creating opportunities in the recycling and waste-processing industries.

Current Challenges in Urban Water and Waste Management

1. **Water Scarcity:** Urban areas in arid and semi-arid regions often experience severe water shortages due to limited resources and rising demand. For instance, cities like Cape Town, South Africa, and Chennai, India, have faced critical water crises, underscoring the urgent need for robust water management systems (Hoekstra & Mekonnen, 2012).
2. **Pollution:** Industrial, agricultural, and domestic wastewater discharges are major sources of urban water pollution. Contaminants such as heavy metals, pesticides, and untreated sewage degrade water quality, posing threats to human health and ecosystems (UNEP, 2016).
3. **Aging Infrastructure:** Outdated water supply and distribution systems result in inefficiencies, leaks, and water loss. Upgrading such infrastructure is often costly, posing challenges for municipalities (UN Water, 2018).
4. **Climate Change:** Altered precipitation patterns, rising temperatures, and extreme weather events further complicate urban water management, necessitating adaptive strategies to cope with droughts and floods (IPCC, 2022).
5. **Waste Generation:** Rapid urbanization has led to an exponential increase in waste, particularly plastics, electronic waste, and organic matter, straining waste management infrastructure (UN-Habitat, 2020).

6. **Improper Waste Disposal:** Landfills and open dumping are still the primary waste disposal methods in many cities, contributing to environmental contamination and health risks. Recycling and composting remain underutilized (OECD, 2018).
7. **Informal Waste Sector:** In developing countries, waste management is often handled by the informal sector, such as waste pickers, who operate without safety measures or formal recognition. Integrating this sector into formal systems can improve efficiency and worker welfare (Wilson et al., 2015).
8. **Public Awareness Deficiency:** Inadequate education on waste segregation, recycling, and reduction results in inefficient waste management. Contaminated recyclables further undermine waste recovery efforts (Kaza et al., 2018).
9. **Financial Constraints:** Many municipalities, particularly in developing nations, lack the funds to invest in necessary infrastructure, such as recycling facilities and composting plants (World Bank, 2018).

INNOVATIVE WATER MANAGEMENT STRATEGIES

Smart Metering and Leak Detection: Smart metering provides real-time data on water usage, allowing utilities and consumers to detect inefficiencies and leaks. Advanced systems can transmit data remotely, minimizing operational costs and water wastage (Kar, 2020). Acoustic sensors and pressure monitors can identify leaks in aging infrastructure, preventing up to 30% of water loss (AWWA, 2021; Liu et al., 2018).

Water-Efficient Appliances and Fixtures: Water-efficient appliances, such as low-flow toilets and water-recycling washing machines, significantly reduce consumption without compromising functionality (Gleick, 2003). Similarly, low-flow faucets and showerheads can cut household water use by up to 20% (EPA, 2021; Zhao et al., 2020).

Public Awareness Campaigns: Educational campaigns like Australia's "Save Water, Save the Future" have demonstrated the power of community engagement in reducing water consumption through behavioral changes (Biel et al., 2017). Workshops, incentives for adopting water-saving devices, and local conservation projects further strengthen public participation (MacRae et al., 2021).

RAINWATER HARVESTING (RWH) AND GREYWATER RECYCLING

Rainwater Harvesting

Rainwater harvesting systems capture and store rainwater for domestic, agricultural, or industrial use. Effective design ensures maximized collection and safe usage. Components include:

- **Catchment Area**: Clean roof surfaces optimize water collection (Koch et al., 2020).
- **Storage Capacity**: Tanks sized according to rainfall and demand.
- **Filtration and Treatment**: Systems for debris removal and water disinfection (Wong et al., 2021).
- **Distribution System**: Pumps or gravity-fed mechanisms transport water efficiently (Gould & Nissen-Peterson, 2019).

Greywater Recycling

Treated greywater from showers, sinks, and washing machines can be reused for irrigation or flushing toilets. This reduces demand on freshwater supplies and decreases wastewater generation, offering a sustainable alternative for urban water management.

System Design Considerations for Rainwater Harvesting

Effective design and planning are critical to ensure that a rainwater harvesting system meets its intended purpose. The following key factors must be considered during the design process:

Catchment Area: The type and size of the surface used for collecting rainwater significantly affect the quantity of water harvested. Larger and cleaner surfaces, such as metal or tiled roofs, are ideal as they maximize water collection and minimize the risk of contamination (Koch et al., 2020).

Storage Capacity: Storage tank size should align with local rainfall patterns, water demand, and available space. Systems can be tailored to a variety of scales, from small residential setups to large municipal facilities designed to serve entire communities.

Water Filtration and Treatment: To ensure safety, filtration systems are essential for removing debris and contaminants. Depending on the intended use, additional treatment methods, such as UV sterilization or chlorination, may be required to make the water potable (Wong et al., 2021).

Distribution System: Water from storage tanks must be efficiently delivered to users, either through gravity-fed systems or pumps integrated into the design. It is equally important to include backflow prevention devices to safeguard municipal water supplies from potential contamination (Gould & Nissen-Peterson, 2019).

Research by Koch et al. (2020) on rainwater harvesting systems in urban areas of Southeast Asia underscores the importance of designing flexible and efficient systems. The study highlighted how well-planned systems could reduce dependency on municipal water supplies and alleviate the effects of seasonal droughts.

Strategies for Implementing Greywater Recycling

Greywater recycling involves the collection, treatment, and reuse of wastewater from sinks, showers, and washing machines for non-potable purposes. It is a promising strategy to reduce fresh water usage and alleviate pressure on urban sewage systems. Despite its benefits, the implementation of greywater recycling faces challenges, including proper treatment requirements and public acceptance (Friedler, 2004).

Benefits of Greywater Recycling in Urban Settings

1. **Water Conservation**: Greywater recycling significantly reduces the demand for fresh water by repurposing wastewater for non-potable uses such as irrigation, landscape watering, and toilet flushing. Studies show that households implementing greywater recycling systems can reduce water consumption by up to 40% (Anwar et al., 2020).
2. **Cost Savings**: By decreasing the reliance on municipal water supplies, greywater recycling can lower water bills, particularly in regions with high water tariffs. This makes it a cost-effective solution for both residential and commercial users over time (Diaper et al., 2020).
3. **Environmental Benefits**: Recycling greywater decreases the volume of wastewater sent to treatment plants, reducing environmental pollution and lessening the strain on urban wastewater treatment infrastructure. This contributes to more sustainable urban water management systems (Zhang & Liao, 2020).

Challenges of Greywater Recycling

1. **Health and Safety Concerns**: Without proper treatment, greywater may harbor pathogens and contaminants, posing health risks. Robust filtration and disinfection systems are necessary to ensure that recycled greywater meets safety standards for its intended uses (Anwar et al., 2020).
2. **Regulatory Barriers**: In many urban areas, the lack of comprehensive legal and regulatory frameworks for greywater recycling hinders its adoption. Establishing clear guidelines and standards for the safe and effective use of greywater is crucial for widespread implementation (Liu & Liao, 2019).
3. **System Costs and Maintenance**: While greywater recycling can result in long-term savings, the initial installation costs and the ongoing expenses associated with maintaining filtration and disinfection systems can deter households and businesses from adopting these systems (Diaper et al., 2020).

FUTURE DIRECTIONS AND INTEGRATION WITH URBAN WATER MANAGEMENT

Research highlights the potential of greywater recycling in urban environments, particularly in water-scarce regions. For example, Anwar et al. (2020) demonstrated that when combined with effective treatment technologies, greywater recycling could contribute significantly to water conservation in residential buildings.

To maximize its effectiveness, greywater recycling must be integrated with broader urban water management strategies. Key steps include:

- **Developing supportive policies and regulations** to promote adoption while ensuring public health and safety.
- **Raising public awareness** about the benefits and safe practices of greywater recycling to enhance acceptance.
- **Investing in cost-efficient technologies** to reduce system installation and maintenance costs.

Complementing Rainwater Harvesting

Greywater recycling and rainwater harvesting together offer transformative potential for urban water conservation. While rainwater harvesting captures and stores fresh water for non-potable uses, greywater recycling provides a continuous source of treated water from daily activities. These strategies can be tailored to local climate conditions, infrastructure, and water demand to optimize their combined impact.

Advanced Wastewater Treatment and Sustainable Urban Drainage Systems (SUDS)

Effective water and wastewater management is a cornerstone of sustainable urban development. With urban populations growing and environmental pressures mounting, advanced wastewater treatment technologies and Sustainable Urban Drainage Systems (SUDS) are becoming vital tools for addressing water resource challenges and managing urban wastewater sustainably. This discussion focuses on the role of advanced wastewater treatment technologies, such as membrane bioreactors (MBRs) and advanced oxidation processes (AOPs), as well as the implementation of SUDS, including green roofs, permeable pavements, and bio-retention systems.

Advanced Wastewater Treatment Technologies

Advanced Wastewater Treatment (AWT) technologies play a critical role in ensuring wastewater meets the required standards for reuse or discharge, particularly in densely populated urban areas. Among the emerging technologies, **Membrane Bioreactors (MBRs)** and **Advanced Oxidation Processes (AOPs)** stand out for their efficiency and effectiveness in treating complex wastewater streams.

1. **Membrane Bioreactors (MBRs):** MBRs combine conventional biological treatment processes, such as activated sludge, with advanced membrane filtration, typically using microfiltration or ultrafiltration membranes. This integration enables the production of high-quality effluent while requiring a smaller physical footprint than traditional wastewater treatment plants (Judd, 2011).

   Key advantages of MBRs include:
   - Enhanced solid-liquid separation.
   - Lower sludge production.
   - High removal efficiency for organic matter, suspended solids, and pathogens.
   - Capacity to treat challenging wastewater streams, such as those containing industrial effluents or high organic loads.
2. For instance, Feng et al. (2019) demonstrated that MBR systems treating municipal wastewater in China achieved high removal rates of contaminants, with improvements in membrane fouling control enhancing energy efficiency. Despite their energy intensity, MBRs are particularly suited for urban areas with limited space for large-scale treatment plants.
3. **Advanced Oxidation Processes (AOPs):** AOPs leverage highly reactive hydroxyl radicals (•OH) to degrade organic pollutants in wastewater. These processes are especially effective for treating recalcitrant compounds, including

pharmaceuticals, pesticides, and industrial chemicals, that resist conventional biological treatment (Barros et al., 2020).

Examples of AOPs include:

- o Ozonation.
- o Photocatalysis.
- o Fenton's reaction.

4. Barros et al. (2020) highlighted the potential of combined AOPs, such as photocatalysis coupled with ozone treatment, for achieving high removal efficiencies at reduced energy costs, making them a promising solution for treating complex wastewater streams.
5. **Decentralized Wastewater Treatment Systems (DEWATS):** DEWATS treat wastewater close to its source, reducing reliance on extensive sewer networks and centralized treatment plants. This approach is especially suitable for rural, peri-urban areas, or small-scale industries. Nair and Goh (2018) found that DEWATS systems, which often integrate biological treatment (e.g., constructed wetlands) with sedimentation and filtration processes, offer:
    - o Cost-effectiveness.
    - o Modular designs that allow for scalability.
    - o Flexibility to adapt to local conditions.
6. Their study underscored the utility of DEWATS for small urban communities, emphasizing its potential to support sustainable wastewater management in less developed areas.

Sustainable Urban Drainage Systems (SUDS)

SUDS are innovative solutions designed to manage surface water runoff, reduce urban flood risks, and enhance water quality. These systems emphasize the use of natural processes and green infrastructure to mimic natural hydrological cycles, supporting sustainable water management in urban areas.

1. **Green Roofs:** Green roofs absorb rainwater, reducing runoff and mitigating urban heat island effects. They also improve water quality by filtering pollutants and provide additional benefits such as enhanced insulation for buildings.
2. **Permeable Pavements:** These pavements allow water to infiltrate through the surface, reducing surface runoff and promoting groundwater recharge. Permeable pavements are particularly effective in urban areas with limited natural infiltration zones.

3. **Bio-Retention Systems:** Bio-retention systems, such as rain gardens and swales, use vegetation and engineered soil to capture and treat stormwater. These systems help remove pollutants, enhance infiltration, and reduce peak flow volumes during heavy rainfall events.

SUDS not only address immediate water management challenges but also contribute to urban biodiversity, aesthetic improvements, and climate resilience.

Integration of Green Infrastructure and Waste Management for Urban Sustainability

Urban sustainability hinges on adopting integrated approaches that address stormwater management, wastewater treatment, and solid waste management. This section discusses the use of **Sustainable Urban Drainage Systems (SUDS)**, innovative waste management strategies, and the role of extended producer responsibility (EPR) in fostering sustainable cities.

Sustainable Urban Drainage Systems (SUDS)

SUDS represent a paradigm shift in urban water management, combining natural processes with engineered solutions to manage surface runoff, reduce flooding, and enhance urban biodiversity. The key components of SUDS include **green roofs**, **permeable pavements**, and **bio-retention systems**, which together contribute to improved water quality and sustainable urban development.

1. **Green Roofs:** Green roofs consist of vegetated layers installed on rooftops to absorb rainwater, reduce stormwater runoff, and mitigate urban heat island effects. Studies show that green roofs can reduce runoff by up to 50%, while simultaneously providing insulation, improving air quality, and supporting biodiversity (Getter & Rowe, 2021). Despite higher initial costs, their long-term benefits—such as energy savings and reduced stormwater infrastructure demands—make them a valuable urban investment.
2. **Permeable Pavements:** Permeable pavements allow water to infiltrate through their surface, reducing runoff and promoting groundwater recharge. Commonly used for sidewalks, parking lots, and driveways, these pavements enhance stormwater management while reducing urban flooding risks (Li et al., 2019).
3. **Bio-Retention Systems:** Bio-retention systems, such as rain gardens, treat stormwater by using soil and plants to remove pollutants, including heavy metals and nutrients. These systems enhance infiltration and support urban greenery, contributing to cleaner waterways and reduced flood risks (Fletcher et al., 2017).

Case Studies of Successful SUDS Implementation

- **London, UK:** The "SuDS for All" initiative transformed urban areas, particularly in the London Borough of Hackney, by integrating green roofs, permeable pavements, and rain gardens. These efforts significantly reduced local flooding and enhanced urban green spaces (Sutcliffe et al., 2018).
- **Melbourne, Australia:** The Melbourne Water "Stormwater Management Strategy" incorporated SUDS into urban planning, leveraging bio-retention systems and permeable pavements to minimize stormwater runoff and improve water quality (Hatt et al., 2016).

These case studies underscore the effectiveness of SUDS in addressing stormwater challenges while delivering co-benefits, such as increased biodiversity and improved urban aesthetics.

Innovative Waste Management Strategies

Urban areas face mounting challenges in managing waste due to growing populations and limited landfill space. To address these issues, cities are adopting innovative strategies, including zero-waste policies, enhanced recycling programs, and Extended Producer Responsibility (EPR).

1. **Zero-Waste Policies and Initiatives:** Zero-waste policies aim to minimize waste generation through comprehensive strategies emphasizing prevention, reuse, and recycling. For instance, Kamikatsu, Japan, reduced landfill waste by over 80% by implementing zero-waste initiatives focused on material recovery and community engagement (Lausten et al., 2020).

Key Features of Zero-Waste Policies

- **Waste Prevention:** Encouraging sustainable consumption and production practices.
- **Product Design:** Promoting durable, reusable, and recyclable products aligned with circular economy principles.
- **Community Engagement:** Educating and incentivizing citizens to adopt waste-reducing behaviors.

2. **Enhancing Recycling Programs and Infrastructure:** Effective recycling requires robust infrastructure and public participation. Cities like San Francisco have achieved high recycling rates by investing in advanced sorting technologies and public education campaigns (Chen et al., 2020).

Focus Areas in Recycling Enhancement

- o **Collection Systems:** Accessible curbside recycling services.
- o **Sorting Facilities:** Advanced technologies to process diverse materials.
- o **Public Education:** Clear guidelines on what and how to recycle.

3. **Extended Producer Responsibility (EPR):** EPR shifts the responsibility for waste management to producers, incentivizing sustainable product design and reducing waste at the source. Successful EPR programs in Europe and North America, such as those in Sweden and Germany, demonstrate their effectiveness in reducing packaging waste and promoting circular economies (Fisher et al., 2019).

Benefits of EPR

- o Encourages innovation in product design for recyclability.
- o Reduces taxpayer burden by financing waste management through producers.
- o Stimulates markets for recycled materials.

Innovative Waste Management Strategies: Organic Waste Management, Technological Innovations, and Policy Frameworks

As urban populations grow, managing waste sustainably has become a significant challenge. Innovative strategies focusing on organic waste management, technological advancements, and robust policy frameworks are essential for addressing these challenges effectively. This discussion explores key approaches within these domains, emphasizing their role in fostering urban sustainability.

Organic Waste Management

Organic waste, which includes food scraps, yard waste, and other biodegradable materials, forms a substantial portion of municipal solid waste. Proper management of this waste reduces landfill dependency, recovers resources, and minimizes environmental impacts. Two primary strategies—composting and anaerobic digestion—are critical in this context.

a. Composting and Anaerobic Digestion
    - **Composting**: This aerobic process converts organic materials into nutrient-rich compost, ideal for improving soil health. It is particularly suitable for urban settings, such as community gardens and urban agriculture, reducing landfill use and methane emissions.
    - **Anaerobic Digestion**: This process decomposes organic waste in the absence of oxygen, generating biogas as a renewable energy source while also producing nutrient-rich digestate for agriculture. Satyro et al. (2020) highlighted that both methods significantly reduce greenhouse gas emissions, enhance resource recovery, and promote urban sustainability.

Urban Agriculture and Community Gardens: Composting supports urban agriculture by providing high-quality compost for soil enrichment, thus promoting food security and reducing waste. Community gardens and urban farms utilizing locally produced compost exemplify sustainable waste-to-resource practices. Cavalett et al. (2021) demonstrated that urban agriculture integrated with composting fosters sustainable food systems, reduces landfill dependency, and enhances community engagement.

Technological Innovations in Waste Management

Technological advancements are revolutionizing waste management systems, enhancing efficiency and sustainability. Two notable innovations are IoT-enabled smart waste bins and waste-to-energy (WTE) technologies.

a. **Smart Waste Bins and IoT:** Smart waste bins equipped with sensors provide real-time data on waste levels, enabling optimized collection routes and reducing unnecessary trips. This approach lowers operational costs and minimizes the carbon footprint of waste collection (Faccio et al., 2011). According to Khan et al. (2021), IoT technologies improve logistics, monitor waste patterns, and streamline urban waste management, especially in high-density areas.
b. **Waste-to-Energy (WTE) Technologies:** WTE technologies such as incineration, pyrolysis, and gasification convert non-recyclable waste into usable energy, reducing landfill use and methane emissions. Pyrolysis and gasification, in particular, offer cleaner alternatives by producing syngas for energy generation while minimizing harmful emissions (Zhou et al., 2019). However, concerns

regarding environmental impacts and public acceptance must be addressed to maximize their potential.

Policy and Regulatory Frameworks

Effective policies and regulations are fundamental in driving sustainable waste management practices. They provide the necessary incentives, frameworks, and enforcement mechanisms to promote waste reduction, recycling, and innovation.

a. Innovative Policies
   - **Extended Producer Responsibility (EPR)**: This policy holds producers accountable for the end-of-life management of their products, incentivizing sustainable product design and recycling. Successful EPR implementations in Europe and North America have improved recycling rates and reduced waste generation (Fisher et al., 2019).
   - **Pay-As-You-Throw (PAYT) Programs**: PAYT systems encourage waste reduction by charging households based on the waste they generate, making recycling more economically attractive.
   - **Landfill Bans on Recyclables**: Prohibiting recyclable materials from landfills promotes recycling and conserves resources.

b. **Impact of Waste Management Regulations:** Case studies highlight the effectiveness of well-implemented waste management policies in promoting urban sustainability. For instance, Mazzanti & Zoboli (2008) demonstrated that strong regulatory frameworks reduce waste generation, improve recycling rates, and lead to better environmental outcomes.

Examples of Successful Waste Management Regulations

Numerous cities and countries have implemented exemplary waste management regulations, achieving significant progress in sustainability. The following examples highlight the effectiveness of such regulations:

Sweden: Comprehensive Waste-to-Energy and Recycling Systems

Sweden has one of the world's most efficient waste management systems, driven by robust policies that promote waste sorting and recycling. The country's waste-to-energy technologies and high recycling rates are supported by stringent regulations incentivizing sustainable practices. Björkqvist et al. (2017) emphasized that Sweden's integrated waste management approach combines extensive recycling initiatives with energy recovery, significantly reducing landfill dependency. This success is attributed to the alignment of government policies, public participation, and advanced waste technologies.

Germany: Extended Producer Responsibility (EPR)

Germany's Packaging Ordinance is a benchmark in waste management, requiring producers to take responsibility for the collection and recycling of packaging materials. This policy has led to a substantial reduction in packaging waste and increased recycling rates. Björkqvist et al. (2017) highlighted the effectiveness of EPR in improving product design and reducing environmental impact, showcasing Germany's leadership in sustainable waste management practices.

Case Studies of Sustainable Cities: Singapore, Copenhagen, and San Francisco

In recent years, several cities have adopted comprehensive strategies to address waste and water management challenges, integrating sustainability into their urban development frameworks. Singapore, Copenhagen, and San Francisco are exemplary in their innovative approaches, emphasizing the importance of policy, technology, and community engagement.

Singapore: Zero Waste Vision and Integrated Water Management

As a resource-constrained city-state, Singapore has developed innovative strategies for water and waste management, setting an example for urban centers globally.

a. **Integrated Water Management and the NEWater Initiative:** Singapore's integrated water management system addresses water scarcity through advanced technologies and urban planning. The NEWater initiative reclaims wastewater using microfiltration, reverse osmosis, and ultraviolet disinfection to produce ultra-pure water for industrial use and potable purposes during droughts. This initiative, along with the Active, Beautiful, Clean Waters (ABC Waters) program, integrates stormwater management with urban development to enhance water security and quality of life (Tan et al., 2019).
b. **Zero-Waste Masterplan:** Singapore aims to become a zero-waste society by 2030. The Zero-Waste Masterplan emphasizes waste reduction at the source, increased recycling, and circular economy principles. Key initiatives include mandatory waste sorting, organic waste composting, and waste-to-energy plants. Public education campaigns play a crucial role in fostering responsible consumption and waste disposal practices. Lee et al. (2020) noted that Singapore's

integrated framework combines policy, infrastructure, and public participation to achieve its sustainability goals.

Copenhagen: Climate Adaptation and Circular Economy

Copenhagen is a global leader in climate adaptation and sustainable urban design, seamlessly integrating water and waste management into its development plans.

a. **Water-Sensitive Urban Design (WSUD):** Copenhagen's Cloudburst Management Plan (CMP) addresses extreme rainfall and climate resilience through green infrastructure such as rain gardens, permeable pavements, and green roofs. This approach mitigates flooding, conserves water, and enhances biodiversity. The city's goal of becoming carbon-neutral by 2025 further underscores its commitment to sustainable urban planning (Bakker & Wester, 2020).
b. **Waste-to-Energy and Recycling:** Copenhagen's waste management system is centered on circular economy principles, prioritizing recycling and energy recovery. The Amager Bakke Waste-to-Energy plant converts household waste into renewable energy, supporting the city's district heating system and reducing reliance on fossil fuels. Policies promoting organic waste segregation and reverse vending machines for recyclables further enhance sustainability. Mikkelsen et al. (2019) emphasized that Copenhagen's waste-to-energy strategy significantly reduces landfill use while contributing to renewable energy production.

San Francisco: Leading the Way in Zero-Waste and Water Conservation

San Francisco has emerged as a global leader in sustainability, focusing on zero-waste initiatives and water conservation to address regional challenges, including waste management and recurring droughts.

a. Zero-Waste Goal and Comprehensive Waste Diversion Strategies

San Francisco has set an ambitious zero-waste goal to divert 100% of its waste from landfills by 2030. Its comprehensive waste management system integrates recycling, composting, and waste reduction practices across households and businesses.

Key initiatives include:

- The **Mandatory Recycling and Composting Ordinance**, which requires all residents and businesses to separate waste into recyclables, compostables, and trash.

- Investments in organic waste infrastructure, such as advanced composting facilities.
- Programs like **Recology**, which streamline waste sorting, and partnerships led by the San Francisco Department of the Environment to encourage sustainable practices among businesses.

Ko et al. (2020) highlighted how San Francisco's policy framework, public engagement, and infrastructure investments have significantly increased recycling rates and minimized landfill dependency, positioning the city as a model for waste diversion.

b. Water Conservation Initiatives and Drought Resilience

Recognizing California's susceptibility to drought, San Francisco has implemented innovative water conservation measures to enhance resilience. These initiatives include:

- Promoting **water-efficient appliances**, rainwater harvesting, and graywater reuse systems.
- Encouraging the adoption of **low-flow fixtures** and water-saving irrigation technologies through campaigns by the San Francisco Public Utilities Commission (SFPUC).
- Developing infrastructure to collect and reuse stormwater and treated wastewater for non-potable purposes, such as irrigation and industrial applications.

Gleick and Cooley (2019) noted the city's proactive approach to drought resilience, emphasizing its emphasis on water-efficient technologies, public awareness campaigns, and sustainable water management practices.

Community and Stakeholder Engagement in Water and Waste Management

Effective community and stakeholder engagement is critical for implementing sustainable water and waste management practices. Grassroots initiatives and public-private partnerships (PPP) have emerged as powerful mechanisms to improve environmental outcomes while building more inclusive and resilient urban systems. This review explores the roles of communities, public-private collaborations, and the associated challenges and opportunities in fostering engagement in water and waste management sectors.

Role of Communities in Water and Waste Management

Communities play a pivotal role in achieving sustainable water and waste management. By participating in decision-making processes, adopting sustainable practices, and driving local innovation, communities ensure the success and longevity of environmental policies and programs.

a. Grassroots Initiatives and Community-Led Projects

Grassroots initiatives often emerge from local residents or community organizations advocating for environmental justice and sustainable practices. These initiatives address localized needs, empower individuals, and foster engagement. Examples include:

- **Community-based recycling programs** that promote waste segregation and reduce landfill dependency.
- **Rainwater harvesting and purification projects** enabling communities to sustainably manage local water resources.
- **Community composting systems** in urban areas, which process organic waste into nutrient-rich compost for urban gardening (Doria et al., 2019).

Such projects enhance local capacity and encourage sustainable living. For instance, Graham & Stinson (2020) highlighted how community composting reduces landfill waste while fostering environmental awareness and cooperation.

b. Education and Awareness Programs

Educational initiatives are essential for empowering communities to adopt sustainable practices. Through workshops, campaigns, and school programs, stakeholders can raise awareness about:

- The importance of reducing water consumption.
- The benefits of waste segregation and recycling.
- Challenges like drought resilience and waste diversion.

Programs like "zero-waste challenges" or "water conservation months" actively engage individuals and drive behavioral change (Schweitzer & Brown, 2021). Public-private collaborations further enhance these efforts by promoting shared responsibility and sustainability goals.

Public-Private Partnerships (PPP)

Public-private partnerships (PPP) provide collaborative frameworks to enhance water and waste management infrastructure. By combining government oversight with private sector expertise, PPPs foster innovation, efficiency, and large-scale investments.

a. Collaborative Projects Between Municipalities and Private Sectors

PPP initiatives leverage the strengths of both sectors to improve services and sustainability outcomes. In water management, these partnerships have modernized:

- **Water treatment plants.**
- **Distribution systems** for efficient delivery.
- **Stormwater management** using advanced technologies (Jiménez & Maestu, 2020).

Similarly, in waste management, PPPs support waste-to-energy projects, reducing landfill dependency while generating renewable energy. These collaborations deliver environmental and economic benefits, addressing urban challenges comprehensively.

b. Successful Examples of PPPs

- **Veolia Environnement (France):** This multinational company partners with cities like London and New York to upgrade water treatment facilities and improve recycling systems.
- **Swachh Bharat Mission (India):** A prominent PPP that enhances sanitation services, promotes waste segregation, and manages solid waste through partnerships between the government and private enterprises (Prasad & Kumar, 2021).

These examples showcase the potential of PPPs to address large-scale challenges by fostering collaboration, leveraging resources, and deploying innovative solutions.

Challenges and Opportunities in Community and Stakeholder Engagement

a. Overcoming Barriers to Engagement

Engaging diverse stakeholders requires addressing challenges such as:

- **Lack of trust** between communities and decision-makers. Building trust through inclusive decision-making ensures marginalized groups feel represented.
- **Cultural and social barriers** that hinder participation in diverse urban settings. Tailored communication strategies promote inclusivity (Doria et al., 2019).
- **Resource limitations** and bureaucratic delays that affect implementation. Proactive planning, cross-sector collaboration, and adequate funding are vital.

Doria et al. (2019) emphasized that overcoming these barriers requires coordinated efforts, including targeted educational campaigns and transparent engagement processes.

b. Leveraging Opportunities for Collaboration and Innovation

Technological and collaborative advancements present significant opportunities for strengthening engagement:

- **Smart technologies and digital platforms** enable real-time data collection, operational monitoring, and decision-making by communities and policy-makers (Graham & Stinson, 2020).
- **Intersectoral collaborations** between environmental organizations, technology firms, and municipalities foster comprehensive solutions, such as circular economy models.

Graham & Stinson (2020) explored how innovative tools like smart sensors and digital dashboards empower communities to actively participate in water and waste management systems.

Future Directions and Recommendations in Water and Waste Management

As urban populations grow and environmental challenges intensify, the future of water and waste management depends on adopting innovative and sustainable approaches. Emerging trends, such as digital technologies, data analytics, and circular economy principles, are transforming how resources are managed. This review highlights these trends, their applications, and future possibilities, offering actionable recommendations for policy makers and practitioners to integrate these solutions into urban planning and governance.

Recommendations for Policy Makers and Practitioners

Urban water and waste management systems are essential for sustainable urban development. By integrating innovative technologies and adopting adaptive, holistic approaches, policy makers and practitioners can improve resource efficiency, strengthen environmental outcomes, and enhance urban resilience.

1. Integrating Innovative Solutions into Urban Planning

Addressing challenges such as urbanization, climate change, and resource scarcity requires policy makers to incorporate advanced technologies and sustainable practices into urban planning. This integration will make cities more efficient, adaptable, and resilient in managing water and waste.

a. **Incorporating Digital Technologies:** Digital technologies, including smart meters, IoT-based sensors, and advanced data analytics, are vital for optimizing water and waste management systems. These tools enable real-time monitoring, predictive analysis, and system optimization, allowing municipalities to:
   - Detect leaks and manage water usage more efficiently.
   - Track waste production and improve recycling rates.
   - Schedule waste collections based on real-time data from IoT sensors in bins, reducing costs and environmental impact (Patel et al., 2020; Dutta et al., 2021).

Urban planning should prioritize digital platforms for data sharing among governments, utilities, and citizens, fostering a smart city approach to resource management. Patel et al. (2020) demonstrated how smart meters and IoT technologies improve water efficiency through demand management and leak detection.

b. **Promoting Circular Economy Principles:** The adoption of circular economy principles can transform urban waste management by prioritizing resource recovery, recycling, and waste reduction. Key strategies include:
   - Encouraging businesses to design products for reuse, recycling, and easy disassembly.
   - Diverting organic waste from landfills through composting and waste-to-energy technologies (Zhang et al., 2020).
   - Implementing Extended Producer Responsibility (EPR) regulations to hold manufacturers accountable for their products' lifecycle.

Local waste sorting programs and investments in recycling and resource recovery infrastructure should also be expanded. Zhang et al. (2020) emphasized the importance of waste-to-energy systems and organic waste recycling in achieving sustainable urban resource management.

2. Building Resilient and Adaptive Urban Infrastructure

Urban infrastructure must be resilient and adaptable to climate change, population growth, and evolving environmental pressures. Policy makers should prioritize sustainable water and waste management solutions that address these challenges effectively.

a. **Designing Climate-Resilient Water Infrastructure:** Urban water systems face risks such as flooding, water scarcity, and temperature fluctuations due to climate change. Climate-resilient infrastructure should include:
   - **Sustainable Urban Drainage Systems (SUDS):** Green roofs, permeable pavements, and bio-retention areas can absorb rainwater, reduce flooding risks, and improve water quality (Hollis et al., 2019).
   - **Decentralized Water Systems:** Rainwater harvesting and on-site wastewater treatment systems can reduce the burden on centralized infrastructure, increase water availability, and decrease environmental impacts (Pereira et al., 2020).

Hollis et al. (2019) highlighted the importance of integrating green infrastructure into urban planning to mitigate urban flooding and drought impacts.

b. **Adaptive Waste Management Systems:** Evolving waste streams and increasing urban complexity require flexible and scalable waste management solutions. These include:
   - **Smart Waste Technologies:** AI-driven sorting systems and robotics can automate waste separation, reducing human intervention and improving efficiency (Rai & Tiwari, 2020).
   - **Waste-to-Energy Plants:** Scaling up these facilities can help cities process non-recyclable waste while generating renewable energy.
   - **Expanded Composting Initiatives:** Promoting composting programs for organic waste supports local food systems and urban agriculture.

Rai & Tiwari (2020) demonstrated the potential of AI and robotics to optimize waste processing, allowing cities to adapt to changing waste streams while enhancing operational efficiency.

Final Thoughts on Achieving Sustainable Urban Water and Waste Management

Achieving sustainable urban water and waste management requires a coordinated, forward-thinking approach from policy makers, urban planners, and practitioners. Key priorities include:

- **Continuous innovation and investment** in smart infrastructure, circular systems, and climate-resilient solutions.
- **Collaboration across sectors,** including public, private, and civil society, to create inclusive, economically viable solutions.
- **Holistic governance frameworks** that integrate technology, sustainability, and community participation.

By embracing efficiency, adaptability, and resource recovery, cities can address the pressures of rapid urbanization, climate change, and resource scarcity. Such strategies will ensure that urban centers remain livable, resilient, and capable of thriving in a complex and uncertain future.

CONCLUSION

Achieving sustainable urban water and waste management requires a holistic approach that integrates innovative strategies, resilient infrastructure, and active community engagement. Cutting-edge waste management solutions, such as zero-waste policies, advanced recycling programs, and extended producer responsibility (EPR), are pivotal in reducing environmental impacts and supporting the transition to a circular economy. Zero-waste initiatives prioritize waste prevention, sustainable design, and active community participation to divert waste from landfills. Enhanced recycling systems improve material recovery and reduce contamination, while EPR encourages producers to adopt sustainable practices and take accountability for their products' lifecycle.

Innovations like composting, anaerobic digestion, smart waste bins, and waste-to-energy technologies further enhance urban sustainability by addressing waste reduction, resource recovery, and energy generation. Policy frameworks, such as pay-as-you-throw programs and EPR, incentivize responsible consumption and waste management behaviors.

Case studies from cities like Singapore, Copenhagen, and San Francisco demonstrate how integrated water systems, circular economy strategies, and robust waste diversion programs can serve as models for sustainable urban development. These cities have shown that adopting innovative and inclusive approaches can significantly improve sustainability, despite challenges such as resource limitations and socio-cultural barriers.

Community and stakeholder engagement remain central to achieving these goals. Grassroots initiatives, public-private partnerships, and education campaigns foster responsible consumption and waste reduction behaviors. Technology and intersectoral collaboration offer opportunities to overcome barriers, build trust, and enhance resource efficiency.

Ultimately, the integration of innovative technologies, circular economy principles, and climate-resilient infrastructure will be essential in addressing the complex challenges of urbanization and climate change. By prioritizing efficiency, adaptability, and sustainability, cities can build resilient systems that ensure long-term environmental health, economic vitality, and social well-being.

REFERENCES

Anwar, R., Mollah, M. H., & Islam, S. (2020). Greywater recycling in urban residential buildings: Current trends and future challenges. *Environmental Science and Pollution Research International, 27*(11), 12241–12256. DOI: 10.1007/s11356-020-07827-5

Anwar, R., Mollah, M. H., & Islam, S. (2020). Greywater recycling in urban residential buildings: Current trends and future challenges. *Environmental Science and Pollution Research International, 27*(11), 12241–12256. DOI: 10.1007/s11356-020-07827-5

Arena, U. (2012). Process and technological aspects of municipal solid waste gasification. *A review.Waste Management (New York, N.Y.), 32*(4), 625–639. DOI: 10.1016/j.wasman.2011.09.025 PMID: 22035903

AWWA. (2021). *The impact of water loss on urban systems and strategies for reducing it.* American Water Works Association.

Bakker, J., & Wester, P. (2020). Water-sensitive urban design in Copenhagen: Strategies for climate adaptation. *Urban Climate, 34*, 100678. DOI: 10.1016/j.uclim.2020.100678

Barros, E. M., Silva, S. R., & Oliveira, D. M. (2020). Advanced oxidation processes for the treatment of pharmaceuticals in wastewater: A review. *Journal of Environmental Management, 260*, 110085. DOI: 10.1016/j.jenvman.2020.110085

Biel, A., Mårtensson, F., & Gärling, T. (2017). Public awareness campaigns and their role in water conservation. *Journal of Environmental Psychology, 51*, 68–75. DOI: 10.1016/j.jenvp.2017.04.003

Björkqvist, S., Elser, C., & Thörne, E. (2017). Sweden's waste management system: A case study in circular economy. *Waste Management (New York, N.Y.), 59*, 1–12. DOI: 10.1016/j.wasman.2017.02.003

Brown, S., Leonard, P., & Saunders, R. (2012). *Urban agriculture and community food security in the United States: Farming from the city center to the urban fringe.* Sustainable Urban Agriculture Coalition.

Cavalett, O., Lima, D., & Silva, M. A. (2021). Urban agriculture and composting: A sustainable approach to waste management in community gardens. *Environmental Science & Policy, 113*, 32–43. DOI: 10.1016/j.envsci.2020.08.004

Chen, M., Zhao, W., & Wang, X. (2020). Enhancing urban recycling through advanced sorting technologies: Lessons from San Francisco. *Journal of Environmental Management, 259*, 110057. DOI: 10.1016/j.jenvman.2019.110057

Diaper, C., Williams, K., & Jackson, T. (2020). Greywater recycling: A review of technologies, benefits, and challenges in urban areas. *Journal of Environmental Management*, *259*, 109613. DOI: 10.1016/j.jenvman.2019.109613

Doria, M., Tedesco, L., & Gagliardi, R. (2019). Overcoming barriers to community engagement in urban water management: Lessons from international case studies. *Water Policy*, *21*(5), 1016–1034. DOI: 10.2166/wp.2019.103

Doria, M., Tedesco, L., & Gagliardi, R. (2019). The role of community-driven approaches in improving urban water management and waste recycling. *Environmental Sustainability*, *11*(3), 217–230. DOI: 10.1016/j.envsus.2019.02.006

Dutta, A., Sharma, A., & Patel, P. (2021). IoT-based smart waste management systems for urban sustainability. *Urban Climate*, *37*, 100845. DOI: 10.1016/j.uclim.2021.100845

EPA. (2021). WaterSense: A guide to water-efficient plumbing products. U.S. Environmental Protection Agency. https://www.epa.gov/watersense

EPA. (2021). WaterSense: A guide to water-efficient plumbing products. U.S. Environmental Protection Agency. https://www.epa.gov/watersense

Faccio, M., Persona, A., & Zanin, G. (2011). Waste collection multi objective model with real time traceability data. *Waste Management (New York, N.Y.)*, *31*(12), 2391–2405. DOI: 10.1016/j.wasman.2011.07.005 PMID: 21821406

Feng, H., Zhou, J., & Li, H. (2019). Membrane bioreactor (MBR) technology for wastewater treatment: Recent advances and future directions. *Environmental Technology*, *40*(8), 1056–1069. DOI: 10.1080/09593330.2018.1564107

Fisher, P. C., Blanchard, D. L., & McCormick, R. A. (2019). The role of extended producer responsibility in reducing municipal waste: Global perspectives and lessons learned. *Waste Management & Research*, *37*(6), 557–568. DOI: 10.1177/0734242X19859448

Fletcher, T. D., Shuster, W., Hunt, W. F., Ashley, R. M., Butler, D., Arthur, S., & Uhl, M. (2015). SUDS, LID, BMPs, WSUD and more–The evolution and application of terminology surrounding urban drainage. *Urban Water Journal*, *12*(7), 525–542. DOI: 10.1080/1573062X.2014.916314

Fletcher, T. D., Shuster, W. D., & Gersonius, B. (2017). *Water sensitive urban design: A case study approach*. Springer.

Friedler, E. (2004). Quality of individual domestic greywater streams and its implications for on-site treatment and reuse possibilities. *Environmental Technology, 25*(9), 997–1008. DOI: 10.1080/09593330.2004.9619393 PMID: 15515266

García, A., & López, R. (2020). Circular economy in urban waste management: Opportunities and challenges. *Waste Management (New York, N.Y.), 102*, 349–359. DOI: 10.1016/j.wasman.2020.01.045

Getter, K. L., & Rowe, D. B. (2021). *Green roofs: A comprehensive guide to the planning, design, and construction of green roofs*. John Wiley & Sons.

Gleick, P. H. (2003). Water use. *Annual Review of Environment and Resources, 28*(1), 275–314. DOI: 10.1146/annurev.energy.28.040202.122849

Gleick, P. H., & Cooley, H. (2019). Water conservation and drought resilience in San Francisco: Policies and outcomes. *Water Resources Research, 55*(9), 7539–7550. DOI: 10.1029/2019WR024585

Gould, J., & Nissen-Peterson, E. (2019). Rainwater catchment systems for domestic water supply: Design, construction, and implementation. CRC Press.

Graham, S., & Stinson, C. (2020). Leveraging digital technologies for community engagement in urban waste management. *Journal of Urban Technology, 27*(4), 35–51. DOI: 10.1080/10630732.2020.1794825

Hatt, B. E., Fletcher, T. D., & Deletic, A. (2016). Water-sensitive urban design: A review of its potential and practice in Melbourne, Australia. *Australian Journal of Water Resources, 20*(3), 259–272. DOI: 10.1080/13241583.2016.1183684

Hoekstra, A. Y., & Mekonnen, M. M. (2012). The water footprint of humanity. *Proceedings of the National Academy of Sciences of the United States of America, 109*(9), 3233–3237. DOI: 10.1073/pnas.1109936109 PMID: 22331890

Hollis, A., Johnson, R., & Zhuang, M. (2019). Green infrastructure in urban water management: Building resilience against climate change. *Environmental Science & Policy, 104*, 66–74. DOI: 10.1016/j.envsci.2019.01.003

Hopewell, J., Dvorak, R., & Kosior, E. (2009). Plastics recycling: Challenges and opportunities. *Philosophical Transactions of the Royal Society of London. Series B, Biological Sciences, 364*(1526), 2115–2126. DOI: 10.1098/rstb.2008.0311 PMID: 19528059

IPCC. (2022). Climate change 2022: Impacts, adaptation, and vulnerability. Intergovernmental Panel on Climate Change. https://www.ipcc.ch/report/ar6/wg2/

Jiménez, B., & Maestu, J. (2020). Public-private partnerships in water management: Enhancing sustainability through collaborative infrastructure. *Water Resources Management, 34*(7), 2291–2302. DOI: 10.1007/s11269-020-02537-4

Judd, S. (2011). *The MBR book: Principles and applications of membrane bioreactors for water and wastewater treatment.* Elsevier.

Kar, S. (2020). Smart water metering and real-time monitoring systems for water management. *Journal of Water Supply: Research & Technology - Aqua, 69*(3), 168–179. DOI: 10.2166/aqua.2020.145

Kar, S. (2020). Smart water metering and real-time monitoring systems for water management. *Journal of Water Supply: Research & Technology - Aqua, 69*(3), 168–179. DOI: 10.2166/aqua.2020.145

Kaza, S., Yao, L., Bhada-Tata, P., & Van Woerden, F. (2018). *What a waste 2.0: A global snapshot of solid waste management to 2050.* World Bank Group. DOI: 10.1596/978-1-4648-1329-0

Kaza, S., Yao, L., Bhada-Tata, P., & Van Woerden, F. (2018). What a waste 2.0: A global snapshot of solid waste management to 2050. World Bank Group. https://openknowledge.worldbank.org/handle/10986/30317

Khalid, A., Arshad, M., Anjum, M., Mahmood, T., & Dawson, L. (2011). The anaerobic digestion of solid organic waste. *Waste Management (New York, N.Y.), 31*(8), 1737–1744. DOI: 10.1016/j.wasman.2011.03.021 PMID: 21530224

Khan, M. Z., Hassan, M., & Yu, Z. (2021). Internet of Things (IoT) in smart waste management: A comprehensive review. *Sustainable Cities and Society, 69*, 102758. DOI: 10.1016/j.scs.2021.102758

Ko, H., Chang, J., & Li, Y. (2020). San Francisco's zero waste program: Policies, practices, and challenges. *Environmental Management, 66*(1), 85–94. DOI: 10.1007/s00267-020-01314-6

Koch, F., Bassetti, L., & Huynh, D. (2020). Performance of rainwater harvesting systems in urban Southeast Asia: Opportunities and challenges. *Water (Basel), 12*(6), 1607. DOI: 10.3390/w12061607

Koch, F., Bassetti, L., & Huynh, D. (2020). Performance of rainwater harvesting systems in urban Southeast Asia: Opportunities and challenges. *Water (Basel), 12*(6), 1607. DOI: 10.3390/w12061607

Lausten, M. A., Aschenbrenner, A., & Hughes, J. R. (2020). Zero-waste initiatives and urban waste management: Lessons from Kamikatsu, Japan. *Waste Management & Research, 38*(10), 1167–1177. DOI: 10.1177/0734242X20962921

Lee, H. K., Tan, B. K., & Soh, Y. W. (2020). The Zero Waste Masterplan of Singapore: Strategies for sustainable waste management. *Environmental Science & Technology, 54*(19), 11872–11881. DOI: 10.1021/acs.est.0c03094 PMID: 32852946

Li, J., Zhao, X., & He, X. (2019). Permeable pavements for stormwater management in urban areas: A review of materials and techniques. *Environmental Science and Pollution Research International, 26*(5), 4680–4691. DOI: 10.1007/s11356-019-04387-9 PMID: 30565107

Li, Z., Boyle, F., & Reynolds, A. (2016). Rainwater harvesting and greywater treatment systems for domestic application in Ireland. *Desalination and Water Treatment, 57*(48-49), 22911–22919. DOI: 10.1080/19443994.2015.1127175

Lifset, R., Atasu, A., & Tojo, N. (2013). Extended producer responsibility. *Journal of Industrial Ecology, 17*(2), 162–166. DOI: 10.1111/jiec.12022

Liu, C., & Liao, Z. (2019). Overcoming barriers to greywater recycling in urban settings: A review of regulations and technological advancements. *Resources, Conservation and Recycling, 142*, 83–92. DOI: 10.1016/j.resconrec.2018.11.018

Liu, C., & Liao, Z. (2019). Overcoming barriers to greywater recycling in urban settings: A review of regulations and technological advancements. *Resources, Conservation and Recycling, 142*, 83–92. DOI: 10.1016/j.resconrec.2018.11.018

Liu, J., Zhang, Q., & Li, Y. (2018). Enhancing water conservation in urban distribution systems: A study of smart water metering and real-time leak detection. *Journal of Water Resources Planning and Management, 144*(12), 04018087. DOI: 10.1061/(ASCE)WR.1943-5452.0001005

MacRae, R., McLeod, A., & Green, J. (2021). Community-based water conservation programs: Impacts on urban sustainability. *Environmental Education Research, 27*(4), 525–539. DOI: 10.1080/13504622.2020.1819331

Massoud, M. A., Tarhini, A., & Nasr, J. A. (2009). Decentralized approaches to wastewater treatment and management: Applicability in developing countries. *Journal of Environmental Management, 90*(1), 652–659. DOI: 10.1016/j.jenvman.2008.07.001 PMID: 18701206

Mikkelsen, P. S., Andersen, R., & Mortensen, T. (2019). Copenhagen's waste management and recycling strategies: Waste-to-energy and circular economy. *Waste Management (New York, N.Y.), 94*, 218–226. DOI: 10.1016/j.wasman.2019.06.026

Mills, J. D., & Boersma, D. L. (2021). Big data and AI in waste management: The future of urban resource optimization. *Journal of Environmental Informatics, 35*(1), 74–89. DOI: 10.1016/j.envinform.2021.05.006

Nair, R., & Goh, K. (2018). Decentralized wastewater treatment systems for small urban communities: Case studies and lessons learned. *Environmental Engineering Science, 35*(4), 240–250. DOI: 10.1089/ees.2017.0324

OECD. (2018). *The state of play on extended producer responsibility*. Organisation for Economic Co-operation and Development.

Patel, S. J., Yang, X., & Roy, P. K. (2020). The role of digital technologies in enhancing urban water management: Smart meters and IoT-based solutions. *Water Resources Management, 34*(8), 2557–2571. DOI: 10.1007/s11269-020-02601-1

Prasad, B., & Kumar, V. (2021). Swachh Bharat Mission: Public-private partnership in waste management and sanitation. *Waste Management & Research, 39*(8), 939–948. DOI: 10.1177/0734242X21102194

Rai, A., & Tiwari, S. (2020). Robotics and AI in urban waste management: Advancing efficiency and adaptability in waste systems. *Waste Management & Research, 38*(1), 13–24. DOI: 10.1177/0734242X20923056

Satyro, J. R., Silva, A. L., & Almeida, S. S. (2020). Managing organic waste through composting and anaerobic digestion: Case studies and sustainable strategies. *Waste Management & Research, 38*(4), 403–414. DOI: 10.1177/0734242X20914401

Schweitzer, A., & Brown, J. (2021). The impact of education and public awareness on sustainable waste management practices. *Journal of Environmental Management, 256*, 109885. DOI: 10.1016/j.jenvman.2020.109885

Sutcliffe, J., Hunter, S., & Sim, N. (2018). The role of green infrastructure in sustainable urban drainage systems: Case studies from London. *Urban Water Journal, 15*(1), 50–60. DOI: 10.1080/1573062X.2017.1280226

Tan, A. L., Wong, H. L., & Yap, S. L. (2019). NEWater: Singapore's success story in water reuse and sustainability. *Desalination and Water Treatment, 146*, 207–217. DOI: 10.5004/dwt.2019.24274

UN-Habitat. (2020). *The state of African cities 2020: Inclusive cities for resilient economies*. United Nations Human Settlements Programme.

UN-Habitat. (2020). The state of African cities 2020: Inclusive cities for resilient economies. United Nations Human Settlements Programme. https://unhabitat.org/the-state-of-african-cities-2020-inclusive-cities-for-resilient-economies

UNEP. (2016). *Global outlook on sustainable consumption and production policies*. United Nations Environment Programme.

UNEP. (2016). Global outlook on sustainable consumption and production policies. United Nations Environment Programme. https://www.unep.org/resources/report/global-outlook-sustainable-consumption-and-production-policies

Water, U. N. (2018). *The United Nations world water development report 2018: Nature-based solutions for water*. United Nations.

Water, U. N. (2018). The United Nations world water development report 2018: Nature-based solutions for water. United Nations. https://www.unwater.org/publications/world-water-development-report-2018/

Wilson, D. C., Velis, C. A., & Cheeseman, C. R. (2015). Role of informal sector recycling in waste management in developing countries. *Habitat International*, *31*(1), 6–15.

Wong, K. S., Ibrahim, R., & Lim, S. (2021). Advances in rainwater harvesting technologies: Challenges and solutions for urban areas. *Environmental Science & Technology*, *55*(13), 8904–8914. DOI: 10.1021/acs.est.1c01677

World Bank. (2018). *Solid waste management in developing countries*. World Bank Group.

World Bank. (2018). Solid waste management in developing countries. World Bank Group. https://openknowledge.worldbank.org/handle/10986/30303

Zhang, L., & Liao, B. (2020). The role of greywater recycling in sustainable urban water management. *Environmental Science & Technology*, *54*(15), 9208–9217. DOI: 10.1021/acs.est.0c02238

Zhang, Y., Wang, X., & He, L. (2020). Circular economy in urban waste management: Concepts, applications, and future perspectives. *Waste Management & Research*, *38*(4), 347–357. DOI: 10.1177/0734242X20910101 PMID: 32116140

Zhao, M., Li, W., & Chen, Z. (2020). The role of water-efficient appliances in urban water conservation. *Water Resources Management*, *34*(2), 561–577. DOI: 10.1007/s11269-019-02439-9

Zhou, Y., Wang, S., & Jiang, C. (2019). Waste-to-energy technologies and their potential in reducing urban waste: A case study of pyrolysis and gasification. *Renewable & Sustainable Energy Reviews*, *109*, 358–368. DOI: 10.1016/j.rser.2019.04.037

Chapter 10
Innovations for Sustainable Coastal Cities:
Strategies for Water and Waste Management

Treesa Sindhu P. Thomas
https://orcid.org/0009-0002-7405-5162
St. Teresa's College, Ernakulam, India

Leena Leon
Rajagiri College of Social Sciences, Kalamasseri, India

ABSTRACT

Coastal cities face unique challenges at the intersection of rapid urbanization and climate change, making sustainable water and waste management critical for their development. Innovative approaches to resource management that harmonize environmental stability, economic resilience, and social well-being by leveraging cutting-edge technologies, adaptive urban planning, and participatory governance models are essential for mitigating these issues. The adoption of holistic strategies, including zero-waste policies, enhanced recycling programs, and extended producer responsibility (EPR) frameworks, as key drivers of environmental impact reduction and resource recovery. The transformative role of community engagement in fostering behavioural change and policy instruments such as pay-as-you-throw initiatives and public-private partnerships are identified as crucial enablers of scalable, sustainable solutions.

DOI: 10.4018/979-8-3373-5278-7.ch010

INTRODUCTION

Coastal urban areas represent some of the most dynamic and essential components of cities, offering valuable resources, driving economic growth, and delivering vital ecological services. However, these regions are uniquely vulnerable to challenges such as rising sea levels, rapid urbanization, environmental degradation, and socio-economic disparities. As coastal cities expand, sustainable development becomes a necessity to safeguard both human populations and the natural environment.

Home to nearly 40% of the global population within 100 kilometers of the coastline (UN-Habitat, 2020), coastal areas are pivotal to the world's economy. These regions serve as hubs for international trade, tourism, fisheries, and transportation, with ports facilitating a significant share of global commerce. Iconic coastal cities such as New York, Tokyo, and Mumbai stand as economic powerhouses, playing central roles in finance, business, and culture.

In addition to economic importance, coastal ecosystems like mangroves, coral reefs, and wetlands provide essential services such as biodiversity conservation, carbon sequestration, and natural disaster mitigation. For example, mangroves buffer against storm surges, while coral reefs support marine biodiversity. These ecosystems also contribute to coastal stabilization, water filtration, and habitat creation, making them indispensable for urban and environmental resilience (Barbier et al., 2019). Tourism further underscores the importance of coastal areas, as millions are drawn to their aesthetic and recreational appeal. The resulting economic benefits highlight the need to balance urban development with ecological preservation to maintain these regions' allure and functionality.

Despite their importance, coastal cities face growing threats from climate change and unsustainable practices. Rising sea levels, extreme weather events, and resource scarcity strain water and waste management systems, crucial pillars of urban sustainability. These challenges directly impact public health, economic stability, and environmental integrity, requiring immediate and innovative solutions.

Water resources in coastal cities are under unprecedented pressure due to natural and human-induced factors. Climate change intensifies extreme weather events such as storms, droughts, and flooding, disrupting water infrastructure and reducing freshwater availability. Saline intrusion from rising sea levels contaminates water supplies, compelling cities to invest in costly desalination systems (UN-Habitat, 2011). Meanwhile, rapid urbanization exacerbates water resource depletion, increases pollution, and undermines natural water systems (Kallis et al., 2020).

Waste management in coastal cities also faces significant challenges, driven by high population densities, industrial activity, and proximity to water bodies. Inefficient systems lead to the accumulation of waste, much of which ends up in oceans, causing marine pollution and ecological harm (Jambeck et al., 2015). Organic waste and industrial byproducts further contribute to greenhouse gas emissions and environmental degradation, necessitating urgent intervention.

To address these interconnected issues, coastal cities are adopting integrated water resource management (IWRM) systems, green infrastructure, and circular economy models. IWRM emphasizes efficient water use, equitable distribution, and environmental protection, while green infrastructure solutions such as rainwater harvesting and wetland restoration enhance flood resilience and groundwater recharge (Gill et al., 2007). Circular economy approaches, including waste-to-energy technologies and improved recycling systems, aim to minimize waste generation and promote resource recovery.

Governance plays a critical role in the success of these strategies. Resilient infrastructure investments, climate adaptation policies, and public-private partnerships are essential for ensuring sustainable water and waste systems. Collaboration among governments, businesses, and communities is vital to designing systems that prioritize both immediate needs and long-term resilience (Kallis et al., 2020).

By integrating innovative practices and fostering collective action, coastal cities can address the dual challenges of water and waste management while enhancing their sustainability. These efforts will not only protect ecosystems and improve quality of life but also contribute to global climate resilience and sustainable urban development.

CHALLENGES IN WATER AND WASTE MANAGEMENT FOR COASTAL CITIES

Despite the importance of these systems, coastal cities face significant challenges in ensuring sustainable water and waste management.

Water Scarcity and Increased Demand: Population growth and industrial development in coastal cities drive rising water demands. Many of these cities are situated in arid or semi-arid regions where freshwater resources are already scarce or overexploited. Climate change further exacerbates water scarcity by altering precipitation patterns and increasing evaporation rates, straining local water supplies (Pereira et al., 2012).

Saltwater Intrusion: Rising sea levels caused by climate change contribute to saltwater intrusion into freshwater aquifers, contaminating critical water sources. This is particularly problematic for cities that rely heavily on groundwater for drinking water. Once aquifers are contaminated, desalination becomes a necessary but expensive solution, often beyond the reach of many cities (UN-Habitat, 2011).

Flooding and Stormwater Management: Coastal cities are increasingly vulnerable to flooding, driven by sea-level rise and extreme weather events such as hurricanes, typhoons, and heavy rainfall. Insufficient stormwater infrastructure exacerbates this issue, leading to urban flooding, water contamination, and infrastructure degradation. Robust stormwater management systems, including green infrastructure and improved drainage networks, are essential to mitigate these risks (Villarini et al., 2019).

Pollution of Water Resources: Untreated sewage, industrial discharges, and agricultural runoff frequently contaminate water sources in coastal cities, degrading ecosystems and posing significant public health risks. This issue is particularly acute in developing regions, where rapid urbanization often outpaces infrastructure development (Kallis et al., 2020). Enhancing wastewater treatment facilities and adopting stringent pollution control measures are vital to addressing this challenge.

CHALLENGES IN WASTE MANAGEMENT IN COASTAL CITIES

Effective waste management in coastal cities is vital for safeguarding public health and preserving the environment. However, these urban areas face complex and multifaceted challenges due to their dense populations, dynamic economies, and proximity to marine ecosystems.

High Waste Generation: Coastal cities are hubs of economic activity and dense human settlement, which results in the generation of vast amounts of waste. Beyond household waste, these cities also grapple with significant volumes of industrial and commercial waste, much of which can be hazardous or difficult to recycle. A considerable portion of this waste ends up in the ocean, contributing to global marine pollution, particularly the escalating crisis of plastic waste (Jambeck et al., 2015). The sheer volume of waste poses a significant burden on local waste management systems.

Inadequate Waste Management Infrastructure: Many coastal cities, especially in developing regions, lack adequate waste management infrastructure. Insufficient waste collection, recycling, and disposal facilities lead to the accumulation of waste in streets, waterways, and open spaces, causing widespread pollution and health hazards (Pillai et al., 2021). Furthermore, the absence of effective waste segregation practices exacerbates the issue, making recycling inefficient and accelerating the overflow of landfills.

Marine Litter and Plastic Pollution: Marine litter, particularly plastic waste, is one of the most pressing waste management challenges for coastal cities. Proximity to shipping routes, industrial zones, and high levels of human activity contribute to the dumping of plastics and other non-biodegradable materials into the ocean

(Jambeck et al., 2015). This pollution harms marine ecosystems, endangers aquatic life, and poses risks to human health through the consumption of contaminated seafood. The lack of stringent policies and comprehensive infrastructure for plastic waste collection and recycling exacerbates this challenge, demanding immediate action to mitigate its impact.

Waste-to-Energy (WTE) Systems and Landfill Overload: To address the mounting waste problem, some coastal cities have adopted waste-to-energy (WTE) technologies, which convert waste into electricity or heat. While WTE systems help reduce landfill dependency and provide alternative energy, they come with significant hurdles. These include high implementation costs, technical complexities, and the risk of increased air pollution if not managed effectively (Pillai et al., 2021). Despite advancements, many coastal cities continue to rely heavily on landfills as their primary waste disposal method. This reliance contributes to environmental degradation, groundwater contamination, and the loss of valuable land that could otherwise be used for urban development or green spaces.

WATER RESOURCE MANAGEMENT: THE BACKBONE OF SUSTAINABILITY IN COASTAL CITIES

Water and waste management are critical pillars of urban sustainability, particularly in coastal cities where the interplay of land, freshwater, and marine ecosystems creates unique challenges and opportunities. These essential systems ensure the availability of clean water, the safe disposal of waste, and the preservation of urban ecosystems, all of which are fundamental to achieving sustainable development goals and improving quality of life.

Effective water management is central to sustaining urban life by providing clean water for domestic consumption, sanitation, agriculture, industry, and ecosystem support. However, coastal cities face mounting pressures on water resources due to population growth, pollution, over-extraction of groundwater, and climate change. These pressures demand innovative, integrated water management strategies, including:

1. Water conservation to reduce consumption and wastage.
2. Rainwater harvesting to supplement urban water supplies.
3. Recycling and reuse of treated wastewater to optimize water resource utilization.
4. Advanced wastewater treatment to prevent pollution and promote environmental health.

These measures collectively play a vital role in addressing the water needs of growing populations while safeguarding environmental integrity.

INNOVATIVE WATER MANAGEMENT STRATEGIES IN COASTAL AREAS

Coastal regions, with their dense populations, vibrant economies, and proximity to water, face unique and pressing challenges in water management. The increasing impacts of climate change, rapid urbanization, and rising freshwater demand make it critical to adopt innovative strategies that ensure water security, sustainability, and resilience. Here are key approaches transforming water management in coastal areas:

1. Integrated Water Resource Management (IWRM)

Integrated Water Resource Management (IWRM) is a holistic approach that optimizes water use while balancing environmental, social, and economic objectives. In coastal cities, IWRM integrates the management of freshwater, wastewater, and stormwater to create efficient and sustainable systems.

IWRM is especially vital in coastal regions, where land and water systems are intricately connected. For example, during heavy rainfall or storms, stormwater can overwhelm drainage systems, causing flooding and water quality issues. By adopting integrated management practices, cities can enhance water supply efficiency and strengthen infrastructure resilience against extreme weather events—a growing threat in the era of climate change (Kallis et al., 2020).

2. Green Infrastructure for Sustainable Water Management

Green infrastructure leverages natural or engineered systems that mimic natural processes to manage water sustainably. Examples include green roofs, rain gardens, permeable pavements, and constructed wetlands. In coastal cities, these systems reduce stormwater runoff, recharge groundwater, and improve water quality. Additionally, green infrastructure mitigates urban heat, enhances biodiversity, and provides recreational spaces. For instance, Singapore's extensive use of rain gardens and rooftop greenery not only reduces flooding but also enhances urban aesthetics and environmental quality (Tan et al., 2018). Green infrastructure offers an eco-friendly, cost-effective alternative to traditional grey infrastructure like concrete flood barriers.

3. Desalination: Transforming Seawater into Freshwater

Desalination is a crucial technology for addressing freshwater scarcity in coastal areas, particularly in arid regions. By removing salt and impurities from seawater, desalination produces potable water for drinking and agricultural use.

Reverse Osmosis (RO): RO is the most widely used desalination method, utilizing pressure to push seawater through semi-permeable membranes that remove salts. While effective, RO is energy-intensive and produces brine, which requires careful disposal to prevent environmental harm.

Forward Osmosis (FO): FO is an emerging technology that relies on natural osmotic pressure gradients, offering lower energy consumption and the potential for integration with renewable energy sources like solar power.

Despite its high costs and environmental challenges, desalination remains a valuable solution, particularly when integrated with other strategies to minimize its ecological footprint (Ghaffour et al., 2013).

4. Advanced Water Treatment and Membrane Filtration

Innovative water treatment technologies, such as membrane filtration (microfiltration, ultrafiltration, and reverse osmosis), are enabling the safe reuse of wastewater. These methods remove pathogens, contaminants, and salts, making reclaimed water suitable for drinking and industrial use.

For instance, Singapore's advanced wastewater treatment plants provide high-quality recycled water for both industrial and potable applications (Leong et al., 2020). Water reuse strategies significantly reduce freshwater demand, cut treatment costs, and minimize environmental discharge. However, substantial infrastructure investment and rigorous monitoring are essential to ensure safety and reliability.

5. Nature-Based Solutions (NbS) for Climate Adaptation

Nature-based solutions (NbS) harness natural ecosystems to address water management challenges while enhancing climate resilience. In coastal areas, NbS includes restoring wetlands, mangroves, coral reefs, and dunes, which provide vital ecosystem services such as coastal protection, water filtration, and carbon sequestration.

Mangroves, for example, act as natural barriers against storm surges and coastal erosion, while filtering pollutants from stormwater. These ecosystems not only safeguard inland areas but also offer habitats for marine life and support biodiversity. NbS provides a sustainable, cost-effective alternative to traditional engineering solutions like seawalls, making them indispensable in climate adaptation strategies (Barbier et al., 2019).

6. Smart Water Management and Technological Innovations

Technological advancements are revolutionizing water management in coastal cities. Smart water systems equipped with sensors, data analytics, and automation enable real-time monitoring of water quality, distribution, and usage.

For instance, smart meters detect leaks and inefficiencies in distribution networks, allowing for quick repairs and reducing water waste. Advanced tools, such as digital platforms that integrate weather patterns and urban data, support long-term planning and informed decision-making for water sustainability (Mauzerall et al., 2017).

Effective water management is vital in addressing the growing challenges faced by coastal areas. Innovative strategies such as smart metering, efficient appliances, rainwater harvesting, and greywater recycling play a crucial role in ensuring sustainability, reducing water wastage, and promoting conservation.

1. Smart Metering and Leak Detection

Smart metering systems provide real-time data on water usage, enabling both utilities and consumers to identify inefficiencies and detect leaks swiftly. Advanced meters transmit data remotely, significantly reducing operational costs and minimizing water loss (Kar, 2020).

Technological advancements, such as acoustic sensors and pressure monitors, are instrumental in identifying leaks in aging water infrastructure. Studies indicate that these systems can prevent up to 30% of water loss, a critical improvement for urban water management in coastal areas (AWWA, 2021; Liu et al., 2018).

2. Water-Efficient Appliances and Fixtures

Water-efficient appliances and fixtures offer a practical and impactful way to reduce water consumption without compromising performance.

***Low-flow Toilets and Washing Machines*:** Advanced appliances, such as low-flow toilets and water-recycling washing machines, reduce water use substantially, benefiting both households and the environment (Gleick, 2003).

Low-flow Faucets and Showerheads: These fixtures can cut household water consumption by up to 20%, making them essential for sustainable water use (EPA, 2021; Zhao et al., 2020).

3. Water Reuse and Recycling

With limited freshwater availability in many coastal regions, water reuse and recycling are pivotal strategies to enhance water supply. Advanced treatment technologies make wastewater suitable for non-potable and, in some cases, potable use, transforming wastewater into a valuable resource.

Rainwater Harvesting (RWH)

Rainwater harvesting captures and stores rainwater for diverse applications, including domestic, agricultural, and industrial purposes. Effective design and system integration maximize collection and ensure safe usage. Key components include:

***Catchment Area*:** Clean roof surfaces optimize water collection efficiency (Koch et al., 2020).

***Storage Capacity*:** Tanks designed based on rainfall patterns and water demand store harvested water effectively.

***Filtration and Treatment*:** Systems such as debris filters and UV sterilization ensure water safety (Wong et al., 2021).

***Distribution Systems*:** Gravity-fed or pump-driven mechanisms deliver water efficiently to users, supported by backflow prevention devices to protect municipal supplies (Gould & Nissen-Peterson, 2019).

Research highlights the benefits of flexible, well-designed RWH systems, which reduce reliance on municipal water and mitigate the impacts of seasonal droughts in urban areas (Koch et al., 2020).

Greywater Recycling

Greywater recycling involves treating wastewater from sinks, showers, and washing machines for reuse in non-potable applications like irrigation and toilet flushing. This approach reduces freshwater demand while alleviating pressure on urban sewage systems. Effective greywater systems incorporate robust treatment processes to meet quality standards and ensure safety. And the benefits include: Water Conservation: Households can cut water consumption by up to 40% through greywater recycling (Anwar et al., 2020). Cost Savings: Reduced dependence on municipal water lowers bills, especially in areas with high water tariffs (Diaper et al., 2020). Environmental Impact: Recycling decreases wastewater discharge, easing the load on treatment plants and reducing pollution (Zhang & Liao, 2020). Despite its benefits, widespread adoption of greywater recycling requires overcoming challenges related to treatment infrastructure, regulatory standards, and public awareness (Friedler, 2004).

Addressing Challenges in Greywater Recycling and Urban Water Management

As water scarcity intensifies in urban and coastal areas, greywater recycling and advanced water management practices emerge as critical solutions. However, these approaches are not without challenges and require holistic strategies for effective implementation.

Health and Safety Concerns**:** Untreated greywater can contain pathogens and harmful contaminants, posing significant health risks. To mitigate these risks, robust filtration and disinfection systems are essential to ensure that recycled greywater meets safety standards for its intended non-potable uses (Anwar et al., 2020).

Regulatory Barriers**:** The absence of comprehensive regulatory frameworks often limits the adoption of greywater recycling in urban settings. Establishing clear guidelines and safety standards for greywater reuse is imperative to build public trust and encourage widespread implementation (Liu & Liao, 2019).

System Costs and Maintenance**:** Although greywater recycling offers long-term cost savings, high initial installation costs and ongoing maintenance expenses, particularly for advanced filtration and disinfection systems, can deter adoption by households and businesses (Diaper et al., 2020).

Advanced Wastewater Treatment Technologies

Emerging technologies, such as Membrane Bioreactors (MBRs) and Advanced Oxidation Processes (AOPs), ensure high-quality wastewater treatment, enabling safe reuse or discharge.

- ***Membrane Bioreactors (MBRs)*****:** Combining biological treatment with membrane filtration, MBRs deliver exceptional effluent quality, efficiently removing pathogens, organic matter, and suspended solids (Judd, 2011).
- ***Advanced Oxidation Processes (AOPs)*****:** AOPs employ chemical methods like ozonation or UV-based systems to degrade complex pollutants, making treated wastewater suitable for reuse.

Studies by Feng et al. (2019) demonstrate the success of MBR systems in China, highlighting high contaminant removal rates and improvements in energy efficiency through membrane fouling control.

INNOVATIVE APPROACHES TO WASTEWATER TREATMENT, URBAN DRAINAGE, AND WASTE MANAGEMENT

As urbanization accelerates, sustainable solutions for wastewater treatment, stormwater management, and waste reduction are critical to fostering resilient and eco-friendly cities. This article explores advanced oxidation processes (AOPs), decentralized wastewater treatment systems (DEWATS), sustainable urban drainage systems (SUDS), and waste management innovations, showcasing their transformative potential.

Advanced Oxidation Processes (AOPs)

AOPs employ highly reactive hydroxyl radicals (•OH) to degrade organic pollutants in wastewater, effectively treating recalcitrant compounds like pharmaceuticals, pesticides, and industrial chemicals that resist conventional biological methods (Barros et al., 2020).

Examples of AOPs:

- Ozonation: Utilizes ozone gas to oxidize contaminants.
- Photocatalysis: Employs light-activated catalysts to degrade pollutants.
- Fenton's Reaction: Combines hydrogen peroxide and iron catalysts to generate hydroxyl radicals.

Barros et al. (2020) highlight the efficacy of combined AOPs, such as photocatalysis paired with ozone treatment, which achieve superior pollutant removal while optimizing energy use. This makes AOPs a promising solution for managing complex wastewater streams in densely populated or industrial areas.

Decentralized Wastewater Treatment Systems (DEWATS)

DEWATS treat wastewater close to its source, reducing dependence on extensive sewer networks and centralized plants. This approach is particularly suitable for rural, peri-urban areas, or small-scale industries.

Key Features of DEWATS:

- Cost-Effectiveness: Low operational costs compared to centralized systems.
- Scalability: Modular designs that can expand based on community needs.

- Adaptability: Tailored to local conditions, often incorporating constructed wetlands, sedimentation, and filtration.

Nair and Goh (2018) emphasized DEWATS' potential for small urban communities, noting its ability to support sustainable wastewater management in underdeveloped regions.

Sustainable Urban Drainage Systems (SUDS)

SUDS integrate green infrastructure and natural processes to manage stormwater runoff, mitigate urban flooding, and enhance water quality. These systems mimic natural hydrological cycles, contributing to resilient and sustainable urban environments.

Key SUDS Components:

- Green Roofs: Absorb rainwater, filter pollutants, and reduce urban heat island effects while providing insulation for buildings.
- Permeable Pavements: Allow water to infiltrate surfaces, reducing runoff and promoting groundwater recharge.
- Bio-Retention Systems: Rain gardens and swales capture and treat stormwater using vegetation and engineered soil, enhancing infiltration and removing pollutants.

SUDS not only address stormwater challenges but also promote urban biodiversity, improve aesthetics, and build climate resilience.

CASE STUDIES OF SUCCESSFUL SUDS IMPLEMENTATION:

- London, UK: The "SuDS for All" initiative in Hackney integrated green roofs, permeable pavements, and rain gardens, reducing flooding and creating vibrant green spaces (Sutcliffe et al., 2018).
- Melbourne, Australia: The "Stormwater Management Strategy" employed bio-retention systems and permeable pavements to minimize runoff and improve water quality (Hatt et al., 2016).

These examples demonstrate the dual benefits of SUDS: mitigating stormwater issues and enhancing urban livability.

Water-Efficient Technologies and Demand Management

Technological Solutions

Water-efficient technologies, such as low-flow faucets, water-saving appliances, and precision irrigation systems, have become integral to reducing freshwater demand in coastal regions. These innovations ensure efficient water usage without compromising functionality.

Demand Management Strategies

Demand management tools like tiered water pricing, public education campaigns, and conservation incentives encourage responsible water use. Initiatives promoting water-efficient landscaping, such as incorporating drought-tolerant plants, have proven effective in reducing irrigation needs in urban coastal areas (Mauzerall et al., 2017).

Policy and Governance for Water Sustainability

Effective water management in coastal areas requires robust policy frameworks and inclusive governance structures. Local governments must:

1. **Promote Conservation Policies:** Implement regulations to conserve water and protect water quality.
2. **Encourage Public Participation:** Engage stakeholders, including private sectors, NGOs, and local communities, in water management planning.
3. **Adopt Integrated Water Management (IWRM):** Develop adaptive strategies that address climate variability and resource scarcity (Pereira et al., 2012).

Educational campaigns, such as Australia's "Save Water, Save the Future," illustrate the power of public engagement in promoting behavioral changes. Workshops, incentives for adopting water-saving devices, and community conservation projects further enhance participation (Biel et al., 2017; MacRae et al., 2021).

WASTE MANAGEMENT: PIONEERING A CIRCULAR ECONOMY

Sustainable waste management is equally vital for coastal cities, which generate vast quantities of municipal solid waste (MSW), industrial byproducts, and hazardous materials. Improper waste disposal leads to severe environmental and health risks, particularly in densely populated areas. Strategies such as waste reduction, recycling, composting, and safe disposal of hazardous materials form the cornerstone of sustainable waste management.

By integrating these approaches into a circular economy framework, coastal cities can minimize waste generation, conserve resources, and reduce reliance on landfills. Moreover, the adoption of circular economy principles fosters economic growth by creating jobs in recycling, composting, and waste-processing industries, driving innovation and resource efficiency.

Innovative Waste Management Strategies

Urban areas face mounting waste management challenges due to growing populations and diminishing landfill capacities. Innovative strategies, including zero-waste policies, enhanced recycling programs, and Extended Producer Responsibility (EPR), offer practical solutions.

1. Zero-Waste Policies and Initiatives

Zero-waste approaches aim to minimize waste generation by emphasizing prevention, reuse, and recycling. Kamikatsu, Japan, reduced landfill waste by over 80% through community-driven zero-waste initiatives, focusing on material recovery and public engagement (Lausten et al., 2020).

Key Strategies:

- Promoting sustainable consumption and production.
- Designing durable, reusable, and recyclable products.
- Educating communities on waste reduction practices.

2. Enhancing Recycling Programs and Infrastructure

Effective recycling relies on robust infrastructure and active public participation. San Francisco achieved high recycling rates through advanced sorting technologies and comprehensive public education (Chen et al., 2020).

Focus Areas:

- Accessible curbside collection systems.
- Advanced sorting facilities for diverse materials.
- Clear public guidelines to improve recycling efficiency.

3. Extended Producer Responsibility (EPR)

EPR shifts waste management responsibilities to producers, incentivizing sustainable product designs and reducing waste at the source. Sweden and Germany have implemented EPR initiatives that reduce packaging waste and promote circular economies (Fisher et al., 2019). Benefits include:

- Encourages innovation in recyclable product designs.
- Reduces taxpayer burdens by financing waste management through producers.
- Stimulates demand for recycled materials.

Integration of Green Infrastructure and Waste Management

A holistic approach that integrates SUDS, innovative wastewater treatment technologies, and waste management strategies is essential for urban sustainability. Collaborative efforts between policymakers, industries, and communities can transform cities into models of environmental resilience and resource efficiency.

By adopting solutions such as AOPs, DEWATS, and SUDS while leveraging zero-waste principles and EPR frameworks, cities can effectively address the challenges of water and waste management, paving the way for a sustainable future.

INNOVATIVE WASTE MANAGEMENT STRATEGIES: ORGANIC WASTE MANAGEMENT, TECHNOLOGICAL ADVANCEMENTS, AND POLICY FRAMEWORKS

As urban populations expand, sustainable waste management has emerged as a critical challenge requiring innovative solutions. Addressing this challenge involves a multidimensional approach encompassing organic waste management, technological innovations, and robust policy frameworks. Together, these strategies foster urban sustainability by reducing environmental impact, recovering resources, and improving system efficiency.

Organic Waste Management

Organic waste, including food scraps, yard debris, and biodegradable materials, constitutes a significant portion of municipal solid waste. Managing this waste sustainably reduces landfill dependency, curtails greenhouse gas emissions, and recovers valuable resources. Two pivotal strategies, composting and anaerobic digestion play vital roles in this domain.

- *Composting*: This aerobic process converts organic materials into nutrient-rich compost, enhancing soil health. It is particularly suited for urban settings, such as community gardens and urban agriculture, where it reduces methane emissions and supports sustainable practices.
- *Anaerobic Digestion*: Decomposing organic waste in oxygen-free environments generates biogas, a renewable energy source, and produces digestate, a nutrient-rich agricultural product. Studies by Satyro et al. (2020) highlight these processes as essential for mitigating greenhouse gas emissions and enhancing urban resource recovery.

Urban Agriculture and Community Gardens

Composting directly supports urban agriculture by providing high-quality organic matter for soil enrichment, thus promoting food security and community engagement. Cavalett et al. (2021) demonstrated that integrating composting into urban agricultural systems not only reduces landfill dependency but also strengthens sustainable food systems.

Technological Innovations in Waste Management

Advancements in technology are revolutionizing waste management, increasing operational efficiency and sustainability. Two standout innovations are **IoT-enabled smart waste bins** and **waste-to-energy (WTE) technologies**.

- *Smart Waste Bins and IoT*: Equipped with sensors, smart bins provide real-time data on waste levels, enabling optimized collection schedules and reducing operational costs. Research by Khan et al. (2021) shows how IoT integration enhances urban waste logistics, particularly in densely populated areas, by reducing carbon footprints and streamlining waste collection.
- *Waste-to-Energy (WTE) Technologies*: Processes like incineration, pyrolysis, and **gasification** convert non-recyclable waste into usable energy. Among these, pyrolysis and gasification are noted for producing syngas with minimal emissions, offering cleaner alternatives to traditional methods (Zhou et al., 2019). These technologies significantly reduce landfill use and contribute to renewable energy production, despite challenges related to environmental impacts and public acceptance.

POLICY AND REGULATORY FRAMEWORKS

Effective waste management policies drive sustainable practices by providing clear incentives, frameworks, and enforcement mechanisms.

Key Policies

- **Extended Producer Responsibility (EPR)**: EPR policies mandate that producers manage the lifecycle of their products, encouraging sustainable design and recycling. Successful implementations in Europe and North America have significantly reduced packaging waste and increased recycling rates (Fisher et al., 2019).
- **Pay-As-You-Throw (PAYT) Programs**: PAYT systems incentivize waste reduction by charging households based on the amount of waste generated, promoting recycling and minimizing waste production.
- **Landfill Bans on Recyclables**: Prohibiting recyclable materials from landfills conserves resources and encourages better waste sorting and recycling practices.

Case Studies in Policy Impact

Studies by Mazzanti and Zoboli (2008) reveal that strong regulatory frameworks can significantly reduce waste generation, improve recycling rates, and enhance environmental outcomes.

Global Leaders in Waste Management

Several countries and cities have implemented innovative waste management strategies that serve as benchmarks for sustainable urban practices.

1. **Sweden**: Sweden's integrated waste management system combines extensive recycling initiatives with advanced WTE technologies, achieving high recycling rates and minimal landfill dependency. Björkqvist et al. (2017) attribute this success to stringent regulations, public participation, and technological innovation.
2. **Germany**: Germany's Packaging Ordinance, a pioneer in EPR policies, has significantly reduced packaging waste while promoting sustainable product design. This approach demonstrates the potential of EPR in fostering circular economies and minimizing environmental impact.

3. **Singapore**: Singapore's Zero-Waste Masterplan emphasizes waste reduction, increased recycling, and circular economy principles. Key initiatives, such as mandatory waste sorting and WTE plants, are complemented by extensive public education campaigns (Lee et al., 2020).
4. **Copenhagen**: Through its Climate Adaptation and Circular Economy strategy, Copenhagen integrates green infrastructure with WTE technologies to enhance resilience against climate change. The Amager Bakke plant exemplifies how WTE can support renewable energy while reducing landfill use (Mikkelsen et al., 2019).
5. **San Francisco**: San Francisco's ambitious zero-waste goal integrates mandatory recycling, advanced composting infrastructure, and extensive public engagement to achieve significant waste diversion rates. Its comprehensive waste management system, supported by the Mandatory Recycling and Composting Ordinance, serves as a global model for sustainable urban practices (Ko et al., 2020).

COMMUNITY AND STAKEHOLDER ENGAGEMENT IN WATER AND WASTE MANAGEMENT

Engaging communities and stakeholders is pivotal to the success of sustainable water and waste management initiatives. Grassroots movements and public-private partnerships (PPPs) have proven to be effective in addressing environmental challenges, fostering inclusive urban systems, and building resilience. This article explores the significant roles of community involvement, collaborative partnerships, and the challenges and opportunities in driving stakeholder engagement for sustainable water and waste management.

The Role of Communities in Water and Waste Management

Communities are central to the success of sustainable water and waste management strategies. Their active participation in decision-making processes, adoption of eco-friendly practices, and promotion of localized innovation ensure the effectiveness and sustainability of environmental policies and programs.

Grassroots Initiatives and Community-Led Projects

Grassroots movements initiated by local residents and organizations often address specific environmental needs while empowering individuals and fostering engagement. Key examples include:

- **Community-Based Recycling Programs:** Promoting waste segregation to reduce landfill dependency.
- **Rainwater Harvesting Projects:** Enabling sustainable local water management.
- **Urban Community Composting Systems:** Transforming organic waste into nutrient-rich compost for gardening and agriculture.

Such initiatives enhance local capacities, foster environmental consciousness, and strengthen social cohesion. For example, *Graham and Stinson (2020)* demonstrated that community composting reduces waste, raises environmental awareness, and builds a sense of collaboration among participants.

Education and Awareness Programs

Educational campaigns are essential for empowering communities to adopt sustainable practices. Workshops, school programs, and awareness drives address issues such as:

- Reducing water consumption.
- Benefits of waste segregation and recycling.
- Strategies for drought resilience and waste diversion.

Interactive programs like "Zero-Waste Challenges" or "Water Conservation Months" inspire behavioral change, while partnerships with private entities amplify the outreach and impact of these campaigns (*Schweitzer & Brown, 2021*).

Public-Private Partnerships (PPPs): A Catalyst for Sustainable Development

Public-private partnerships (PPPs) combine governmental oversight with private sector expertise, fostering innovation, enhancing efficiency, and mobilizing resources for large-scale environmental projects.

Collaborative Frameworks for Water and Waste Management

PPPs play a transformative role in modernizing infrastructure and integrating advanced technologies into water and waste management systems. Examples include:

- Upgraded water treatment facilities and distribution systems for efficient delivery.

- Advanced stormwater management solutions.
- Waste-to-energy projects that reduce landfill dependency while generating renewable energy (*Jiménez & Maestu, 2020*).

Successful PPP Examples

- **Veolia Environnement (France):** Partnerships with cities like London and New York to modernize water treatment and recycling systems.
- **Swachh Bharat Mission (India):** A flagship PPP enhancing sanitation, promoting waste segregation, and managing solid waste through government-private collaborations (*Prasad & Kumar, 2021*).

These examples highlight the ability of PPPs to deliver environmental and economic benefits by fostering collaboration, leveraging resources, and driving innovation.

Challenges and Opportunities in Stakeholder Engagement

Barriers to Effective Engagement

Despite its importance, stakeholder engagement faces numerous challenges:

- **Lack of Trust:** Gaps between decision-makers and communities hinder collaboration. Inclusive decision-making can bridge this divide.
- **Cultural and Social Barriers:** Diverse urban settings require tailored communication strategies to promote inclusivity (*Doria et al., 2019*).
- **Resource Constraints:** Limited funding and bureaucratic delays can stall implementation. Proactive planning and cross-sector collaboration are critical to overcoming these hurdles.

Opportunities for Innovation and Collaboration

Emerging technologies and intersectoral collaborations offer significant potential to enhance stakeholder engagement:

- **Smart Technologies:** Real-time data collection and monitoring using IoT-enabled tools empower communities and streamline operations (*Graham & Stinson, 2020*).
- **Circular Economy Models:** Collaborations among environmental organizations, technology firms, and municipalities drive holistic and sustainable solutions.

As *Graham and Stinson (2020)* noted, innovative tools like smart sensors and digital dashboards enable communities to actively participate in decision-making processes, fostering ownership and accountability.

FUTURE DIRECTIONS AND RECOMMENDATIONS IN WATER AND WASTE MANAGEMENT

As urban populations continue to grow and environmental challenges escalate, the future of water and waste management hinges on the adoption of innovative, sustainable approaches. Emerging trends such as digital technologies, data analytics, and circular economy principles are revolutionizing resource management. This review explores these trends, their applications, and future possibilities, offering actionable recommendations for policymakers and practitioners to incorporate these solutions into urban planning and governance.

Recommendations for Policymakers and Practitioners

Urban water and waste management systems are critical for achieving sustainable urban development. By embracing cutting-edge technologies and adopting adaptive, holistic approaches, policymakers and practitioners can enhance resource efficiency, improve environmental outcomes, and bolster urban resilience.

1. Integrating Innovative Solutions into Urban Planning

To tackle the challenges of urbanization, climate change, and resource scarcity, policymakers must prioritize advanced technologies and sustainable practices within urban planning frameworks. This will ensure cities are more efficient, adaptable, and resilient in managing water and waste.

a. Incorporating Digital Technologies

Digital innovations such as smart meters, IoT-based sensors, and advanced data analytics are essential tools for optimizing water and waste management systems. These technologies provide real-time monitoring, predictive analysis, and system optimization, enabling municipalities to:

- Detect leaks and manage water usage more effectively.
- Monitor waste production and enhance recycling efforts.

- Optimize waste collection schedules with real-time data from IoT-enabled sensors, reducing operational costs and environmental impact (*Patel et al., 2020;Dutta et al., 2021*).

Urban planning should integrate digital platforms that facilitate data sharing between governments, utilities, and citizens, fostering a smart city model for resource management. For example, *Patel et al. (2020)* demonstrated how smart meters and IoT technologies can improve water efficiency through demand management and leak detection.

b. Promoting Circular Economy Principles

Adopting circular economy principles is key to transforming urban waste management, prioritizing resource recovery, recycling, and waste reduction. Key strategies include:

- Encouraging businesses to design products for reuse, recycling, and disassembly.
- Diverting organic waste from landfills through composting and waste-to-energy solutions (*Zhang et al., 2020*).
- Implementing Extended Producer Responsibility (EPR) regulations to hold manufacturers accountable for the lifecycle of their products.

Local waste sorting initiatives and investments in recycling infrastructure should be expanded to reinforce circular economy principles. *Zhang et al. (2020)* stressed the role of waste-to-energy systems and organic waste recycling in achieving sustainable urban resource management.

2. Building Resilient and Adaptive Urban Infrastructure

As cities confront climate change, population growth, and evolving environmental pressures, it is imperative that urban infrastructure be designed to be resilient and adaptive. Policymakers must prioritize sustainable water and waste management solutions that address these complex challenges.

Designing Climate-Resilient Water Infrastructure: Urban water systems face increasing risks such as flooding, water scarcity, and temperature fluctuations due to climate change. Climate-resilient infrastructure should include:

Sustainable Urban Drainage Systems (SUDS): Green roofs, permeable pavements, and bio-retention areas that absorb rainwater, reduce flooding, and improve water quality (*Hollis et al., 2019*).

Decentralized Water Systems: Rainwater harvesting and on-site wastewater treatment systems that reduce reliance on centralized infrastructure, enhance water availability, and minimize environmental impacts (*Pereira et al., 2020*).

Integrating green infrastructure into urban planning can mitigate the impacts of urban flooding and droughts, as emphasized by *Hollis et al. (2019)*.

Adaptive Waste Management Systems: To manage the evolving complexity of waste streams, cities need flexible and scalable waste management solutions, including:

Smart Waste Technologies: AI-driven sorting systems and robotics that automate waste separation, reducing human intervention and improving operational efficiency (*Rai & Tiwari, 2020*).

Waste-to-Energy Plants: Expanding these facilities to process non-recyclable waste while generating renewable energy.

Composting Initiatives: Promoting composting programs for organic waste to support urban agriculture and local food systems.

Rai & Tiwari (2020) demonstrated the potential of AI and robotics to optimize waste processing, enabling cities to adapt to changing waste streams and improve operational efficiency.

Final Thoughts on Achieving Sustainable Urban Water and Waste Management

Achieving sustainable urban water and waste management requires a coordinated, forward-looking approach from policymakers, urban planners, and practitioners. Key priorities include:

- Ongoing innovation and investment in smart infrastructure, circular systems, and climate-resilient solutions.
- Cross-sector collaboration between public, private, and civil society actors to develop inclusive and economically viable solutions.
- Holistic governance frameworks that integrate technology, sustainability, and community participation.

By embracing resource efficiency, adaptability, and recovery strategies, cities can effectively navigate the pressures of rapid urbanization, climate change, and resource scarcity. These strategies will ensure that urban centers remain livable, resilient, and capable of thriving in an increasingly complex and uncertain future.

CONCLUSION

Achieving sustainable urban water and waste management necessitates a comprehensive approach that combines innovative strategies, resilient infrastructure, and active community involvement. Cutting-edge waste management solutions, including zero-waste policies, advanced recycling programs, and Extended Producer Responsibility (EPR), play a crucial role in minimizing environmental impacts and facilitating the shift toward a circular economy. Zero-waste initiatives emphasize waste prevention, sustainable design, and strong community engagement, significantly reducing landfill waste. Enhanced recycling systems improve material recovery and minimize contamination, while EPR encourages manufacturers to adopt sustainable practices and take responsibility for the entire lifecycle of their products.

Innovations such as composting, anaerobic digestion, smart waste bins, and waste-to-energy technologies further contribute to urban sustainability by promoting waste reduction, resource recovery, and energy generation. Policy frameworks like pay-as-you-throw and EPR programs incentivize responsible consumption and waste management behaviors.

Case studies from forward-thinking cities like Singapore, Copenhagen, and San Francisco highlight how integrated water systems, circular economy strategies, and robust waste diversion initiatives can serve as exemplary models for sustainable urban development. These cities demonstrate that by adopting innovative and inclusive approaches, sustainability can be achieved despite challenges such as resource constraints and socio-cultural barriers.

Community and stakeholder engagement remains at the heart of these efforts. Grassroots initiatives, public-private partnerships, and education campaigns are instrumental in fostering responsible consumption and waste reduction practices. Additionally, technological advancements and intersectoral collaboration offer significant opportunities to overcome challenges, build trust, and enhance resource efficiency.

Ultimately, the integration of innovative technologies, circular economy principles, and climate-resilient infrastructure will be crucial in addressing the complex challenges posed by urbanization and climate change. By focusing on efficiency, adaptability, and sustainability, cities can develop resilient systems that ensure long-term environmental health, economic prosperity, and social well-being.

REFERENCES

Al-Anzi, B., Zubari, W., & Al-Rashidi, A. (2020). *Desalination technologies and sustainability in the coastal regions of the Arabian Gulf: Advances and challenges*. Springer., DOI: 10.1007/978-3-030-36938-5

Anwar, R., Mollah, M. H., & Islam, S. (2020). Greywater recycling in urban residential buildings: Current trends and future challenges. *Environmental Science and Pollution Research International, 27*(11), 12241–12256. DOI: 10.1007/s11356-020-07827-5

Arena, U. (2012). Process and technological aspects of municipal solid waste gasification. *A review.Waste Management (New York, N.Y.), 32*(4), 625–639. DOI: 10.1016/j.wasman.2011.09.025 PMID: 22035903

AWWA. (2021). *The impact of water loss on urban systems and strategies for reducing it*. American Water Works Association.

Bakker, J., & Wester, P. (2020). Water-sensitive urban design in Copenhagen: Strategies for climate adaptation. *Urban Climate, 34*, 100678. DOI: 10.1016/j.uclim.2020.100678

Barbier, E. B., Hacker, S. D., Kennedy, C., Koch, E. W., & Silliman, B. R. (2019). The value of estuarine and coastal ecosystem services. *Ecological Monographs, 89*(3), e01333. DOI: 10.1002/ecm.1333

Barros, E. M., Silva, S. R., & Oliveira, D. M. (2020). Advanced oxidation processes for the treatment of pharmaceuticals in wastewater: A review. *Journal of Environmental Management, 260*, 110085. DOI: 10.1016/j.jenvman.2020.110085

Biel, A., Mårtensson, F., & Gärling, T. (2017). Public awareness campaigns and their role in water conservation. *Journal of Environmental Psychology, 51*, 68–75. DOI: 10.1016/j.jenvp.2017.04.003

Björkqvist, S., Elser, C., & Thörne, E. (2017). Sweden's waste management system: A case study in circular economy. *Waste Management (New York, N.Y.), 59*, 1–12. DOI: 10.1016/j.wasman.2017.02.003

Brown, S., Leonard, P., & Saunders, R. (2012). *Urban agriculture and community food security in the United States: Farming from the city center to the urban fringe*. Sustainable Urban Agriculture Coalition.

Cavalett, O., Lima, D., & Silva, M. A. (2021). Urban agriculture and composting: A sustainable approach to waste management in community gardens. *Environmental Science & Policy, 113*, 32–43. DOI: 10.1016/j.envsci.2020.08.004

Chen, M., Zhao, W., & Wang, X. (2020). Enhancing urban recycling through advanced sorting technologies: Lessons from San Francisco. *Journal of Environmental Management, 259*, 110057. DOI: 10.1016/j.jenvman.2019.110057

Diaper, C., Williams, K., & Jackson, T. (2020). Greywater recycling: A review of technologies, benefits, and challenges in urban areas. *Journal of Environmental Management, 259*, 109613. DOI: 10.1016/j.jenvman.2019.109613

Doria, M., Tedesco, L., & Gagliardi, R. (2019). Overcoming barriers to community engagement in urban water management: Lessons from international case studies. *Water Policy, 21*(5), 1016–1034. DOI: 10.2166/wp.2019.103

Doria, M., Tedesco, L., & Gagliardi, R. (2019). The role of community-driven approaches in improving urban water management and waste recycling. *Environmental Sustainability, 11*(3), 217–230. DOI: 10.1016/j.envsus.2019.02.006

Dutta, A., Sharma, A., & Patel, P. (2021). IoT-based smart waste management systems for urban sustainability. *Urban Climate, 37*, 100845. DOI: 10.1016/j.uclim.2021.100845

EPA. (2021). WaterSense: A guide to water-efficient plumbing products. U.S. Environmental Protection Agency. https://www.epa.gov/watersense

Faccio, M., Persona, A., & Zanin, G. (2011). Waste collection multi objective model with real time traceability data. *Waste Management (New York, N.Y.), 31*(12), 2391–2405. DOI: 10.1016/j.wasman.2011.07.005 PMID: 21821406

Feng, H., Zhou, J., & Li, H. (2019). Membrane bioreactor (MBR) technology for wastewater treatment: Recent advances and future directions. *Environmental Technology, 40*(8), 1056–1069. DOI: 10.1080/09593330.2018.1564107

Fisher, P. C., Blanchard, D. L., & McCormick, R. A. (2019). The role of extended producer responsibility in reducing municipal waste: Global perspectives and lessons learned. *Waste Management & Research, 37*(6), 557–568. DOI: 10.1177/0734242X19859448

Fletcher, T. D., Shuster, W., Hunt, W. F., Ashley, R. M., Butler, D., Arthur, S., & Uhl, M. (2015). SUDS, LID, BMPs, WSUD and more–The evolution and application of terminology surrounding urban drainage. *Urban Water Journal, 12*(7), 525–542. DOI: 10.1080/1573062X.2014.916314

Fletcher, T. D., Shuster, W. D., & Gersonius, B. (2017). *Water sensitive urban design: A case study approach*. Springer.

Friedler, E. (2004). Quality of individual domestic greywater streams and its implications for on-site treatment and reuse possibilities. *Environmental Technology*, *25*(9), 997–1008. DOI: 10.1080/09593330.2004.9619393 PMID: 15515266

García, A., & López, R. (2020). Circular economy in urban waste management: Opportunities and challenges. *Waste Management (New York, N.Y.)*, *102*, 349–359. DOI: 10.1016/j.wasman.2020.01.045

Getter, K. L., & Rowe, D. B. (2021). *Green roofs: A comprehensive guide to the planning, design, and construction of green roofs*. John Wiley & Sons.

Gill, S. E., Handley, J. F., Ennos, A. R., & Pauleit, S. (2007). Adapting cities for climate change: The role of the green infrastructure. *Built Environment*, *33*(1), 115–133. DOI: 10.2148/benv.33.1.115

Gill, S. E., Handley, J. F., Ennos, A. R., & Pauleit, S. (2007). Adapting cities for climate change: The role of green infrastructure. *Built Environment*, *33*(1), 115–133. DOI: 10.2148/benv.33.1.115

Gleick, P. H. (2003). Water use. *Annual Review of Environment and Resources*, *28*(1), 275–314. DOI: 10.1146/annurev.energy.28.040202.122849

Gleick, P. H., & Cooley, H. (2019). Water conservation and drought resilience in San Francisco: Policies and outcomes. *Water Resources Research*, *55*(9), 7539–7550. DOI: 10.1029/2019WR024585

Gould, J., & Nissen-Peterson, E. (2019). Rainwater catchment systems for domestic water supply: Design, construction, and implementation. CRC Press.

Graham, S., & Stinson, C. (2020). Leveraging digital technologies for community engagement in urban waste management. *Journal of Urban Technology*, *27*(4), 35–51. DOI: 10.1080/10630732.2020.1794825

Hatt, B. E., Fletcher, T. D., & Deletic, A. (2016). Water-sensitive urban design: A review of its potential and practice in Melbourne, Australia. *Australian Journal of Water Resources*, *20*(3), 259–272. DOI: 10.1080/13241583.2016.1183684

Hoekstra, A. Y., & Mekonnen, M. M. (2012). The water footprint of humanity. *Proceedings of the National Academy of Sciences of the United States of America*, *109*(9), 3233–3237. DOI: 10.1073/pnas.1109936109 PMID: 22331890

Hollis, A., Johnson, R., & Zhuang, M. (2019). Green infrastructure in urban water management: Building resilience against climate change. *Environmental Science & Policy*, *104*, 66–74. DOI: 10.1016/j.envsci.2019.01.003

Hopewell, J., Dvorak, R., & Kosior, E. (2009). Plastics recycling: Challenges and opportunities. *Philosophical Transactions of the Royal Society of London. Series B, Biological Sciences*, *364*(1526), 2115–2126. DOI: 10.1098/rstb.2008.0311 PMID: 19528059

IPCC. (2022). Climate change 2022: Impacts, adaptation, and vulnerability. Intergovernmental Panel on Climate Change. https://www.ipcc.ch/report/ar6/wg2/

Jambeck, J. R., Geyer, R., Wilcox, C., Siegler, T. R., Perryman, M., Andrady, A., Narayan, R., & Law, K. L. (2015). Plastic waste inputs from land into the ocean. *Science*, *347*(6223), 768–771. DOI: 10.1126/science.1260352 PMID: 25678662

Jiménez, B., & Maestu, J. (2020). Public-private partnerships in water management: Enhancing sustainability through collaborative infrastructure. *Water Resources Management*, *34*(7), 2291–2302. DOI: 10.1007/s11269-020-02537-4

Judd, S. (2011). *The MBR book: Principles and applications of membrane bioreactors for water and wastewater treatment*. Elsevier.

Kallis, G., Varela-Ortega, C., & González-Gómez, T. (2020). Coastal water management under climate change: Integrated approaches and strategies for adaptation. *Global Environmental Change*, *64*, 102147. DOI: 10.1016/j.gloenvcha.2020.102147

Kar, S. (2020). Smart water metering and real-time monitoring systems for water management. *Journal of Water Supply: Research & Technology - Aqua*, *69*(3), 168–179. DOI: 10.2166/aqua.2020.145

Kaza, S., Yao, L., Bhada-Tata, P., & Van Woerden, F. (2018). What a waste 2.0: A global snapshot of solid waste management to 2050. World Bank Group. https://openknowledge.worldbank.org/handle/10986/30317

Khalid, A., Arshad, M., Anjum, M., Mahmood, T., & Dawson, L. (2011). The anaerobic digestion of solid organic waste. *Waste Management (New York, N.Y.)*, *31*(8), 1737–1744. DOI: 10.1016/j.wasman.2011.03.021 PMID: 21530224

Khan, M. Z., Hassan, M., & Yu, Z. (2021). Internet of Things (IoT) in smart waste management: A comprehensive review. *Sustainable Cities and Society*, *69*, 102758. DOI: 10.1016/j.scs.2021.102758

Ko, H., Chang, J., & Li, Y. (2020). San Francisco's zero waste program: Policies, practices, and challenges. *Environmental Management*, *66*(1), 85–94. DOI: 10.1007/s00267-020-01314-6

Koch, F., Bassetti, L., & Huynh, D. (2020). Performance of rainwater harvesting systems in urban Southeast Asia: Opportunities and challenges. *Water (Basel), 12*(6), 1607. DOI: 10.3390/w12061607

Lausten, M. A., Aschenbrenner, A., & Hughes, J. R. (2020). Zero-waste initiatives and urban waste management: Lessons from Kamikatsu, Japan. *Waste Management & Research, 38*(10), 1167–1177. DOI: 10.1177/0734242X20962921

Lee, H. K., Tan, B. K., & Soh, Y. W. (2020). The Zero Waste Masterplan of Singapore: Strategies for sustainable waste management. *Environmental Science & Technology, 54*(19), 11872–11881. DOI: 10.1021/acs.est.0c03094 PMID: 32852946

Li, J., Zhao, X., & He, X. (2019). Permeable pavements for stormwater management in urban areas: A review of materials and techniques. *Environmental Science and Pollution Research International, 26*(5), 4680–4691. DOI: 10.1007/s11356-019-04387-9 PMID: 30565107

Li, Z., Boyle, F., & Reynolds, A. (2016). Rainwater harvesting and greywater treatment systems for domestic application in Ireland. *Desalination and Water Treatment, 57*(48-49), 22911–22919. DOI: 10.1080/19443994.2015.1127175

Lifset, R., Atasu, A., & Tojo, N. (2013). Extended producer responsibility. *Journal of Industrial Ecology, 17*(2), 162–166. DOI: 10.1111/jiec.12022

Liu, C., & Liao, Z. (2019). Overcoming barriers to greywater recycling in urban settings: A review of regulations and technological advancements. *Resources, Conservation and Recycling, 142*, 83–92. DOI: 10.1016/j.resconrec.2018.11.018

Liu, J., Zhang, Q., & Li, Y. (2018). Enhancing water conservation in urban distribution systems: A study of smart water metering and real-time leak detection. *Journal of Water Resources Planning and Management, 144*(12), 04018087. DOI: 10.1061/(ASCE)WR.1943-5452.0001005

MacRae, R., McLeod, A., & Green, J. (2021). Community-based water conservation programs: Impacts on urban sustainability. *Environmental Education Research, 27*(4), 525–539. DOI: 10.1080/13504622.2020.1819331

Massoud, M. A., Tarhini, A., & Nasr, J. A. (2009). Decentralized approaches to wastewater treatment and management: Applicability in developing countries. *Journal of Environmental Management, 90*(1), 652–659. DOI: 10.1016/j.jenvman.2008.07.001 PMID: 18701206

Mauzerall, D. L., Wang, J., & Vigdor, N. (2017). A global assessment of urban water efficiency and conservation technologies. *Environmental Science & Technology, 51*(7), 3989–3998. DOI: 10.1021/acs.est.6b06090

Mikkelsen, P. S., Andersen, R., & Mortensen, T. (2019). Copenhagen's waste management and recycling strategies: Waste-to-energy and circular economy. *Waste Management (New York, N.Y.)*, *94*, 218–226. DOI: 10.1016/j.wasman.2019.06.026

Miller, R. E., Thomas, R. L., & Lehman, P. (2017). *Green infrastructure and water management: Exploring the potential of urban ecosystems in coastal cities*. Wiley.

Mills, J. D., & Boersma, D. L. (2021). Big data and AI in waste management: The future of urban resource optimization. *Journal of Environmental Informatics*, *35*(1), 74–89. DOI: 10.1016/j.envinform.2021.05.006

Nair, R., & Goh, K. (2018). Decentralized wastewater treatment systems for small urban communities: Case studies and lessons learned. *Environmental Engineering Science*, *35*(4), 240–250. DOI: 10.1089/ees.2017.0324

OECD. (2018). *The state of play on extended producer responsibility*. Organisation for Economic Co-operation and Development., DOI: 10.1787/9789264290478-

Patel, S. J., Yang, X., & Roy, P. K. (2020). The role of digital technologies in enhancing urban water management: Smart meters and IoT-based solutions. *Water Resources Management*, *34*(8), 2557–2571. DOI: 10.1007/s11269-020-02601-1

Pereira, L. S., Cordery, I., & Iacovides, I. (2012). *Handbook of land and water management*. Springer., DOI: 10.1007/978-1-4614-6435-8

Pillai, M., Smith, D., & Desai, R. (2021). Waste-to-energy technologies for sustainable waste management. *Renewable & Sustainable Energy Reviews*, *135*, 110186. DOI: 10.1016/j.rser.2020.110186

Prasad, B., & Kumar, V. (2021). Swachh Bharat Mission: Public-private partnership in waste management and sanitation. *Waste Management & Research*, *39*(8), 939–948. DOI: 10.1177/0734242X21102194

Rai, A., & Tiwari, S. (2020). Robotics and AI in urban waste management: Advancing efficiency and adaptability in waste systems. *Waste Management & Research*, *38*(1), 13–24. DOI: 10.1177/0734242X20923056

Satyro, J. R., Silva, A. L., & Almeida, S. S. (2020). Managing organic waste through composting and anaerobic digestion: Case studies and sustainable strategies. *Waste Management & Research*, *38*(4), 403–414. DOI: 10.1177/0734242X20914401

Schoen, R. M., Dzwonczyk, M. K., & Williams, A. D. (2019). Smart water management: Technologies and applications. *Water Resources Management*, *33*(4), 1103–1121. DOI: 10.1007/s11269-019-02248-9

Schweitzer, A., & Brown, J. (2021). The impact of education and public awareness on sustainable waste management practices. *Journal of Environmental Management, 256*, 109885. DOI: 10.1016/j.jenvman.2020.109885

Sutcliffe, J., Hunter, S., & Sim, N. (2018). The role of green infrastructure in sustainable urban drainage systems: Case studies from London. *Urban Water Journal, 15*(1), 50–60. DOI: 10.1080/1573062X.2017.1280226

Swan, W., & Blanchard, M. (2015). *Waste management in coastal cities: Challenges and solutions*. Routledge.

Tan, A. L., Wong, H. L., & Yap, S. L. (2019). NEWater: Singapore's success story in water reuse and sustainability. *Desalination and Water Treatment, 146*, 207–217. DOI: 10.5004/dwt.2019.24274

Tan, C. P., Tan, T. Y., & Ho, C. S. (2018). *Urban green infrastructure in Singapore: Integrating climate adaptation into water management*. Routledge.

Tan, Y. Y., Ang, H. M., & Kuan, W. S. (2018). Smart water management and sustainable water usage in Singapore. *Water Science and Technology: Water Supply, 18*(5), 1463–1471. DOI: 10.2166/ws.2018.104

UN-Habitat. (2011). *Global report on human settlements 2011: Cities and climate change: Global report on human settlements*. United Nations Human Settlements Programme.

UN-Habitat. (2020). The state of African cities 2020: Inclusive cities for resilient economies. United Nations Human Settlements Programme. https://unhabitat.org/the-state-of-african-cities-2020-inclusive-cities-for-resilient-economies

UNEP. (2016). Global outlook on sustainable consumption and production policies. United Nations Environment Programme. https://www.unep.org/resources/report/global-outlook-sustainable-consumption-and-production-policies

Villarini, G., Vecchi, G. A., & Smith, J. A. (2019). The role of climate change in the occurrence and intensity of extreme weather events. *Journal of Climate, 32*(24), 1–15. DOI: 10.1175/JCLI-D-19-0047.1

Water, U. N. (2018). The United Nations world water development report 2018: Nature-based solutions for water. United Nations. https://www.unwater.org/publications/world-water-development-report-2018/

Wilson, D. C., Velis, C. A., & Cheeseman, C. R. (2015). Role of informal sector recycling in waste management in developing countries. *Habitat International, 31*(1), 6–15. DOI: 10.1016/j.habitatint.2005.06.004

Wong, K. S., Ibrahim, R., & Lim, S. (2021). Advances in rainwater harvesting technologies: Challenges and solutions for urban areas. *Environmental Science & Technology*, *55*(13), 8904–8914. DOI: 10.1021/acs.est.1c01677

Wong, K. S., Ibrahim, R., & Lim, S. (2021). Advances in rainwater harvesting technologies: Challenges and solutions for urban areas. *Environmental Science & Technology*, *55*(13), 8904–8914. DOI: 10.1021/acs.est.1c01677

World Bank. (2018). Solid waste management in developing countries. World Bank Group. https://openknowledge.worldbank.org/handle/10986/30303

Zhang, L., & Liao, B. (2020). The role of greywater recycling in sustainable urban water management. *Environmental Science & Technology*, *54*(15), 9208–9217. DOI: 10.1021/acs.est.0c02238

Zhang, Y., Bi, J., & Huang, X. (2020). Circular economy and sustainable waste management in coastal cities: A review of challenges and opportunities. *Environmental Science and Pollution Research International*, *27*(9), 9143–9155. DOI: 10.1007/s11356-020-07768-3

Zhang, Y., Wang, X., & He, L. (2020). Circular economy in urban waste management: Concepts, applications, and future perspectives. *Waste Management & Research*, *38*(4), 347–357. DOI: 10.1177/0734242X20910101 PMID: 32116140

Zhao, M., Li, W., & Chen, Z. (2020). The role of water-efficient appliances in urban water conservation. *Water Resources Management*, *34*(2), 561–577. DOI: 10.1007/s11269-019-02439-9

Zhou, Y., Wang, S., & Jiang, C. (2019). Waste-to-energy technologies and their potential in reducing urban waste: A case study of pyrolysis and gasification. *Renewable & Sustainable Energy Reviews*, *109*, 358–368. DOI: 10.1016/j.rser.2019.04.037

Zhou, Z., Zhang, L., & Li, C. (2021). *Desalination: Principles, technologies, and environmental considerations*. Elsevier., DOI: 10.1016/B978-0-12-819138-1.00001-3

KEY TERMS AND DEFINITIONS

Circular economy: is therefore an economic model designed and intended to reduce, reusing, and making most effective use of products by reusing in a productive manner by giving materials and products for remanufacturing and recycling. Whereas the traditional linear economy depends on the "take, make, dispose" strategy, circular economies want an uninterrupted cycle where materials and products remain in a continuous flow. This approach reduces environmental impact, conserves natural resources, and fosters economic growth by creating new business opportunities.

Green Infrastructure: a network of natural and semi-natural systems that provide environmental, social, and economic benefits by incorporating natural processes into urban environments. Examples of green infrastructure include green roofs, permeable pavements, rain gardens, urban forests, and wetlands. These systems help manage stormwater, reduce flooding, improve air and water quality, enhance biodiversity, and provide recreational spaces for communities. Green infrastructure solutions are sustainable and cost-effective, offering a resilient alternative to traditional gray infrastructure.

Greywater Reuse: It is the treatment and reuse of wastewater produced from household activities like bathing, washing dishes, and laundry except toilet waste. Greywater can be treated and reused for non-potable purposes, such as irrigation, toilet flushing, and landscape watering. By reusing greywater, households can reduce their overall water consumption, lower water bills, and decrease the demand on municipal water supply systems. Greywater reuse also reduces the volume of wastewater discharged to sewer systems and into the environment.

Rainwater Harvesting: The process of accumulation, storage, and utilization of rainwater for irrigation and toilet flushing and even for potable supply after proper treatment. Rainwater harvesting captures the rainwater from the roofs or surface runoff, directing it down the gutters and the downspouts into the storage tanks or cisterns. Rainwater harvesting would decrease reliance on the municipal water supply, reduce stormwater runoff, and furnish an alternative source of freshwater in areas with limited or no freshwater sources.

Recycling Technologies: refer to the many methods and processes used in converting waste materials into new reusable products. These technologies involve mechanical recycling, chemical recycling, and biological recycling. Mechanical recycling deals with the physical processing of waste materials, such as plastics and metals, to form new products. Chemical recycling breaks down waste materials at the molecular level to generate raw materials that can be used in making new items. Biological recycling employs natural processes such as composting and anaerobic digestion in order to decompose organic waste, yielding useful products like compost and biogas.

Smart Water Systems: advanced water management systems that utilize digital technologies, such as IoT (Internet of Things) sensors, data analytics, and artificial intelligence, to monitor, manage, and optimize water resources. These systems enable real-time data collection and analysis, allowing for more efficient water distribution, leak detection, quality monitoring, and demand management. Smart water systems enhance the sustainability and resilience of water infrastructure by providing precise control and timely information to water managers and users.

Sustainable Urban Drainage Systems (SUDS): a variety of water management techniques that aim to replicate natural drainage processes and control surface runoff in urban environments. SUDS aims to diminish flooding, improve water quality, and increase the amenity and biodiversity of urban environments. SUDS techniques involve permeable pavements, swales, detention basins, green roofs, and rain gardens. It mitigates the effects of urbanization on the natural water cycle and promotes a more resilient and sustainable urban environment by slowing down and treating runoff at its source.

Chapter 11
Evaluation and Management of Coastal Erosion Using DSAs and Creation of a Spatial Data Infrastructure in Kribi, Cameroon

Samy Borel Obah Mengue
National Advanced School of Publics Works, Yaoundé, Cameroon

Bikie Gerald Anicet
https://orcid.org/0009-0003-2068-3494
National Advanced School of Publics Works, Yaoundé, Cameroon

Dongmo Hile Bertrand
National Advanced School of Publics Works, Yaoundé, Cameroon

Elime Boubouama Aime
National Advanced School of Engineering,Yaounde, Cameroon

Sakina Mehdioui
Geoscience Laboratory, Faculty of Sciences, Ibn Tofail University, Morocco

Luminita Diaconu
https://orcid.org/0000-0002-6736-6344
Academy of Economic Studies of Moldova, Moldova

Mohamed El Bakkali
Faculty of Sciences, Ibn Tofail University, Kenitra, Morocco

Redouane Kaiss
https://orcid.org/0009-0008-3049-2532
Faculty of Economics and Management, Hassan 1st University, Morocco

Yassine Mouniane
https://orcid.org/0000-0002-6037-5301
Faculty of Sciences, Ibn Tofail University, Morocco

DOI: 10.4018/979-8-3373-5278-7.ch011

Serkan Derici
https://orcid.org/0000-0003-2581-6770
Araştırma Görevlisi, Nevşehir Hacı Bektaş Veli University, Turkey

ABSTRACT

In recent years, the Kribi coast in Cameroon has experienced significant coastal erosion, resulting in the submersion of land parcels, including those with land titles. This study aims to create a spatial data infrastructure (IDoGek) to monitor and manage coastal changes from 2015 to 2023, enabling better planning and sustainable growth in the region. Using DSAS (ArcGIS plug-in), historical data and remote sensing analyses revealed a loss of over 7 hectares of land, with erosion rates averaging 1.1 meters per year. By 2034, up to 11 hectares could be lost without intervention, as some areas may retreat by 10.97 meters. This approach maps land use changes, tracks coastline positions, and assesses the kinetics of coastal erosion. The results emphasize the need for high-resolution imagery, offshore tide gauges, and integration of remote sensing with in-situ data to enhance monitoring and develop effective solutions for the sustainable management of Kribi's vulnerable coastal zones.

1. INTRODUCTION

Coastal and marine areas are among the most dynamic and multifarious ecosystems and are consequently subjected to a multiplicity of anthropogenic and natural pressures (Wang et al., 2024). Cameroon Ordinance 74-2, July 6, 1974, establishes the domain regime, which includes the prescription that a 50 m strip should remain at sea level (high tide). Such provision leads to on-going concerns regarding parcels of land which are partly or totally submerged. A holistic management of such environments would only be possible if good knowledge exists on their functioning and evolution over space and time (Zhou et al., 2018; Htay et al., 2023; Yan et al., 2024). In this case, the establishment of a Spatial Data Infrastructure (SDI) may be a powerful tool in the collection, integration, analysis, and sharing of geospatial information related to marine and coastal ecosystems (Strain et al., 2006; Jabbour et al., 2019; Aretouyap et al., 2020; Vaitis et al., 2022).

The coastline of Cameroon in the Gulf of Guinea houses a number of different marine and coastal ecosystems like mangroves, seagrasses, coral reefs, and wetlands (Ukwe et al., 2006; Feka & Ajonina, 2011). These ecosystems contribute various vital ecosystem services to local communities, namely fishery resources, coastal erosion prevention, and climate regulation (Barbier et al., 2011; Spalding et al., 2014). These environments are under serious pressures; especially they are threatened with regards to economic activities development (fishing, aquaculture, tourism), urbanization, and climate change (Ondoa et al., 2015). The creation of a SDI for the management and space-time control of changes along the Cameroonian coastline is very strategic in that regard. This will help consolidate, standardize, and enhance existing geospatial data while allowing for new data to be collected and subsequently combined into the analysis. It would thus contribute to a better understanding of ecosystem dynamics, identifying the most vulnerable areas, and making informed decisions for the sustainable management of these environments (Politi et al., 2019; Kouamou Njifen et al., 2023; Lin & Pan, 2023).

This study aims to design and implement such a spatial data infrastructure for the Cameroonian coastline. It has a general objective: to contribute to a better understanding (current and future) of the dynamics of the Kribi coastline, to better assess and locate future impacts over a 20-year horizon and propose appropriate solutions. It includes several specific objectives:

- Conduct an assessment of existing geospatial data and identify the geographic information needs for the management of marine and coastal ecosystems;
- Map land use at an earlier date (2015) and a recent date (2023) in the area of interest;
- Map the positions of the coastline at different dates (2015-2018-2020-2023) along the study section;
- Design the technical and organizational architecture of the SDI, relying on international standards and norms;
- Develop the key components of the SDI, including the geospatial database, analysis, and visualization tools;
- Evaluate the impact of the SDI on the management and monitoring of marine and coastal ecosystems along the Cameroonian coastline.

Numerous studies have explored the impact of climate change and urbanization on coastal dynamics. Sustainable management of coastal urban areas relies on mitigation and adaptation strategies such as the use of green infrastructure and the implementation of urban resilience policies. The integration of spatial data infrastructure (SDI) can improve monitoring and decision-making for better coastal risk management. Coastal erosion has major implications for sustainable urban development. In addition to threatening coastal ecosystems, it has a direct impact on urban infrastructures, habitats and the economic activities of coastal populations. Managing the risks associated with coastal erosion is therefore an essential component of sustainable urban development, requiring an integrated approach combining geospatial tools, territorial planning and adaptation strategies.

This study (beyond being a case study) generally illustrates the contribution of remote sensing and GIS tools in monitoring coastal risks and managing coastlines. The ability to quantify and geolocate the retreat of the coastline (as well as the future impacts on various coastal occupations: urban areas, cultivated land, beaches, etc.), combined with the creation of an SDI for its monitoring, will provide a concrete and pressing argument for establishing a proactive and rational management plan for this section in a specific manner. It aims to contribute to a better understanding and sustainable management of Cameroon's coastal and marine ecosystems.

2. SITE PRESENTATION

The coastal zone of Cameroon consists of several distinct topographical units: The coastal plain (This is a stretch of lowlands that opens onto the Atlantic Ocean. It extends to a maximum depth of 360 km and is bordered by the western highlands and the southern plateau of Cameroon); The Ndian Basin (Located between Mount Cameroon and the Rumpi Hills, this basin is drained by the Ndian and Mémé rivers and ends in the Atlantic through the swamps of Rio Del Rey); The Douala Basin (Separated from the Ndian Basin by Mount Cameroon, this basin consists of vast areas where the Wouri, Sanaga, Nyong, and Dibamba rivers flow into long estuaries); The Kribi region (This region has a higher elevation and a very rugged topography. It is watered by the Ntem, Lobé, Kienké, and Lokoundjé rivers); Mount Cameroon (At 4,095 meters high, it is the highest point in the coastal area and one of the tallest peaks in Africa. It is located at the northern end of the Cameroon Highlands chain) (Sighomnou, 2004; Vicat et al., 1998).

The coastal plain of Cameroon is traversed by many rivers that flow into the Atlantic Ocean. The main rivers in this region are:

- The Manyu or Cross River;
- The Ndian and Mémé;
- The Wouri, Sanaga, Nyong, and Dibamba in the Douala Basin;
- The Ntem, Lobé, Kienké, and Lokoundjé in the Kribi region.

These rivers form long estuaries at their mouths, creating a mangrove zone along the coast. The hydrographic network of the coastal plain is very dense and plays an important role in hydroelectricity and traditional fishing. However, these rivers are only navigable in certain sections. In addition to the rivers, the coastal zone also hosts many wetlands, such as the swamps of Rio Del Rey. This entire hydrographic network makes the coastal plain the second richest region in water in Cameroon, after the Congo Basin (Segalen et al., 1967; Mbevo Fendoung et al., 2022).

The climate and weather conditions in this part of the country are quite varied or at least very unstable in recent years, due to several factors such as climate change causes like deforestation, greenhouse gas emissions, etc. In this area, an equatorial climate of the Guinean type prevails, characterized by the existence of four seasons; namely, two dry seasons (from December to March for the major season and from July to August for the minor season) and two rainy seasons (from September to November for the major season and from April to June for the minor season). The average daily duration of sunshine varies from 8 to 10 hours throughout the year (Ebodé, 2020; MINEPDED/PNUD, 2009).

The average daily maximum temperatures range from 25°C to 33°C, while the average daily minimum temperatures are between 15°C and 22°C. The coastal zone of Cameroon is characterized by an equatorial and sub-equatorial climate, with high rainfall and stable temperatures throughout the year. There is a long rainy season in the area from March to October, whereas the rest of the year sneaks in a dry season from November to February. With rainfall averaging from 1,500 to 2,000 mm per year, the temperature is relatively constant; roughly an averaged 22°C per year, maximums, hovering about 28°C during the day. Humidity is high, with an ail-time mean around 84%. The dew point stalls around 24°C. Winds are moderate with good visibility-counting for around 13 km. Also, the UV index is quite high, up to 4 out of 11, thereby requiring precautions for outdoor activities. Kribi is a coastal town in the littoral region in southern Cameroon. Kribi is known for its beaches, port, and biodiversity. An overview of the main socio-economic characteristics of Kribi is provided below (Suchel, 1987; Mena et al., 2016; Ngoran & Xue, 2017).

2.1. Demographics

Kribi is found in southern Cameroon, along the coast. Various sources have approximated its population at 30,000-50,000, made up of various peoples, notably the Bantu, Beti and Sawa, who each bring considerable cultural heritage to the area (MINHDU, 2014). Over the years, the city has undergone rapid demographic growth, or stimulated by major factors. Charmingly, the other advantage of Kribi is its economic appeal through growth in the major infrastructures: deepwater port, which has a strategic role in the trade and industrialization of the country. This has attracted many workers and investors, thus increasing the size of its urban population (Kevin et al. 2022). This aspect is further vitalized by the tourism sector, with its picturesque beaches, verdant flora and fauna, and emblematic spots, most particularly the Lobé waterfalls, which attract thousands of visitors annually; these have furthered the pioneering of new populations in search of economic opportunities in hotel, restaurant, and tourist services. At the same time, the rapid urbanization of the town comes with serious challenges, mainly concerning land management, access to basic services, and environment conservation. For those challenges, local and national authorities are implementing a series of projects to ensure sustainable and harmonious development of Kribi, in which economic growth goes hand in hand with good living conditions for the people (Evariste et al., 2018).

2.2. Economy

Tourism is an essential economic driver, attracting national and international visitors. Fishing is an important traditional activity, contributing to food supply and the local economy. Seafood products are a significant source of income for many families (Moudingo et al., 2020). Agriculture, although less dominant, is still present, with crops such as cocoa, coffee, and tropical fruits. The port of Kribi facilitates trade and import/export activities, supporting industrial activities, particularly in the timber and construction materials sectors (Ntamak-Nida et al., 2008; Mbevo Fendoung et al., 2022).

2.3. Infrastructure

Kribi is accessible by road and has a deep-water port, which is crucial for maritime trade. Road infrastructure is under development, improving access to other regions. Basic infrastructure, such as drinking water, electricity, and sanitation facilities, is being improved, although challenges remain in some areas (Segalen et al., 1967; Mbevo Fendoung et al., 2022).

2.4. Education and Health

Kribi has several primary and secondary educational institutions. Access to education is improving, but disparities persist. The city has hospitals and health centers, but access to healthcare can be limited in surrounding rural areas.

2.5. Culture and Heritage

Kribi is rich in culture, with festivals and events celebrating local traditions. Local cuisine, which includes seafood dishes, is also a key element of its appeal. The city is close to national parks and nature reserves, contributing to biodiversity and attracting ecotourists.

The Cameroonian coast is shared by three regions of the country: the South region, the Littoral region, and the South-West region. This area stretches 590 km along the Atlantic Ocean. The coastal region extends along the border with Nigeria to the northwest, narrows at the foot of Mount Cameroon, widens in the Douala sedimentary basin, and continues up to the border with Equatorial Guinea (Mbaha & Etoundi, 2021).

Figure 1. Geographical presentation of the Cameroon Coast

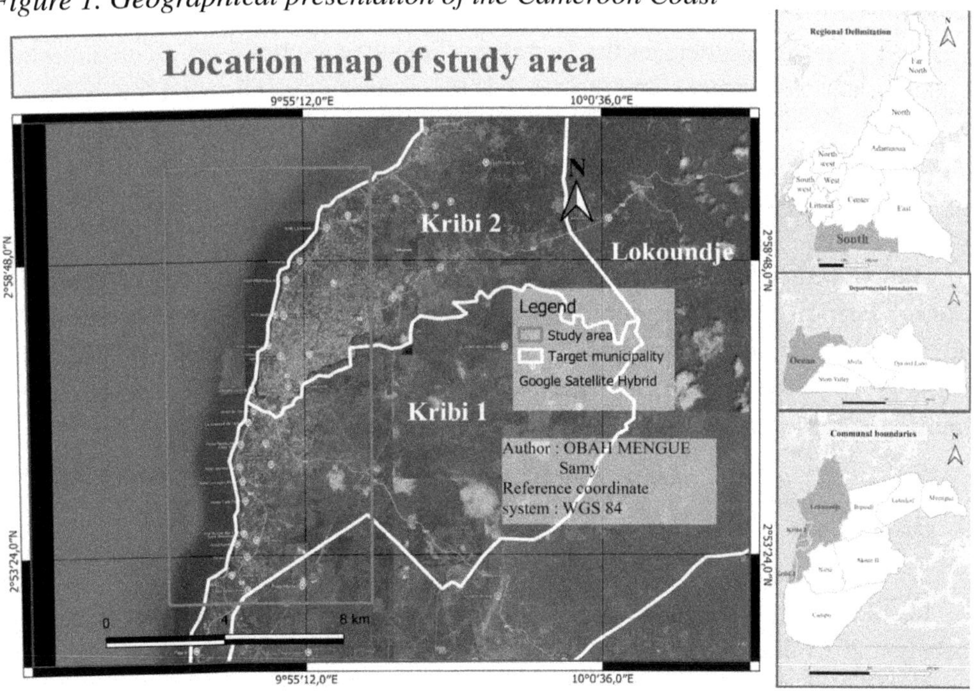

3. METHODOLOGY

The methodology used in this paper is based on a technique widely tested for processing geographic information in coastal dynamics. Essentially, the adopted procedure, as per the availability of data, comprises the literature review; consultations with resource persons and institutions specialized in the production of geospatial information; data collection; data production; data processing and analysis; and the presentation of results accompanied by discussions.

3.1. Data Collection

Through satellite images and various administrative institutions, historical data were collected. The DSAS (Digital Shoreline Analysis System) software was used to analyze the evolution of the shoreline. DSAS is a tool for analysis, coastal problems management, and coastal erosion risk analyses. Its ambition is to assess shoreline variation over time on the basis of geospatial data. In turn, this can be utilized for assessments of coastal erosion impact and better management decisions. The users will digitize the shoreline from the historical images and draw polygons for each shoreline at different times. The DSAS analyzes shoreline shift by applying transects, which are lines that intersect the shoreline. The distance between given shoreline variants, at each transect along the coastal area, is measured, allowing for erosion or accretion rates to be calculated for different time frames (Himmelstoss et al., 2021).

3.2. Development of the IDS

Announcement of the IDS In order to ease analysis and sharing of results, a central custom-made geospatial data information system (IDS) was coded using advanced programming languages such as Python, JavaScript, and HTML. The name given to the platform is IDoGek; it optimally visualizes data and performs multi-structural risk analyses for better decision-making.

4. RESULTS

Results exhibit that more than 7 hectares of land of human interest were lost between 2015 and 2023, according to the analyses. Statistics highlight a slowed contraction of the shores between 2018 and 2020; forecasts, however, indicate a retreat of 10.97 m by the year 2034 if no measures are taken.

4.1. Digitalized Shorelines

This phase involved presenting the shorelines, also referred to as coastal lines, from different years, namely those of 2014, 2018, 2020, and 2023, obtained through remote sensing methods, which form the basis of all the results obtained.

Figure 2. Dimension lines for different years

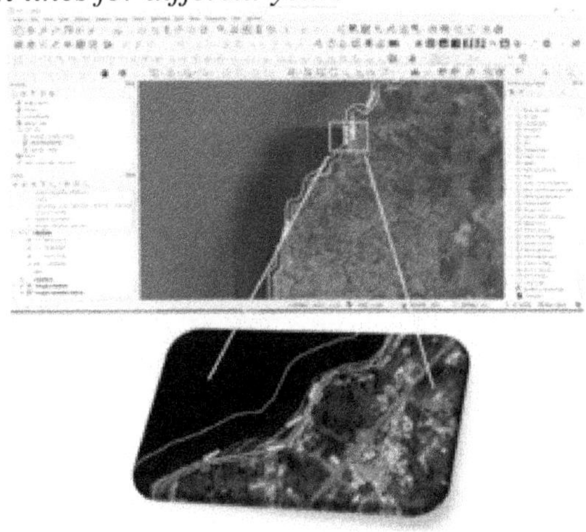

4.2. Land Use Mapping

The land use maps for 2015 and 2023 show significant changes in land use, with a decrease in usable coastal areas. A clearer view of the land use in this area is presented as follows:

4.2.1. In 2015

Figure 3. Land use map 2015

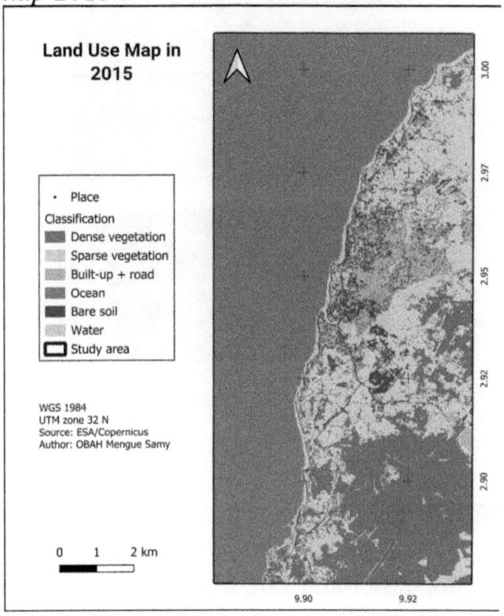

4.2.2. In 2019

Figure 4. Land use map 2019

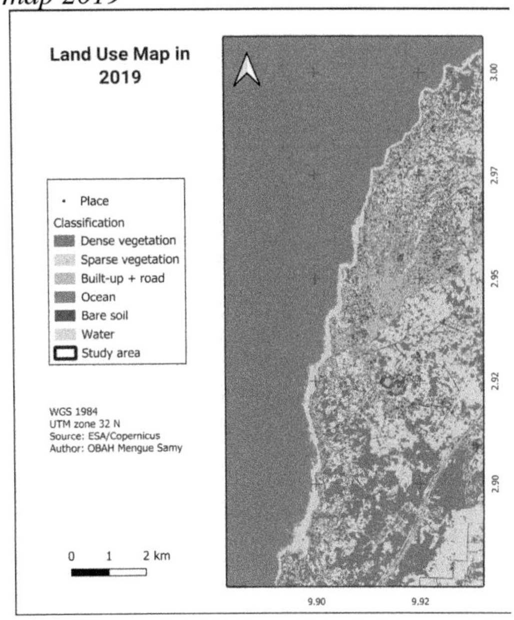

4.2.3. In 2023

Figure 5. Land use map 2023

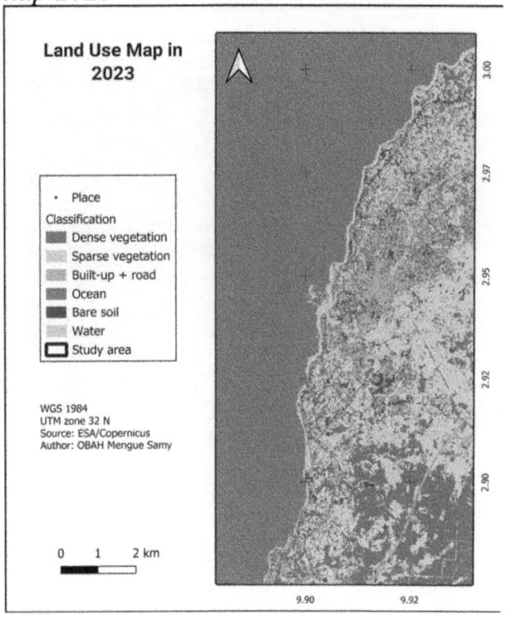

4.3. Presentation of the Platform

Our platform, named IDoGek, which stands for "Geospatial Data Infrastructure of Kribi," is presented as follows:

4.3.1. A Homepage Featuring Its Logo

Figure 6. Screenshot of home page

4.3.2. Creation of an Account

Account creation is done through authentication after filling in all the available fields.

Figure 7. Screenshot illustrating the fields

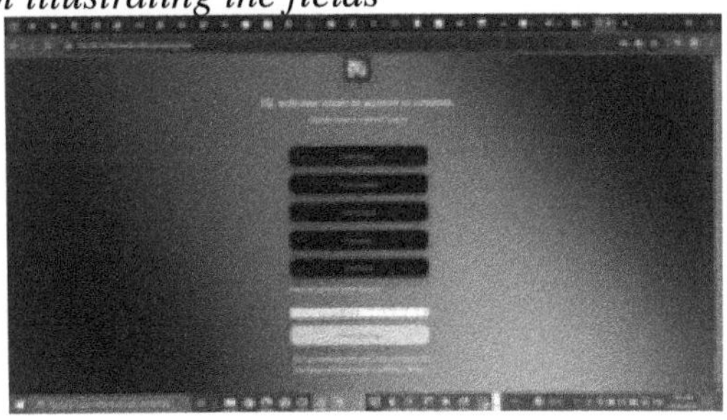

4.3.3. Working Interface

This is the very core of the platform, where the different layers and analyses are conducted, providing access to various features and visualization.

Figure 8. Tool palette

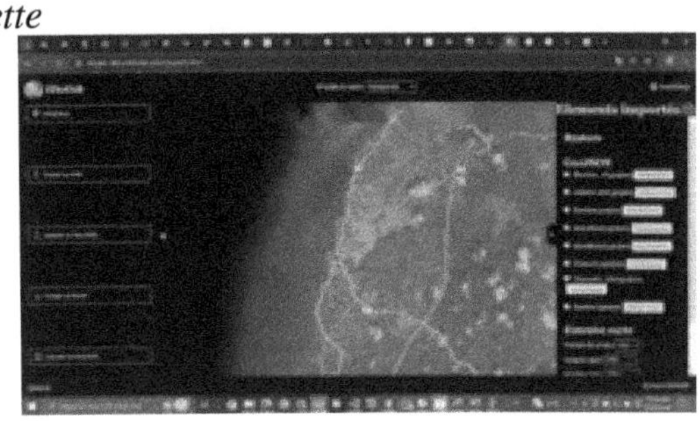

The image of the following table provides a summary of the different variables calculated by the DSAS.

Figure 9. Screen table showing the different values calculated

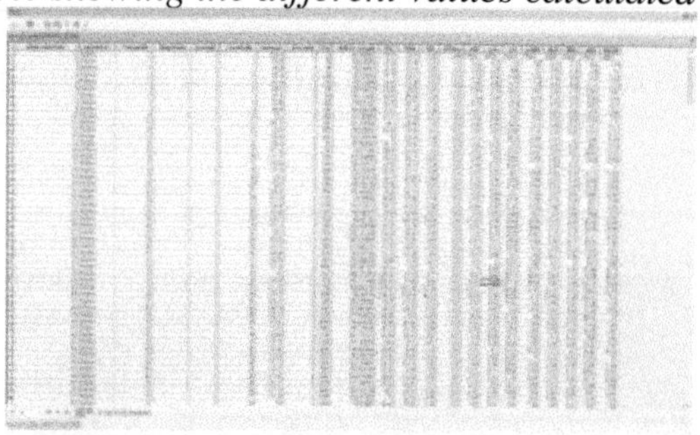

From this table, we derive the graph showing the variation of the shoreline as follows:

Figure 10. Diagram of shoreline evolution

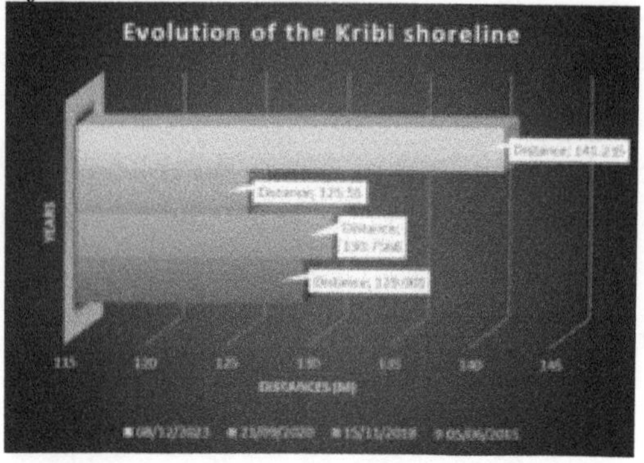

5. DISCUSSION

Based on the results obtained, we observe:

5.1. For the Land Use Maps

A clearer observation for better interpretation is provided by the following graph.

5.1.1. In 2015

Table 1. Quantifying land use in 2015

Class Name	Area (Km²)	Hectare	Pourcentage
Ocean	47,4	4740	47,09
Water	6,595	659,5	6,55
Batis	15,263	1526,3	15,16
Vegetation dense	22,177	2217,7	22,03
Vegetation claire	6,428	642,8	6,39
Bare soil	2,79	279	2,77
Total	100,653	100665,3	100,00

5.1.2. In 2019

Table 2. Quantifying land use in 2019

Class Name	Area (Km²)	Hectare	Pourcentage
Ocean	48,198	4819,8	47,89
Water	4,673	467,3	4,64
Batis	18,604	1860,4	18,48
Vegetation dense	20,951	2095,1	20,82
Vegetation claire	6,273	627,3	6,23
Bare soil	1,954	195,4	1,94
Total	100,653	10065,3	100,00

5.1.3. In 2023

Table 3. Quantifying land use in 2023

Class Name	Area (Km²)	Hectare	Pourcentage
Ocean	48,421	4842,10	48,11
Water	6,041	604,10	6,00
Batis	11,933	1193,30	11,86
Vegetation dense	24,744	2474,40	24,58
Vegetation claire	7,622	762,20	7,57
Bare soil	1,891	189,10	1,88
Total	100,652	10065,20	100,00

From 2015 to 2019, there was an increase in the percentage of the area occupied by the ocean, rising from 47.09% to 47.89%, which represents a progression of 0.8%. This indicates an advancement of the shoreline towards the mainland. From 2019 to 2023, there was also an increase from 47.89% to 48.11%, a progression of 0.22%. An explanation for this regression compared to the previous period is the significant reduction in anthropogenic activities in this area.

This graph presents the different positions of the shoreline over time based on its distance from a baseline that we used during the calculations. It shows an erosion of -1.75m between 2015 and 2018; an accretion of 5.2m between 2018 and 2020, which could be attributed to climatic factors such as wind, precipitation, and even anthropogenic actions, as human presence was significantly reduced during this period due to the COVID-19 pandemic. There was a severe erosion of -12.23m between 2020 and 2023, which may be explained by the significant climate change the world has faced over the last two years and the resumption of anthropogenic activities in the area. Thus, over the period from 2015 to 2023, there is an average change rate of -8.78m, representing a progression speed of -1.1m/year.

By analogy to the western coast of Cameroon, during its second study period from 2005 to 2008, they recorded an erosion situation along almost the entire length of the coastline. The average speed was 3.5m/year for an erosion distance of 10.5m.

The retreat of the shoreline has resulted in considerable losses along the coast. Certain economically significant land use areas (beaches, cultivated areas) as well as human settlements have been destroyed or inundated by water during high tide episodes. The GIS tool has allowed for a more or less precise assessment of the extent of these losses for the period from 2014 to 2021. Indeed, by overlaying the

2021 coastline with the 2014 classification, we were able to perform a statistical analysis of these land use losses.

Based on the average erosion speed in this area (-1.1m/year), we can somewhat predict what the position of the shoreline in these areas would be in 2034, as well as the losses in land use if it does not change significantly by then, especially if no actions are taken to reverse the erosive trend.

The projection of the 2034 shoreline was carried out using ArcGIS by performing a "buffer." We considered the average of "1.1m/year" for the critical shoreline areas (increasing erosion) and the number of years until 2034 (10 years) to find the fictitious projection of the shoreline. Therefore, the projected distance (10.97m) of the shoreline in critical areas was determined using the basic speed formula (speed = Distance = Speed * Time).

This same formula was also used for the western coast of Cameroon, considering the same criteria. With an average speed of 1.09m/year, they recorded that by 2028, the coastal section of the western coast of Cameroon would retreat by 10.90m. The figure 11 below shows the visualization in our developed IDS of the different layers.

Figure 11. Visualization of the coastline in 2034

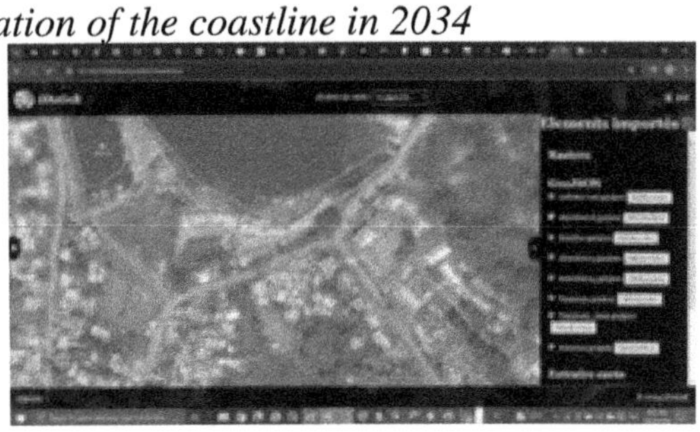

6. CONCLUSION

The creation of the IDS for the spatio-temporal monitoring and management of KRIBI was initiated to measure the evolution of the shoreline in order to estimate land use losses and anticipate damage that may occur in the coming years, through data visualization via a platform, with the aim of encouraging local authorities to implement a sustainable management plan for this tourist coastal strip. The data produced and processed in this study show that this coastal strip is highly dynamic. Indeed, during this periodic analysis, 81% of this coastal section is subject to erosion. A land area loss will be 1.2%. The projection of the shoreline in certain critical areas demonstrated that by 2034, the shoreline is expected to retreat by 10.97 m toward the mainland, and without any protective intervention, the loss of land use area would be much more considerable (estimated at 171,566 hectares). Hotel infrastructures, beaches, urban areas, fishing camps, and some roads would be destroyed or inundated in the coming years. To prevent this from becoming a reality and to mitigate the risks of coastal erosion, it would be prudent for local authorities (mayors and officials from the Ministry of Tourism, MINDHU) to truly recognize this reality and be informed about the existence of these critical coastal areas.

This work has allowed for the establishment of an IDS to monitor the evolution of the KRIBI shoreline; however, it deserves to be complemented by further field investigations, including: the perceptions of local residents on coastal erosion, the analysis of the causes of shoreline dynamics, and the weight of each anthropogenic factor that most influences this coastal dynamic. To significantly reduce the risks associated with coastal erosion phenomena in the coastal areas of Cameroon, specifically along the KRIBI shoreline, and to allow for the restoration of the natural dynamics of the coasts, it is necessary to:

- Establish a unit within the Ministry of Environment to participate in the development and implementation of sustainable management plans for the coastal areas of Cameroon;
- Conduct continuous monitoring of the state of the coasts through regular studies of the shoreline position and install a tide gauge at the coasts to monitor sea level changes so that management strategies can take into account variations resulting from climate change;
- Prohibit certain anthropogenic actions (such as sand extraction for construction and private development, destruction of vegetation that stabilizes sandy sediments along the coast, and vehicle traffic along the beaches) that accelerate coastal degradation.

REFERENCES

Aretouyap, Z., Billa, L., Jones, M., & Richter, G. (2020). Geospatial and statistical interpretation of lineaments: Salinity intrusion in the Kribi-Campo coastland of Cameroon. *Advances in Space Research*, *66*(4), 844–853. DOI: 10.1016/j.asr.2020.05.002

Barbier, E. B., Hacker, S. D., Kennedy, C., Koch, E. W., Stier, A. C., & Silliman, B. R. (2011). The value of estuarine and coastal ecosystem services. *Ecological Monographs*, *81*(2), 169–193. DOI: 10.1890/10-1510.1

Ebodé, V. B. (2020). Variabilité Hydropluviométrique en Afrique Centrale Occidentale Forestière: Entre Analyse des Fluctuations Observées, Modélisation Hydrologique Prédictive et Recherche des Facteurs Explicatifs. Ph.D. Thesis, University of Yaounde 1, Yaoundé, Cameroon.

Evariste, F. F., Jean, S. D., Victor, K., & Claudia, M. (2018). Assessing climate change vulnerability and local adaptation strategies in adjacent communities of the Kribi-Campo coastal ecosystems, South Cameroon. *Urban Climate*, *24*, 1037–1051. DOI: 10.1016/j.uclim.2017.12.007

Feka, N. Z., & Ajonina, G. N. (2011). Drivers causing decline of mangrove in West-Central Africa: A review. *The International Journal of Biodiversity Science, Ecosystem Services & Management*, *7*(3), 217–230. DOI: 10.1080/21513732.2011.634436

Himmelstoss, E. A., Henderson, R. E., Kratzmann, M. G., & Farris, A. S. (2021). Digital Shoreline Analysis System (DSAS) Version 5.1 User Guide. US Geological Survey. https://pubs.usgs.gov/publication/ofr20211091

Htay, T., Røskaft, E., Ringsby, T. H., & Ranke, P. S. (2023). Spatio-temporal variation in avian taxonomic, functional, and phylogenetic diversity and its relevance for conservation in a wetland ecosystem in Myanmar. *Biodiversity and Conservation*, *32*(8-9), 2841–2867. DOI: 10.1007/s10531-023-02632-9

Jabbour, C., Rey-Valette, H., Maurel, P., & Salles, J. M. (2019). Spatial data infrastructure management: A two-sided market approach for strategic reflections. *International Journal of Information Management*, *45*, 69–82. DOI: 10.1016/j.ijinfomgt.2018.10.022

Kevin, S. T. W., Shidiki, A. A., & Tchamba, M. N. (2022). Coastal risk management in a context of climate change: A case study of kribi town of the south region of Cameroon. *Journal of Geoscience and Environment Protection*, *10*(5), 111–124. DOI: 10.4236/gep.2022.105009

Kouamou Njifen, S. R., Enyegue A Nyam, F. M., Bikoro, M. B. A., Ngouokouo, A. T., & Tabod, C. T. (2023). Modeling groundwater potential zones in the Kribi-Campo region, South Cameroon using geospatial techniques and statistical models. *Modeling Earth Systems and Environment*, *9*(1), 913–935. DOI: 10.1007/s40808-022-01544-x

Lin, B., & Pan, F. (2023). Applications of DGT in coastal sediments: Monitoring and biogeochemical study of trace metals and oxyanions. *Trends in Environmental Analytical Chemistry*, *39*, e00207. DOI: 10.1016/j.teac.2023.e00207

Mbaha, J. P., & Etoundi, M. L. B. A. (2021). *Construire une Ville Portuaire Stratégique et Émergente à L'horizon 2035*. Espace Géograph. Soc.

Mbevo Fendoung, P., Tchindjang, M., & Hubert-Ferrari, A. (2022). Weakening of Coastlines and Coastal Erosion in the Gulf of Guinea: The Case of the Kribi Coast in Cameroon. *Land (Basel)*, *11*(9), 1557. DOI: 10.3390/land11091557

Mena, M. S., Tchawa, P., Amougou, J. A., & Tchotsoua, M. (2016). Les changements climatiques à travers la modification du régime pluviométrique dans la région de Kribi (1935–2006). *Rev. Ivoire. Sci. Technol.*, *28*, 389–407.

MINEPDED/PNUD. (2009). Révision /Opérationnalisation Du PNGE Vers Un Programme Environnemental (PE), Volume I: Diagnostic De La Situation De L'Environnement Au Cameroun; Ministère de L'environnement et de la Protection de la Nature: Yaoundé, Cameroon, 203p.

MINHDU. (2014). *Plan Directeur D'Urbanisme De La Ville De Kribi. Rapport Justificatif.* Communauté Urbaine de Kribi.

Moudingo, J. H., Ajonina, G., Kemajou, J., Wassouni, A., Bitomo, M., Assengze, A., & Tomedi, M. (2020). Sylvio-socioeconomic study of urban mangrove patches and challenges: case of Kribi, Cameroon. In *Biotechnological Utilization of Mangrove Resources* (pp. 79–115). Elsevier., DOI: 10.1016/B978-0-12-819532-1.00004-4

Ngoran, S. D., & Xue, X. (2017). Public sector governance in Cameroon: A valuable opportunity or fatal aberration from the Kribi Campo integrated coastal management? *Ocean and Coastal Management*, *138*, 83–92. DOI: 10.1016/j.ocecoaman.2017.01.006

Ntamak-Nida, M. J., Baudin, F., Schnyder, J., Makong, J. C., Komguem, P. B., & Abolo, G. M. (2008). Depositional environments and characterisation of the organic matter of the Lower Mundeck Formation (Barremian? –Aptian) of the Kribi-Campo sub-basin (South Cameroon): Implications for petroleum exploration. *Journal of African Earth Sciences*, *51*(4), 207–219. DOI: 10.1016/j.jafrearsci.2008.01.006

Ondoa, G. A., Onguéné, R., Eyango, M. T., Duhaut, T., Mama, C., Angnuureng, D. B., & Almar, R. (2018). Assessment of the Evolution of Cameroon Coastline: An Overview from 1986 to 2015. *Journal of Coastal Research, 81*(10081), 122–129. DOI: 10.2112/SI81-016.1

Politi, E., Paterson, S. K., Scarrott, R., Tuohy, E., O'Mahony, C., & Cámaro-García, W. C. A. (2019). Earth observation applications for coastal sustainability: Potential and challenges for implementation. *Anthropocene Coasts*, *2*(1), 306–329. DOI: 10.1139/anc-2018-0015

Segalen, P. (1967). Les Sols et la Géomorphologie du Cameroun. Cahiers ORSTOM. *Série Pédologie.*, *1967*, 137–187.

Sighomnou, D. (2004). nalyse Et Redéfinition Des Régimes Climatiques Et Hydrologiques Du Cameroun: Perspectives D'Evolution Des Ressources En Eau. Ph.D. Thesis, University of Yaoundé I, Yaoundé, Cameroon.

Spalding, M. D., Ruffo, S., Lacambra, C., Meliane, I., Hale, L. Z., Shepard, C. C., & Beck, M. W. (2014). The role of ecosystems in coastal protection: Adapting to climate change and coastal hazards. *Ocean and Coastal Management*, *90*, 50–57. DOI: 10.1016/j.ocecoaman.2013.09.007

Strain, L., Rajabifard, A., & Williamson, I. (2006). Marine administration and spatial data infrastructure. *Marine Policy*, *30*(4), 431–441. DOI: 10.1016/j.marpol.2005.03.005

Suchel, J. B. (1987). Rainfall patterns and regimes rainfall in Cameroon. *Doc. Geogr. Trop.*, *5*, 287.

Ukwe, C. N., Ibe, C. A., & Sherman, K. (2006). A sixteen-country mobilization for sustainable fisheries in the Guinea Current Large Marine Ecosystem. *Ocean and Coastal Management*, *49*(7-8), 385–412. DOI: 10.1016/j.ocecoaman.2006.04.006

Vaitis, M., Kopsachilis, V., Tataris, G., Michalakis, V. I., & Pavlogeorgatos, G. (2022). The development of a spatial data infrastructure to support marine spatial planning in Greece. *Ocean and Coastal Management*, *218*, 106025. DOI: 10.1016/j.ocecoaman.2022.106025

Vicat, J.P., Pouclet, A., Nsifa, E. (1998). Les Dolérites du Groupe du Ntem (Sud Cameroun) et des Régions Voisines (Centrafrique, Gabon, Congo, Bas Zaïre): Caractéristiques Géochimiques et Place dans L'évolution du Craton du Congo au Protérozoïque. Géologie Environ. Cameroun Collect. GEOCAM. 305–324.

Wang, Z., Liu, X., Wang, X., Wang, H., Sun, Y., Zhang, J., & Li, H. (2024). Contrasting benthic bacterial and fungal communities in two temperate coastal areas affected by different levels of anthropogenic pressure. *Marine Environmental Research, 198*, 106501. DOI: 10.1016/j.marenvres.2024.106501 PMID: 38615486

Yan, Z., Guo, X., Zhao, Z., & Tang, L. (2024). Achieving fine-grained urban flood perception and spatio-temporal evolution analysis based on social media. *Sustainable Cities and Society, 101*, 105077. DOI: 10.1016/j.scs.2023.105077

Zhou, D., Tian, Y., & Jiang, G. (2018). Spatio-temporal investigation of the interactive relationship between urbanization and ecosystem services: Case study of the Jingjinji urban agglomeration, China. *Ecological Indicators, 95*, 152–164. DOI: 10.1016/j.ecolind.2018.07.007

Compilation of References

Aamer, A. M., Al-Awlaqi, M. A., Affia, I., Arumsari, S., & Mandahawi, N. (2021). The internet of things in the food supply chain: Adoption challenges. *Benchmarking*, *28*(8), 2521–2541. DOI: 10.1108/BIJ-07-2020-0371

Abdul Rahman, A., Mohamed Zaid, S., & Mohammad Shuhaimi, N. D. A. (2022). Effects of Green Roof in Reducing Surface Temperature and Addressing Urban Heat Island in Tropical Climate of Malaysia. *Journal of Design and Built Environment*, *22*(2), 1–20. DOI: 10.22452/jdbe.vol22no2.1

Abdullah, M. N., & Yusof, N. A. (2023). Challenges in retrofitting energy efficiency in heritage buildings: A Malaysian perspective. *Journal of Architectural Conservation*, *29*(1), 20–35.

Abidin, N. Z., & Zainal, R. (2019). Heritage protection regulations in Malaysia: Challenges for adaptive reuse. *Journal of Cultural Heritage Management and Sustainable Development*, *9*(3), 245–258. DOI: 10.1108/JCHMSD-09-2018-0065

Abioye, S. O., Oyedele, L. O., Akanbi, L., Ajayi, A., Davila Delgado, J. M., Bilal, M., Akinade, O. O., & Ahmed, A. (2021). Artificial intelligence in the construction industry: A review of present status, opportunities and future challenges. *Journal of Building Engineering*, *44*, 103299. DOI: 10.1016/j.jobe.2021.103299

Aceyus Team. (2021, June 8). Four benefits of real-time monitoring in the modern contact center. Aceyus. https://www.aceyus.com/four-benefits-of-real-time-monitoring-in-the-modern-contact-center/

Achtnicht, M., & Madlener, R. (2014). Factors influencing German house owners' preferences on energy retrofits. *Energy Policy*, *68*, 254–263. DOI: 10.1016/j.enpol.2014.01.006

Adams, W. C. (2015). Conducting semi-structured interviews. In Newcomer, K., Hatry, H., & Wholey, J. (Eds.), *Handbook of practical program evaluation* (pp. 492–505). Jossey-Bass. DOI: 10.1002/9781119171386.ch19

Ahamad, R., Rahman, A., & Tan, S. P. (2022). Sustainable heritage: The role of energy efficiency in adaptive reuse. *International Journal of Heritage Studies, 28*(4), 295–310.

Ahlqvist, T., & Rizzo, I. (2023). Community involvement in urban heritage: The case of adaptive reuse projects. *Journal of Urban Management, 12*(1), 45–58.

Ahluwalia, I. J. (2017). Urban Governance in India. *Journal of Urban Affairs, 41*(1), 83–102. https://www.tandfonline.com/doi/full/10.1080/07352166.2016.1271614. DOI: 10.1080/07352166.2016.1271614

Ahmad, F., & Din, N. A. M. (2021). Financial challenges in heritage building conservation: The Malaysian context. *International Journal of Built Environment and Sustainability, 8*(1), 35–44. DOI: 10.11113/ijbes.v8.n1.494

Ahmad, M., & Rahman, A. (2022). Community engagement in heritage conservation: A pathway to sustainable practices. *International Journal of Heritage Studies, 28*(5), 399–412.

Ahmad, N., & Din, S. H. (2021). Economic challenges in heritage building conservation: A case for financial incentives. *Heritage Conservation Journal, 7*(2), 65–79.

Ahmad, N., & Rasli, N. (2021). Digital tools in heritage conservation: Bridging the gap between tradition and technology. *Journal of Sustainable Architecture, 15*(1), 34–50.

Ahmad, N., Salleh, M. F., & Ibrahim, R. (2021). Community participation in heritage conservation: A Malaysian perspective. *Journal of Heritage Conservation, 8*(2), 34–45.

Ahmad, S., & Hussain, N. (2021). The role of digital technology in heritage building conservation. *Journal of Heritage Management, 10*(2), 123–134.

Ahmad, S., & Hussain, N. (2021). The use of original materials in heritage building conservation: Challenges and approaches. *Journal of Heritage Management, 10*(2), 123–134.

Ahmad, Y. (2019). The challenges of preserving heritage buildings in Malaysia. *International Journal of Heritage Studies, 25*(3), 265–280.

Aikivuori, A. (2006). Periods and demand for private sector housing refurbishment. *Construction Management and Economics, 14*(1), 3–12. DOI: 10.1080/014461996000000002

Ajith, P. S. (2016). A Study on the Solid Waste Collection, Treatment and Disposal System of Municipalities in Kerala. *JETIR* 3(12), 702-712. https://www.jetir.org/papers/JETIR1701510.pdf

Ajslev, J. Z. N., & Nimb, I. E. E. (2022). Virtual design and construction for occupational safety and health purposes–A review on current gaps and directions for research and practice. *Safety Science, 155*, 105876. DOI: 10.1016/j.ssci.2022.105876

Akinsete, E., Koundouri, P., Kartala, X., Englezos, N., Lautze, J., Yihdego, Z., Gibson, J., Scholz, G., van Bers, C., & Sodoge, J. (2022). Sustainable WEF Nexus Management: A Conceptual Framework to Integrate models of Social, Economic, Policy, and Institutional Developments. *Frontiers in Water, 4*, 727772. https://www.frontiersin.org/journals/water/articles/10.3389/frwa.2022.727772/full. DOI: 10.3389/frwa.2022.727772

Aladayleh, K. J., & Aladaileh, M. J. (2024). Applying Analytical Hierarchy Process (AHP) to BIM-Based Risk Management for Optimal Performance in Construction Projects. *Buildings (Basel, Switzerland), 14*(11), 3632. DOI: 10.3390/buildings14113632

Al-Amleh, K. I. (2020). A study into the adoption of Internet of Things – IoT technologies within contractors in Dubai, United Arab Emirates [British University]. https://bspace.buid.ac.ae/bitstream/handle/1234/1811/20172453.pdf?sequence=3&isAllowed=y

Alamri, A. W. (2019). *Effectiveness of qualitative research methods: Interviews and diaries*. University of Jeddah.

Al-Anzi, B., Zubari, W., & Al-Rashidi, A. (2020). *Desalination technologies and sustainability in the coastal regions of the Arabian Gulf: Advances and challenges*. Springer., DOI: 10.1007/978-3-030-36938-5

Alberini, A., Banfi, S., & Ramseier, C. (2013). Energy efficiency investments in the home: Swiss homeowners and expectations about future energy prices. *The Energy Journal (Cambridge, Mass.), 34*(1), 49–86. DOI: 10.5547/01956574.34.1.3

Aleksandrova, E., Vinogradova, V., & Tokunova, G. (2019). Integration of digital technologies in the field of construction in the Russian Federation. *Engineering Management in Production and Services, 11*(3), 40–49. DOI: 10.2478/emj-2019-0019

Alexa, W., & David, H. (2020). Implementing HBIM on conservation heritage projects: Lessons from renovation case studies.

Alexenberg, M. (2011). *The future of art in a post-digital age*. Intellect Books. DOI: 10.2307/j.ctv36xvpkc

Ali, M., Ismail, K. M., Hashim, K. S. H. Y., Suhaimi, S., & Mustafa, M. H. (2018). Historic Building Information Modelling (HBIM) for Malaysian Construction Industry. Planning Malaysia Journal, 16(7).

Alibaba, H. (2019). Green roof benefits, opportunities and challenges. *International Journal of Civil and Structural Engineering Research*, *7*(2), 106–112.

Ali, Z., Nor, N. F. M., & Ismail, F. (2020). Heritage Building Reuse in Malaysia: A Case Study of Kuala Lumpur. *Journal of Design and Built Environment*, *20*(2), 55–66.

Allison, L. (2024, September 5). All projects worth RM10mil and above to use construction software BIM. *The Star*. https://www.thestar.com.my/news/nation/2024/09/04/all-projects-worth-rm10mil-and-above-to-use-construction-software-bim-says-ahmad-maslan

Almeida, S. L. G. D. (2014). Retrofitting and refurbishment processes of heritage buildings: application to three case studies (Doctoral dissertation).

Al-Qawasmi, J., & Hadjri, K. (2007). Architecture in the digital age: The effect of digital media on design, production and evaluation of the built environment. *Open House International*.

Alshahrani, A., & Aoyama, Y. (2023). Augmented and virtual reality in architectural design: Engagement, visualization, and education. *Architectural Engineering and Design Management*, *19*(2), 108–122. DOI: 10.1080/17452007.2022.2146554

Alshawabkeh, Y., Baik, A., & Miky, Y. (2024). HBIM for Conservation of Built Heritage. *ISPRS International Journal of Geo-Information*, *13*(7), 231. DOI: 10.3390/ijgi13070231

Alvarez, R., & Duarte, F. (2021). *Urban play: Make believe, technology, and urban space*. The MIT Press.

Amato, A., Andreoli, M., & Rovai, M. (2021). Adaptive reuse of a historic building by introducing new functions: A scenario evaluation based on participatory MCA applied to a former Carthusian monastery in Tuscany, Italy. *Sustainability (Basel)*, *13*(4), 2335. DOI: 10.3390/su13042335

American Institute of Architects. (2022). Building reuse: A proven climate and economic strategy.

American Institute of Architects. (2023). The importance of building reuse in achieving carbon reduction targets.

Amin, N., Ahamad, R., & Lim, L. Y. (2022). Challenges in preserving Malaysia's built heritage amidst urban development. *Journal of Heritage Management*, *7*(1), 45–60.

Anand, P. B. (1999). Waste Management in Madras Revisited. *Environment and Urbanization, 11*(2), 161–176. https://journals.sagepub.com/doi/abs/10.1177/095624789901100214. DOI: 10.1177/095624789901100214

Andenæs, E., Engebø, A., Time, B., Lohne, J., Torp, O., & Kvande, T. (2020). Perspectives on quality risk in the building process of blue-green roofs in Norway. *Buildings (Basel, Switzerland), 10*(10), 10. DOI: 10.3390/buildings10100189

Andenæs, E., Time, B., Muthanna, T., Asphaug, S., & Kvande, T. (2021). Risk reduction framework for blue-Green roofs. *Buildings (Basel, Switzerland), 11*(5), 5. DOI: 10.3390/buildings11050185

Antonopoulou, S., & Bryan, P. (2017). *BIM for heritage: developing a historic building information model*. Historic England.

Anwar, R., Mollah, M. H., & Islam, S. (2020). Greywater recycling in urban residential buildings: Current trends and future challenges. *Environmental Science and Pollution Research International, 27*(11), 12241–12256. DOI: 10.1007/s11356-020-07827-5

Aravind, R. Yadikiumarani, Meghna, K., R, D., & Naregowda, H. (2021). IoT based real time data monitoring for industry. Journal of Engineering Research & Technology (IJERT) NCCDS, 09(12). https://www.ijert.org/iot-based-real-time-data-monitoring-for-industry

Arayici, Y., Counsell, J., Mahdjoubi, L., Nagy, G. A., Hawas, S., & Dweidar, K. (Eds.). (2017). *Heritage building information modelling*. Taylor & Francis. DOI: 10.4324/9781315628011

Architects, Q. W. A. (2022). Sri Peraptih Palace: An exhibition space in the metaverse. Retrieved from www.qwaarchitects.com

Arena, U. (2012). Process and technological aspects of municipal solid waste gasification. *A review.Waste Management (New York, N.Y.), 32*(4), 625–639. DOI: 10.1016/j.wasman.2011.09.025 PMID: 22035903

Aretouyap, Z., Billa, L., Jones, M., & Richter, G. (2020). Geospatial and statistical interpretation of lineaments: Salinity intrusion in the Kribi-Campo coastland of Cameroon. *Advances in Space Research, 66*(4), 844–853. DOI: 10.1016/j.asr.2020.05.002

Arowoiya, V. A., Oke, A. E., Aigbavboa, C. O., & Aliu, J. (2020). An appraisal of the adoption Internet of Things (IoT) elements for sustainable construction. *Journal of Engineering. Defence Technology, 18*(5), 1193–1208. DOI: 10.1108/JEDT-10-2019-0270

Arplan. (2024). *Introduction to HBIM: Fundamentals and Applications*. Retrieved from. https://arplan.com.co/

Arslan, V., Ulubeyli, S., & Kazaz, A. (2019). The use of Internet of Things in the construction industry. https://www.researchgate.net/publication/337890965

Arya, V., & Rejuna, C. A. (2024). Unveiling Kerala's Distinct Urbanization: A Comparative Analysis within India. *The Indian Economic Journal*, •••, 1–16. https://journals.sagepub.com/doi/full/10.1177/00194662241278066

ASHRAE. (2019). *ASHRAE terminology. ASHRAE searchable terminology tool. American Association of Heating, Refrigeration and Air-conditioning Engineers*. ASHRAE.

Asite. (2021, January, 20). Why we should be paying more attention to advanced building materials. [Online blog post]. https://www.asite.com/blogs/why-we-should-be-paying-more-attention-to-advanced-building-materials

Aslam, M., Gao, Z., & Smith, G. (2021). Integrated implementation of Virtual Design and Construction (VDC) and lean project delivery system (LPDS). *Journal of Building Engineering*, *39*, 102252. DOI: 10.1016/j.jobe.2021.102252

Asmat, B. I. (2013). *Cooling potential of passive roof treatments on the concrete flat roof in Penang, Malaysia*. Ph.D. Thesis. Universiti Sains Malaysia, Penang, Malaysia.

Assan, L. (2018). *Potensi internet saling berhubung (internet of things) dalam industri pembinaan di Malaysia* [Ijazah Sarjana Muda Ukur Bahan, UNIVERSITI TEKNOLOGI MALAYSIA]. https://bic.utm.my/files/2018/08/LAROMI-ASSAN.pdf

Athmani, W., Sriti, L., Dabaieh, M., & Younsi, Z. (2022). The potential of using passive cooling roof techniques to improve thermal performance and energy efficiency of residential buildings in hot arid regions. *Buildings (Basel, Switzerland)*, *13*(1), 21. DOI: 10.3390/buildings13010021

Aung, T. H. S. (2022). 10 - Trendy advanced construction materials to know (2022). [Online blog post]. https://www.novatr.com/blog/advanced-construction-materials

Awolusi, I., Marks, E., & Hallowell, M. (2018). Wearable technology for personalized construction safety monitoring and trending: Review of applicable devices. *Automation in Construction*, *85*, 96–106. DOI: 10.1016/j.autcon.2017.10.010

AWWA. (2021). *The impact of water loss on urban systems and strategies for reducing it*. American Water Works Association.

Azira, S., Rahim, A., Noor Hayati, I., Nurul Syala, A. L., Nurulhuda, A., Mohd Amir Shazwan, H., & Sangeetha, A. (2022). Historic Building Information Modelling (HBIM): The Application of Digitalization in Conserving Negeri Sembilan Traditional Malay House. *INTI JOURNAL, 2022*(10), 1–7.

Azis, I. A. A., Khaidzir, K. A. M., & Rashid, A. M. (2022). Identifying Guidelines Limitation Of Historic Interior In Adaptive Re-Use Of Heritage Building In Malaysia. *Journal of Tourism Hospitality and Environment Management, 7*(27), 69–76. DOI: 10.35631/JTHEM.727006

Azizi, N. Z. M., Razak, A. A., Din, M. A. M., & Nasir, N. M. (2016). Recurring Issues in Historic Building Conservation. *Procedia: Social and Behavioral Sciences, 222*, 587–595. DOI: 10.1016/j.sbspro.2016.05.217

Baccega, E., Bortoloni, M., Bottarelli, M., Dino, G. E., & Zannoni, G. (2022). Improving the air permeability of ventilated roofs. *Future Cities and Environment, 8*(1), 6. DOI: 10.5334/fce.144

Baharuddin, M. N., Bahardin, N. F., Harun, S. N., Abd Manap, M. D., & Ab Rashid, M. S. (2022). ASSESSING CRITICAL RISK FACTORS FOR HERITAGE CONSERVATION PROJECTS IN COMPLIANCE WITH NATIONAL HERITAGE ACT 2005 (ACT 645). *PLANNING MALAYSIA, 20.*

Baik, A. H. (2020). *Heritage building information modelling for implementing UNESCO procedures: Challenges, potentialities, and issues*. Routledge. DOI: 10.1201/9781003036548

Bailey, M. (2019). Snowball sampling in business oral history: Accessing and analyzing professional networks in the Australian property industry. *Enterprise and Society, 20*(1), 74–88. DOI: 10.1017/eso.2018.110

Bakker, J., & Wester, P. (2020). Water-sensitive urban design in Copenhagen: Strategies for climate adaptation. *Urban Climate, 34*, 100678. DOI: 10.1016/j.uclim.2020.100678

Banfi, F. (2021). The evolution of interactivity, immersion and interoperability in HBIM: Digital model uses, VR and AR for built cultural heritage. *ISPRS International Journal of Geo-Information, 10*(10), 685. DOI: 10.3390/ijgi10100685

Barber, B. (2013). *If Mayors Ruled the World: Dysfunctional Nations and Rising Cities*. Yale University Press.

Barbier, E. B., Hacker, S. D., Kennedy, C., Koch, E. W., & Silliman, B. R. (2019). The value of estuarine and coastal ecosystem services. *Ecological Monographs, 89*(3), e01333. DOI: 10.1002/ecm.1333

Barrios, S., Bertinelli, L., & Stribl, E. (2006). Climate Change and Rural-Urban Migration: The Case of Sub-Saharan Africa. *Journal of Urban Economics, 60*(3), 357–371. https://www.sciencedirect.com/science/article/abs/pii/S0094119006000398. DOI: 10.1016/j.jue.2006.04.005

Barros, E. M., Silva, S. R., & Oliveira, D. M. (2020). Advanced oxidation processes for the treatment of pharmaceuticals in wastewater: A review. *Journal of Environmental Management, 260*, 110085. DOI: 10.1016/j.jenvman.2020.110085

Bazeley, P., & Jackson, K. (2013). *Qualitative data analysis with NVivo* (2nd ed.). SAGE Publications.

Bello, S. A., Oyedele, L. O., Akinade, O. O., Bilal, M., Delgado, J. M. D., Akanbi, L. A., Ajayi, A. O., & Owolabi, H. A. (2021). Cloud computing in construction industry: Use cases, benefits and challenges. *Automation in Construction, 122*, 103441. DOI: 10.1016/j.autcon.2020.103441

Benedikt, M. (1991). *Cyberspace: First steps* (Benedikt, M., Ed.). MIT Press.

Beniwal, G., & Singhrova, A. (2022). A systematic literature review on IoT gateways. *Journal of King Saud University. Computer and Information Sciences, 34*(10), 9543–9563. DOI: 10.1016/j.jksuci.2021.11.007

Benny, V. (2022). Practices and Challenges of Household Plastic Waste Disposal: An Evaluation of Waste Management System in Kerala. *Orissa Journal of Commerce* 43(2), 27-44. https://ojcoca.org/volume43_issue2/OJC_2022_43_2_03.pdf

Berawi, M. A., Sunardi, A., & Ichsan, M. (2019). Chief-Screen 1.0 as the Internet of Things platform in project monitoring & controlling to improve project schedule performance. *Procedia Computer Science, 161*, 1249–1257. DOI: 10.1016/j.procs.2019.11.239

BERNAMA. (2024, June, 07). *Govt ready to adopt BIM for development projects - PM Anwar.* https://thesun.my/malaysia-news/govt-ready-to-adopt-bim-for-development-projects-pm-anwar-IC12546781

Bertelsen, O. W., & Vines, J. (2023). Virtual architecture: Designing for the metaverse. *International Journal of Architectural Computing, 21*(1), 45–60.

Biagini, C., Capone, P., Donato, V., & Facchini, N. (2016). Towards the BIM implementation for historical building restoration sites. *Automation in Construction, 71*, 74–86. DOI: 10.1016/j.autcon.2016.03.003

Biel, A., Mårtensson, F., & Gärling, T. (2017). Public awareness campaigns and their role in water conservation. *Journal of Environmental Psychology*, *51*, 68–75. DOI: 10.1016/j.jenvp.2017.04.003

Bielefeld, B. (2013). *Basics: Architectural design*. Birkhäuser.

Biňas, M. (2019). Advanced applications of IoT: low power. https://bletvaska.gitbooks.io/advanced-iot-applications/content/en/chapter-1/introduction.html

Björkqvist, S., Elser, C., & Thörne, E. (2017). Sweden's waste management system: A case study in circular economy. *Waste Management (New York, N.Y.)*, *59*, 1–12. DOI: 10.1016/j.wasman.2017.02.003

Blavier, C. L. S., Huerto-Cardenas, H. E., Aste, N., Del Pero, C., Leonforte, F., & Della Torre, S. (2023). Adaptive measures for preserving heritage buildings in the face of climate change: A review. *Building and Environment*, *245*, 110832. DOI: 10.1016/j.buildenv.2023.110832

Bortoloni, M., Bottarelli, M., & Piva, S. (2017). Summer thermal performance of ventilated roofs with tiled coverings. *Journal of Physics: Conference Series*, *796*, 012023. DOI: 10.1088/1742-6596/796/1/012023

Bottani, E., D'Ambrogio, A., & Marini, M. (2021). Digital transformation in the construction industry: A review of innovations and their impact on energy efficiency. *Sustainable Cities and Society*, *70*, 102928. DOI: 10.1016/j.scs.2021.102928

Bourdelais, P. (1987). Transitions from Agricultural to Industrial Societies: Some Introductory Remarks. In Schröder, W. H., & Jarausch, K. H. (Eds.), *Quantitative History of Society and Economy: Some International Studies* (pp. 113–119). Scripta Mercaturae Verl.

Braun, V., & Clarke, V. (2019). Reflecting on reflexive thematic analysis. *Qualitative Research in Psychology*, *16*(2), 297–309.

Braun, V., & Clarke, V. (2019). Reflecting on reflexive thematic analysis. *Qualitative Research in Sport, Exercise and Health*, *11*(4), 589–597. DOI: 10.1080/2159676X.2019.1628806

Braun, V., Clarke, V., Hayfield, N., & Terry, G. (2019). Thematic analysis. In Liamputtong, P. (Ed.), *Handbook of research methods in health social sciences* (pp. 843–860). Springer., DOI: 10.1007/978-981-10-5251-4_103

Brinkmann, S. (2018). The interview. In Denzin, N. K., & Lincoln, Y. S. (Eds.), *The SAGE handbook of qualitative research* (5th ed., pp. 576–599). SAGE.

Brown, A., & Dixon, M. (2017). Design thinking for social equity. *Journal of Architectural Education*.

Browne, J., & Russell, S. (2003). The significance of sample size in qualitative research. *Qualitative Inquiry*, *9*(2), 129–148.

Brown, S., Leonard, P., & Saunders, R. (2012). *Urban agriculture and community food security in the United States: Farming from the city center to the urban fringe*. Sustainable Urban Agriculture Coalition.

Brown, T. (2008). Design thinking. *Harvard Business Review*. PMID: 18605031

Bruno, N., & Roncella, R. (2019). HBIM for conservation: A new proposal for information modeling. *Remote Sensing (Basel)*, *11*(15), 1751. DOI: 10.3390/rs11151751

Bruno, S., De Fino, M., & Fatiguso, F. (2018). Historic Building Information Modelling: Performance assessment for diagnosis-aided information modelling and management. *Automation in Construction*, *86*, 256–276. DOI: 10.1016/j.autcon.2017.11.009

Bullen, P. A., & Love, P. E. D. (2019). Adaptive reuse of heritage buildings: Sustainability, and social and economic considerations. *International Journal of Sustainable Built Environment*, *8*(1), 87–95.

Calati, M., Doretti, L., Gatto, C., & Mancin, S. (2021). 3D CFD Simulation of a New Ventilated Roof. *International High Performance Buildings Conference*, Purdue, Paper 358.

Canter, D. (1977). *The psychology of place*. The Architectural Press.

Cao, D., Li, H., & Wang, G. (2017). Impacts of building information modeling (BIM) implementation on design and construction performance: A resource dependence theory perspective. *Frontiers of Engineering Management*, *4*(1), 20–34. DOI: 10.15302/J-FEM-2017010

Capone, M., & Lanzara, E. (2019). Scan-to-BIM vs 3D ideal model HBIM: Parametric tools to study domes geometry. *The International Archives of the Photogrammetry, Remote Sensing and Spatial Information Sciences*, *42*(W9), 219–226. DOI: 10.5194/isprs-archives-XLII-2-W9-219-2019

Carvalho, J. P., Bragança, L., & Mateus, R. (2019). Optimising building sustainability assessment using BIM. *Automation in Construction*, *102*, 170–182. DOI: 10.1016/j.autcon.2019.02.021

Cascone, S. (2019). Green roof design: State of the art on technology and materials. *Sustainability (Basel)*, *11*(11), 3020. DOI: 10.3390/su11113020

Castleberry, A., & Nolen, A. (2018). Thematic analysis of qualitative research data: Is it as easy as it sounds? *Currents in Pharmacy Teaching & Learning, 10*(6), 807–815. DOI: 10.1016/j.cptl.2018.03.019 PMID: 30025784

Cavalett, O., Lima, D., & Silva, M. A. (2021). Urban agriculture and composting: A sustainable approach to waste management in community gardens. *Environmental Science & Policy, 113*, 32–43. DOI: 10.1016/j.envsci.2020.08.004

Cavallaro, C., Borsellino, V., & Gueli, A. (2023). Digital twin technology and its application in adaptive reuse of historic buildings. *Journal of Cultural Heritage, 54*, 191–203. DOI: 10.1016/j.culher.2023.04.004

Celoza, A., Leite, F., & de Oliveira, D. P. (2021). Impact of BIM-related contract factors on project performance. *Journal of Legal Affairs and Dispute Resolution in Engineering and Construction, 13*(3), 04521011. DOI: 10.1061/(ASCE)LA.1943-4170.0000478

Chai, C., Chan, S., Xiong, L., Lim, B., & Shan, J. (2023) Digital tools adoption towards construction industry revolution. *Journal of Engineering Science and Technology, 11(2022),* 231-243. https://jestec.taylors.edu.my/Special%20Issue%20THINK%20SPACE%202022/STAAUH%202022_19.pdf

Chakrabarti, P. G. D. (2001). Urban Crisis in India: New Initiatives for Sustainable Cities. *Development in Practice, 11*(2-3), 260–272. https://ideas.repec.org/a/taf/cdipxx/v11y2001i2-3p260-272.html. DOI: 10.1080/09614520120056397

Chanal, P. M., & Kakkasageri, M. S. (2020). Security and privacy in IoT: A survey. *Wireless Personal Communications, 115*(2), 1667–1693. DOI: 10.1007/s11277-020-07649-9

Changeworks, E. W. H. (2008). Energy Heritage. A Guide to Improving Energy Efficiency in Traditional and Historic Homes.

Charlton, J., Kelly, K., Greenwood, D., & Moreton, L. (2021). The complexities of managing historic buildings with BIM. *Engineering, Construction, and Architectural Management, 28*(2), 570–583. DOI: 10.1108/ECAM-11-2019-0621

Charmaz, K. (2014). *Constructing grounded theory* (2nd ed.). SAGE Publications.

Chathukulam, J. (2024). Climate-Resilient Governance: Nurturing Eco-friendly Public Services at Grassroots in Kerala, India. *Unisia* 42(2), art.21. https://journal.uii.ac.id/Unisia/article/view/37651

Cheng, L., & Tan, H. (2022). Integrating renewable energy into historic buildings: Case studies and best practices. *Heritage Management, 10*(2), 112–127.

Chen, J.-H., Ha, N. T. T., Tai, H.-W., & Chang, C.-A. (2020). The willingness to adopt the Internet of Things (IoT) conception in Taiwan's construction industry. *Journal of Civil Engineering and Management*, *26*(6), 534–550. DOI: 10.3846/jcem.2020.12639

Chen, M., Zhao, W., & Wang, X. (2020). Enhancing urban recycling through advanced sorting technologies: Lessons from San Francisco. *Journal of Environmental Management*, *259*, 110057. DOI: 10.1016/j.jenvman.2019.110057

Chen, X., Huang, H., Liu, Y., Li, J., & Liu, M. (2022). Robot for automatic waste sorting on construction sites. *Automation in Construction*, *141*, 104387. DOI: 10.1016/j.autcon.2022.104387

Cherail, K. (1994). Haul Your own Garbage. *Down to Earth*, *3*(8), 10. https://www.downtoearth.org.in/environment/haul-your-own-garbage-32442

Chougan, M., Al-Kheetan, M. J., & Ghaffar, S. H. (2023). Additive Manufacturing and the Construction Industry. In *Disrupting Buildings* (pp. 97–109). Palgrave Macmillan., DOI: 10.1007/978-3-031-32309-6_7

Chowdhury, P. K. R., Weaver, J. E., & Bhaduri, B. L. (2019). Electricity Consumption Patterns Within Cities: Application of a Data-Driven Settlement Characterization Method. *International Journal of Digital Earth*, *13*(1), 119–135. https://www.tandfonline.com/doi/full/10.1080/17538947.2018.1556355. DOI: 10.1080/17538947.2018.1556355

Ciampi, M., Leccese, F., & Tuoni, G. (2005). Energy analysis of ventilated and micro-ventilated roofs. *Solar Energy*, *79*(2), 183–192. DOI: 10.1016/j.solener.2004.08.014

CIDB. (2015). *Construction Industry Transformation Programme 2016-2020. Construction Industry Development Board*. CIDB.

CIDB. (2021). *Construction 4.0 Strategic Plan 2021-2025. Construction Industry Development Board*. CIDB.

Clarke, V., & Braun, V. (2017). Thematic analysis. *The Journal of Positive Psychology*, *12*(3), 297–298. DOI: 10.1080/17439760.2016.1262613

Cochran, L. T., & O'Kane, J. M. (1977). Urbanization-Industrialization and the Theory of Demographic Transition. *Pacific Sociological Review* 20(1), 113.134. https://www.jstor.org/stable/1388908

Corbin, J., & Strauss, A. (2008). *Basics of qualitative research: Techniques and procedures for developing grounded theory* (3rd ed.). SAGE Publications. DOI: 10.4135/9781452230153

Cousins, S. (2018, May, 8). *How 3D building scanning transforms construction projects*. https://www.navvis.com/blog/54-how-3d-scanning-is-being-used-to-transform-construction-projects

Creswell, J. W., & Poth, C. N. (2018). *Qualitative Inquiry and Research Design: Choosing Among Five Approaches*. Sage Publications.

da Cruz, N. F., Rode, P., & McQuarrie, M. (2019). New Urban Governance: A Review of Current Themes and Future Priorities. *Journal of Urban Affairs*, *41*(1), 1–19. https://www.tandfonline.com/doi/full/10.1080/07352166.2018.1499416. DOI: 10.1080/07352166.2018.1499416

da Marta, D., Deichmann, U., & Wang, H. G. (2007). Determinants of City Growth in Brazil. *Journal of Urban Economics*, *62*(2), 252–272. https://ideas.repec.org/a/eee/juecon/v62y2007i2p252-272.html. DOI: 10.1016/j.jue.2006.08.010

Dabaieh, M., Wanas, O., Hegazy, M. A., & Johansson, E. (2015). Reducing cooling demands in a hot dry climate: A simulation study for non-insulated passive cool roof thermal performance in residential buildings. *Energy and Building*, *89*, 142–152. DOI: 10.1016/j.enbuild.2014.12.034

Davey, K. (1993). *Elements of Urban Management*. World Bank. DOI: 10.1596/0-8213-2424-1

David, L. O., Nwulu, N. I., Aigbavboa, C. O., & Adepoju, O. O. (2022). Integrating Fourth Industrial Revolution (4IR) Technologies into the Water, Energy, and Food Nexus for Sustainable Resource Security. *Clean Technologies and Environmental Policy*, *26*(3), 803–820. https://pure.uj.ac.za/en/publications/integrating-fourth-industrial-revolution-4ir-technologies-into-th. DOI: 10.1007/s10098-023-02651-8

Davis, J. C., & Henderson, J. V. (2003). Evidence on the Political Economy of the Urbanization Process. *Journal of Urban Economics*, *53*(1), 98–125. https://www.sciencedirect.com/science/article/abs/pii/S0094119002005041. DOI: 10.1016/S0094-1190(02)00504-1

de Vass, T., Shee, H., & Miah, S. (2021). IoT in supply chain management: opportunities and challenges for businesses in early industry 4.0 context. *Operations and Supply Chain Management: An International Journal*, 148–161. DOI: 10.31387/oscm0450293

DeJonckheere, M., & Vaughn, L. M. (2019). Semistructured interviewing in primary care research: A balance of relationship and rigor. *Family Medicine and Community Health*, *7*(2), e000057. DOI: 10.1136/fmch-2018-000057 PMID: 32148704

Deng, F., & Smyth, H. (2014). Nature of firm performance in construction. *Journal of Construction Engineering and Management, 140*(2), 04013040. DOI: 10.1061/(ASCE)CO.1943-7862.0000778

Designboom. (2023). *Zaha Hadid explores the future of architecture in the metaverse with Metrotopia venture.* Retrieved from https://www.designboom.com/architecture/zaha-hadid-future-architecture-metaverse-metrotopia-venture-06-01-2023/

Designingbuildings. (2024), Blue roof. Retrieved October 15, 2024, from https://www.designingbuildings.co.uk

Dewi, C. P., & Paramitha, P. A. (2021). An analysis of the efficiency of green roofs on cooling energy demand in residential building. *Proceedings of the 3rd International Conference of Green Civil and Environmental Engineering (GCEE 2021)*, Malang, Indonesia, n.d DOI: 10.1063/5.0072616

Di Biccari, C., Calcerano, F., D'Uffizi, F., Esposito, A., Campari, M., & Gigliarelli, E. (2022). Building information modeling and building performance simulation interoperability: State-of-the-art and trends in current literature. *Advanced Engineering Informatics, 54*, 101753. DOI: 10.1016/j.aei.2022.101753

Diaper, C., Williams, K., & Jackson, T. (2020). Greywater recycling: A review of technologies, benefits, and challenges in urban areas. *Journal of Environmental Management, 259*, 109613. DOI: 10.1016/j.jenvman.2019.109613

Diara, F. (2022). HBIM open source: A review. *ISPRS International Journal of Geo-Information, 11*(9), 472. DOI: 10.3390/ijgi11090472

Dinesh, B. (2022, August, 9). *IoT is changing the construction industry*. https://www.datasciencecentral.com/iot-in-construction-changing-the-construction-industry/

DJMCADCoordination. (2021, August, 26). The ROI of 3D scanning: Is it worth it? [Online blog post]. https://caddjm.com/is-3d-scanning-worth-it/

Dore, C., & Murphy, M. (2012, September). Integration of Historic Building Information Modeling (HBIM) and 3D GIS for recording and managing cultural heritage sites. In 2012 18th International conference on virtual systems and multimedia (pp. 369-376). IEEE.

Dore, C., & Murphy, M. (2015). Historic building information modelling (HBIM). In *Handbook of research on emerging digital tools for architectural surveying, modeling, and representation* (pp. 233-273). IGI Global Scientific Publishing.

Doria, M., Tedesco, L., & Gagliardi, R. (2019). Overcoming barriers to community engagement in urban water management: Lessons from international case studies. *Water Policy, 21*(5), 1016–1034. DOI: 10.2166/wp.2019.103

Doria, M., Tedesco, L., & Gagliardi, R. (2019). The role of community-driven approaches in improving urban water management and waste recycling. *Environmental Sustainability, 11*(3), 217–230. DOI: 10.1016/j.envsus.2019.02.006

Dristy, N. F., Parvez, M., & Rahman, N. (2020). Potential measures of cool roof treatment for different building types in urban context of Bangladesh. *5th International Conference on Advances in Civil Engineering (ICACE-2020).* CUET, Chattogram, Bangladesh, n.d

Du, L., Feng, Y., Lu, W., Kong, L., & Yang, Z. (2020). Evolutionary game analysis of stakeholders' decision-making behaviours in construction and demolition waste management. *Environmental Impact Assessment Review, 84*, 106428. DOI: 10.1016/j.eiar.2020.106408

Dutta, A., & Jinsart, W. (2020). Waste Generation and Management Status in the Fast-Expanding Indian Cities: A Review. *Journal of the Air & Waste Management Association, 70*(5), 491–503. https://www.tandfonline.com/doi/pdf/10.1080/10962247.2020.1738285. DOI: 10.1080/10962247.2020.1738285 PMID: 32150518

Dutta, A., Sharma, A., & Patel, P. (2021). IoT-based smart waste management systems for urban sustainability. *Urban Climate, 37*, 100845. DOI: 10.1016/j.uclim.2021.100845

Eames, M., Dixon, T., Lannon, S., Hunt, M., Laurentis, C., Marvin, S., Hodson, M., Guthrie, P., Georgiadou, M. (2014). Retrofit 2050: Critical Challenges for Urban Transitions.

Ebodé, V. B. (2020). Variabilité Hydropluviométrique en Afrique Centrale Occidentale Forestière: Entre Analyse des Fluctuations Observées, Modélisation Hydrologique Prédictive et Recherche des Facteurs Explicatifs. Ph.D. Thesis, University of Yaounde 1, Yaoundé, Cameroon.

Egbelakin, T., & Wilkinson, S. (2008). Factors affecting motivation for improved seismic retrofit implementation. In *Australian Earthquake Engineering Conference (AEES).*

Elghaish, F., Hosseini, M. R., Matarneh, S., Talebi, S., Wu, S., Martek, I., Poshdar, M., & Ghodrati, N. (2021). Blockchain and the 'Internet of Things' for the construction industry: Research trends and opportunities. *Automation in Construction, 132*, 104144. DOI: 10.1016/j.autcon.2021.103942

Elghaish, F., Matarneh, S., Talebi, S., Kagioglou, M., Hosseini, M. R., & Abrishami, S. (2021). Toward digitalization in the construction industry with immersive and drones technologies: A critical literature review. *Smart and Sustainable Built Environment, 10*(3), 347–373. DOI: 10.1108/SASBE-06-2020-0077

Elnabawi, M. H., Alhumaidi, A., Osman, B., & Alshehhi, R. (2022). Cool roofs in hot climates: A conceptual review of modelling methods and limitations. *Buildings (Basel, Switzerland), 12*(11), 1968. DOI: 10.3390/buildings12111968

Elshafey, A., Saar, C. C., Aminudin, E. B., Gheisari, M., & Usmani, A. (2020). Technology acceptance model for Augmented Reality and Building Information Modeling integration in the construction industry. *Journal of Information Technology in Construction, 25*, 161–172. DOI: 10.36680/j.itcon.2020.010

Ensafi, M., Thabet, W., Devito, S., & Lewis, A. (2021). Field testing of mixed reality (MR) technologies for quality control of as-built models at project handover: A case study. *EPiC Series in Built Environment, 2*, 246–254. DOI: 10.29007/4wrj

EPA. (2021). WaterSense: A guide to water-efficient plumbing products. U.S. Environmental Protection Agency. https://www.epa.gov/watersense

Espinosa-Fernández, A., Echarri-Iribarren, V., & Sáez, C. A. (2020). Water-covered roof versus inverted flat roof on the mediterranean coast: A comparative study of thermal and energy behavior. *Applied Sciences (Basel, Switzerland), 10*(7), 2288. DOI: 10.3390/app10072288

Evariste, F. F., Jean, S. D., Victor, K., & Claudia, M. (2018). Assessing climate change vulnerability and local adaptation strategies in adjacent communities of the Kribi-Campo coastal ecosystems, South Cameroon. *Urban Climate, 24*, 1037–1051. DOI: 10.1016/j.uclim.2017.12.007

Ewing, R., & Thomas, R. (2021). Architectural innovation in augmented and virtual realities. *Journal of Digital Architecture and Design, 12*(3), 78–92.

Faccio, M., Persona, A., & Zanin, G. (2011). Waste collection multi objective model with real time traceability data. *Waste Management (New York, N.Y.), 31*(12), 2391–2405. DOI: 10.1016/j.wasman.2011.07.005 PMID: 21821406

Falahat, M. (2006). Sense of place and the factors shaping it. *HONAR-HA-YE-ZIBA, 26*, 57–66.

Fan, Y. V., Jiang, P., Hemzal, M., & Klemeš, J. J. (2021). An update of COVID-19 influence on waste management. *The Science of the Total Environment, 754*, 142014. DOI: 10.1016/j.scitotenv.2020.142014 PMID: 32920389

Farhadi, F. (2023, July, 13). *Block chain in Construction: Ultimate Guide for 2024*. https://neuroject.com/blockchain-in-construction/

Feka, N. Z., & Ajonina, G. N. (2011). Drivers causing decline of mangrove in West-Central Africa: A review. *The International Journal of Biodiversity Science, Ecosystem Services & Management*, *7*(3), 217–230. DOI: 10.1080/21513732.2011.634436

Feng, H., Zhou, J., & Li, H. (2019). Membrane bioreactor (MBR) technology for wastewater treatment: Recent advances and future directions. *Environmental Technology*, *40*(8), 1056–1069. DOI: 10.1080/09593330.2018.1564107

Fernandez, M. A. S. (2019). *Implementation of smart devices in the construction industry* [Faculty of Science and Engineering, University of Wolverhampton]. https://wlv.openrepository.com/bitstream/handle/2436/622965/Silverio_Fernandez_PhD_Thesis.pdf?sequence=1&isAllowed=y

Fields, G. (1999). City Systems, Urban History, and Economic Modernity: Urbanization and the Transition from Agrarian to Industrial Society. *Berkeley Planning Journal*, *13*, 102–128. https://escholarship.org/uc/item/7242m3b1

Finlay, L. (2017). Engaging with the lived body: Reflections on the embodied nature of research. In *The Sage Handbook of Qualitative Research* (pp. 202–221). Sage Publications.

Fisher, P. C., Blanchard, D. L., & McCormick, R. A. (2019). The role of extended producer responsibility in reducing municipal waste: Global perspectives and lessons learned. *Waste Management & Research*, *37*(6), 557–568. DOI: 10.1177/0734242X19859448

Flanagan, M. (2009). *Critical play: Radical game design*. MIT Press. DOI: 10.7551/mitpress/7678.001.0001

Fletcher, T. D., Shuster, W. D., & Gersonius, B. (2017). *Water sensitive urban design: A case study approach*. Springer.

Fletcher, T. D., Shuster, W., Hunt, W. F., Ashley, R. M., Butler, D., Arthur, S., & Uhl, M. (2015). SUDS, LID, BMPs, WSUD and more–The evolution and application of terminology surrounding urban drainage. *Urban Water Journal*, *12*(7), 525–542. DOI: 10.1080/1573062X.2014.916314

Flick, U. (2014). *An introduction to qualitative research* (5th ed.). SAGE Publications.

Fosse, R., Ballard, G., & Fischer, M. (2017). Virtual Design and Construction: Aligning BIM and Lean in Practice, *25th Annual Conference of the International Group for Lean Construction*, 499-506. DOI: 10.24928/2017/0159

Friedler, E. (2004). Quality of individual domestic greywater streams and its implications for on-site treatment and reuse possibilities. *Environmental Technology*, *25*(9), 997–1008. DOI: 10.1080/09593330.2004.9619393 PMID: 15515266

Friese, S. (2019). *Qualitative data analysis with ATLAS.ti* (3rd ed.). SAGE Publications.

Fu, Y. (2020). *Study on suitability evaluation system of low energy consumption technologies for rural housing envelope in southern Jiangsu province.* Doctoral dissertation, Southeast University, Nanjing, China.

Gagliano, A., Detommaso, M., Nocera, F., & Evola, G. (2015). A multi-criteria methodology for comparing the energy and environmental behavior of cool, green and traditional roofs. *Building and Environment*, *90*, 71–81. DOI: 10.1016/j.buildenv.2015.02.043

Gamil, Y., Abdullah, A., Abd, M., Rahman, I., & Asad, M. M. (2020). Internet of things in construction industry revolution 4.0. *Journal of Engineering. Defence Technology*, *18*(5), 1091–1102. DOI: 10.1108/JEDT-06-2019-0164

Gamil, Y., & Rahman, I. A. R. (2019). Awareness and challenges of building information modelling (BIM) implementation in the Yemen construction industry. *Journal of Engineering. Defence Technology*, *17*(5), 1077–1084.

Gan, V. J. L., Wong, H. K., Tse, K. T., Cheng, J. C. P., Lo, I. M. C., & Chan, C. M. (2019). Simulation-based evolutionary optimization for energy-efficient layout plan design of high-rise residential buildings. *Journal of Cleaner Production*, *231*, 1375–1388. DOI: 10.1016/j.jclepro.2019.05.324

García, A., & López, R. (2020). Circular economy in urban waste management: Opportunities and challenges. *Waste Management (New York, N.Y.)*, *102*, 349–359. DOI: 10.1016/j.wasman.2020.01.045

García, D., Melero, L., & Martorell, I. (2022). Data-driven sustainability: The role of big data analytics in the renovation of historical buildings. *Energy and Building*, *267*, 112145. DOI: 10.1016/j.enbuild.2022.112145

García-Valldecabres, J., Galiano-Garrigós, A., Meseguer, L. C., & González, M. C. L. (2021). HBIM work methodology applied to preventive maintenance: A state-of-the-art review. *Building Information Modelling (BIM) in Design. Construction and Operations IV*, *205*, 157.

Garg, B. S., & Raut, A. V. (2015). Adarsh Gram: A Gandhian Dream of Gram Swaraj. *Indian Journal of Community Medicine : Official Publication of Indian Association of Preventive & Social Medicine*, *40*(1), 1–4. https://pmc.ncbi.nlm.nih.gov/articles/PMC4317974/. DOI: 10.4103/0970-0218.149260 PMID: 25657504

Gbadamosi, A. Q., Oyedele, L., Mahamadu, A. M., Kusimo, H., & Olawale, O. (2019). *The role of internet of things in* delivering *smart construction* [Paper presentation]. CIB world building congress. Hong Kong, China.

Gbadamosi, A.-Q., Oyedele, L., Mahamadu, A.-M., Kusimo, H., & Olawale, O. (2019). The role of Internet of Things in delivering smart construction. In CIB World Building Congress. https://uwe-repository.worktribe.com/preview/1492592/Gbadamosi%20et

Gerber, A., & Götz, U. (2019). *Architectonics of game spaces*. Transcript.

Getter, K. L., & Rowe, D. B. (2021). *Green roofs: A comprehensive guide to the planning, design, and construction of green roofs*. John Wiley & Sons.

Ghosh, A., Edwards, D. J., & Hosseini, M. R. (2020). Patterns and trends in Internet of Things (IoT) research: Future applications in the construction industry. *Engineering, Construction, and Architectural Management*, *28*(2), 477–501. DOI: 10.1108/ECAM-04-2020-0271

Gibberd, J. (2020). Green building technologies. In Yang, P. (Ed.), (pp. 482–510). Practice, Progress, and Proficiency in Sustainability.

Gigliarelli, E., Calcerano, F., & Cessari, L. (2017). Heritage Bim, numerical simulation and decision support systems: An integrated approach for historical buildings retrofit. *Energy Procedia*, *133*, 135–144. DOI: 10.1016/j.egypro.2017.09.379

Gigliarelli, E., Calcerano, F., Martinelli, L., Artopoulos, G., Thravalou, S., & Alexandrou, K. (2022). *Methodology for the energy renovation of heritage buildings using BIM. Guidelines for the development of an Energy Efficient Heritage Building Information Model (EE-HBIM)* (Gigliarelli, E., Calcerano, F., & Martinelli, L., Eds.). Consiglio Nazionale delle Ricerche.

Gillis, A. S. (2022, March). What is the Internet of Things (IoT)? Tech Target. https://www.techtarget.com/iotagenda/definition/Internet-of-Things-IoT

Gill, S. E., Handley, J. F., Ennos, A. R., & Pauleit, S. (2007). Adapting cities for climate change: The role of the green infrastructure. *Built Environment*, *33*(1), 115–133. DOI: 10.2148/benv.33.1.115

Glaeser, E. L. (2014). A World of Cities: The Causes and Consequences of Urbanization in Poorer Countries. *Journal of the European Economic Association, 12*(5), 1154–1199. https://academic.oup.com/jeea/article-abstract/12/5/1154/2319678?redirectedFrom=fulltext. DOI: 10.1111/jeea.12100

Gleick, P. H. (2003). Water use. *Annual Review of Environment and Resources, 28*(1), 275–314. DOI: 10.1146/annurev.energy.28.040202.122849

Gleick, P. H., & Cooley, H. (2019). Water conservation and drought resilience in San Francisco: Policies and outcomes. *Water Resources Research, 55*(9), 7539–7550. DOI: 10.1029/2019WR024585

GoCodes. (2022). *7 Positive impacts of construction technology on businesses.* https://gocodes.com/positive-impacts-construction-technology/?srsltid=AfmBOoqED4nehztCrB2DXHk8mZoCIqYfKe3hlZlZlo1pdOZGXXcYQRDb

Goodhue, D. L. (1998). Development and measurement validity of a task-technology fit instrument for user evaluations of information system. *Decision Sciences, 29*(1), 105–138. DOI: 10.1111/j.1540-5915.1998.tb01346.x

Goodhue, D. L., Klein, B. D., & March, S. T. (2000). User evaluations of IS as surrogates for objective performance. *Information & Management, 38*(2), 87–101. DOI: 10.1016/S0378-7206(00)00057-4

Goodhue, D. L., & Thompson, R. L. (1995). Task-technology fit and individual performance. *Management Information Systems Quarterly, 19*(2), 213–236. DOI: 10.2307/249689

Gore, C. D., & Gopakumar, G. (2015). Infrastructure and Metropolitan Reorganization: An Exploration of the Relationship in Africa and India. *Journal of Urban Affairs, 37*(5), 548–567. https://www.tandfonline.com/doi/full/10.1111/juaf.12180. DOI: 10.1111/juaf.12180

Górecki, J., & Núñez-Cacho, P. (2022). Decision-making problems in construction projects executed under the principles of sustainable development—Bridge construction case. *Applied Sciences (Basel, Switzerland), 12*(12), 6132. Advance online publication. DOI: 10.3390/app12126132

Gould, J., & Nissen-Peterson, E. (2019). Rainwater catchment systems for domestic water supply: Design, construction, and implementation. CRC Press.

Govt of Kerala. (2022). *Report of the Comptroller and Auditor General of India on Waste Management in Urban Local Bodies.* Govt of Kerala.

Graham, S., & Stinson, C. (2020). Leveraging digital technologies for community engagement in urban waste management. *Journal of Urban Technology*, *27*(4), 35–51. DOI: 10.1080/10630732.2020.1794825

Grant, C., & Osanloo, A. (2014). Understanding, selecting, and integrating a theoretical framework in dissertation research: Creating the blueprint for your "house". *Administrative issues journal, 4*(2), 4.

Guba, E. G., & Lincoln, Y. S. (2020). Fourth Generation Evaluation. Sage Publications. • Abidin, S. Z., & Hasan, N. A. (2020). Environmental Impact of Adaptive Reuse of Heritage Buildings in Malaysia. *Sustainability*, *12*(5), 1856.

Guest, G., Bunce, A., & Johnson, L. (2006). How many interviews are enough? An experiment with data saturation and variability. *Field Methods*, *18*(1), 59–82. DOI: 10.1177/1525822X05279903

Gupta, N., & Tiwari, G. (2016). Review of passive heating/cooling systems of buildings. *Energy Science & Engineering*, *4*(5), 4. DOI: 10.1002/ese3.129

Halim, U. L., & Tambi, N. (2021). Awareness of community on the conservation of heritage buildings in George Town, Penang. *Planning Malaysia, 19.*

Halim, M. I., Rusuli, M. S. C., & Yaziz, M. F. A. (2021). Attitudes, awareness, readiness and barriers towards the Internet of Things adoption among the construction industry in East Coast Malaysia. *Journal of Sustainable Management Studies*, *2*(1), 7–15. http://myscholar.umk.edu.my/bitstream/123456789/2573/1/Che%20Rusuli%20et%20al._Published.pdf

Hamdi, A. (2023). Discreet architecture in the metaverse. School of Architecture, Urban Planning and Construction Engineering Building and Architectural Engineering.

Hamzah, N., Mohamed, S., & Rahim, A. H. A. (2019). The financial viability of adaptive reuse projects in Malaysia. *Journal of Building Performance*, *10*(2), 25–34. DOI: 10.21894/jbp.2019.10.2.3

Han, D., Kalantari, M., & Rajabifard, A. (2021). Building information modeling (BIM) for construction and demolition waste management in Australia: A research agenda. *Sustainability (Basel)*, *13*(23), 12983. DOI: 10.3390/su132312983

hanthi R. (2023, July, 20). *Leveraging big data analytics for construction success.* https://www.contractormag.com/technology/article/21269911/leveraging-big-data-analytics-for-construction-success

Hao, J., Chen, Z., Zhang, Z., & Loehlein, G. (2021). Quantifying construction waste reduction through the application of prefabrication: A case study in Anhui, China. *Environmental Science and Pollution Research International*, *28*(19), 24499–24510. DOI: 10.1007/s11356-020-09026-2 PMID: 32358748

Harikrishnan, G. (2014). Solid Waste Management: A Comparative Study between Kerala and Tamil Nadu. *IJRSSH* 4(4), 29-37. https://www.ijrssh.com/admin/upload/05%20Harikrishnan.pdf

Harputlugil, T., & de Wilde, P. (2021). The interaction between humans and buildings for energy efficiency: A critical review. *Energy Research & Social Science*, *71*, 101828. DOI: 10.1016/j.erss.2020.101828

Harun, S. N. (2021). Pemuliharaan Bangunan Warisan. Dewan Bahasa dan Pustaka, ISBN 978-983-49-1955-9.

Hasan, R., Yusof, N., & Idris, M. (2022). The role of local narratives in adaptive reuse: Enhancing community acceptance in heritage projects. *Journal of Urban Heritage*, *24*(3), 227–241.

Hashim, S. R., Mohamed, M., & Azhar, H. (2022). The role of 3D modeling in heritage conservation: A case study of Malaysian historical sites. *International Journal of Heritage Studies*, *28*(3), 213–227.

Hassan, R., Qamar, F., Hasan, M. K., Aman, A. H. M., & Ahmed, A. S. (2020). Internet of Things and its applications: A comprehensive survey. *Symmetry*, *12*(10), 1674. DOI: 10.3390/sym12101674

Hassan, S. A., & Husain, S. I. (2020). The effect of green roof on improving the environmental efficiency for the residential buildings in Baghdad city. *Materials Science and Engineering*, •••, 928.

Hatt, B. E., Fletcher, T. D., & Deletic, A. (2016). Water-sensitive urban design: A review of its potential and practice in Melbourne, Australia. *Australian Journal of Water Resources*, *20*(3), 259–272. DOI: 10.1080/13241583.2016.1183684

Haughton, G. (1997). Developing Sustainable Urban Development Models. *Cities (London, England)*, *14*(4), 189–195. https://www.sciencedirect.com/science/article/abs/pii/S0264275197000024. DOI: 10.1016/S0264-2751(97)00002-4

Heaton, J., Parlikad, A. K., Owens, D., & Pawsey, N. (2019). BIM as an Enabler for Digital Transformation. In International Conference on Smart Infrastructure and Construction 2019 (ICSIC) Driving data-informed decision-making (pp. 49-54). ICE Publishing. DOI: 10.1680/icsic.64669.049

He, C., Liu, Z., Wu, J., Pan, X., Fang, Z., Li, J., & Bryan, B. (2021). Future Global Urban Water Scarcity and Potential Solutions. *Nature Communications*, *12*(1), 4667. https://www.nature.com/articles/s41467-021-25026-3. DOI: 10.1038/s41467-021-25026-3 PMID: 34344898

Henderson, J. V., Nigmatulina, D., & Kriticos, S. (2021). Measuring Urban Economic Density. *Journal of Urban Economics*, *125*, 103188. https://ideas.repec.org/a/eee/juecon/v125y2021ics0094119019300658.html. DOI: 10.1016/j.jue.2019.103188

Henze, A. (2022, February, 8). *Augmented reality in construction: Benefits and limitations*. https://www.teamviewer.com/en/insights/augmented-reality-in-construction/

Himmelstoss, E. A., Henderson, R. E., Kratzmann, M. G., & Farris, A. S. (2021). Digital Shoreline Analysis System (DSAS) Version 5.1 User Guide. US Geological Survey. https://pubs.usgs.gov/publication/ofr20211091

Historic England. (2015). Planning responsible retrofit of traditional buildings.

Historic England. (2024). New advice on adapting historic buildings for energy and carbon efficiency.

Historic Scotland. (2021). Guide to Energy Retrofit of Traditional Buildings. https://www.historicenvironment.scot/archives-and research/publications/publication/?publicationId=47c9f2eb-1ade-4a76-a775-add0008972f3

Hoekstra, A. Y., & Mekonnen, M. M. (2012). The water footprint of humanity. *Proceedings of the National Academy of Sciences of the United States of America*, *109*(9), 3233–3237. DOI: 10.1073/pnas.1109936109 PMID: 22331890

Hohenberg, P. M. (1990). *The City: Agent or Product of Urbanization. Urbanization in History*. Clarendon Press.

Holgate, S., & Baur, M. (2011). *A companion to Hegel*. Blackwell Publishing Ltd. DOI: 10.1002/9781444397161

Hollis, A., Johnson, R., & Zhuang, M. (2019). Green infrastructure in urban water management: Building resilience against climate change. *Environmental Science & Policy*, *104*, 66–74. DOI: 10.1016/j.envsci.2019.01.003

Holloway, I., & Wheeler, S. (2010). *Qualitative research in nursing and healthcare* (3rd ed.). Wiley-Blackwell.

Holzmann, V., & Lechiara, M. (2022). Artificial intelligence in construction Projects: An explorative study of professionals' expectations. *European Journal of Business Management and Research*, *7*(3), 151–162. DOI: 10.24018/ejbmr.2022.7.3.1432

Hong, C. C., & Mansor, N. S. (2023). Internet of Things (IoT): Real-time monitoring for decision making among the Malaysian contractors. *Journal of Advanced Research in Applied Sciences and Engineering Technology, 32*(3), 455–470. DOI: 10.37934/araset.32.3.455470

Hong, T., Yu, N., Mao, Z., & Zhang, S. (2021). Government-Driven Urbanization and its Impact on Regional Economic Growth in China. *Cities (London, England), 117*, 103299. https://www.sciencedirect.com/science/article/abs/pii/S0264275121001992?via%3Dihub. DOI: 10.1016/j.cities.2021.103299

Hopewell, J., Dvorak, R., & Kosior, E. (2009). Plastics recycling: Challenges and opportunities. *Philosophical Transactions of the Royal Society of London. Series B, Biological Sciences, 364*(1526), 2115–2126. DOI: 10.1098/rstb.2008.0311 PMID: 19528059

Hosagrahar, J. (2017). *Culture: Urban Future; Global Report on Culture for Sustainable Urban Development*. UNESCO.

Hou, J., & Li, B. (2020). The evolutionary game for collaborative innovation of the IoT industry under government leadership in China: An IoT infrastructure perspective. *Sustainability (Basel), 12*(9), 3750. DOI: 10.3390/su12093648

Htay, T., Røskaft, E., Ringsby, T. H., & Ranke, P. S. (2023). Spatio-temporal variation in avian taxonomic, functional, and phylogenetic diversity and its relevance for conservation in a wetland ecosystem in Myanmar. *Biodiversity and Conservation, 32*(8-9), 2841–2867. DOI: 10.1007/s10531-023-02632-9

Huang, B., Wang, X., Kua, H., Geng, Y., Bleischwitz, R., & Ren, J. (2018). Construction and demolition waste management in China through the 3R principle. *Resources, Conservation and Recycling, 129*, 36–44. DOI: 10.1016/j.resconrec.2017.09.029

Hull, J., & Bryan, P. (2019). *Bim for heritage: Developing the asset information model*. Historic England.

Hwang, B. G., & Bao Yeo, Z. (2011). Perception on benefits of construction waste management in the Singapore construction industry. *Engineering, Construction, and Architectural Management, 18*(4), 394–406. DOI: 10.1108/09699981111145835

Hwang, Y., & Park, K. S. (2020). Adaptive reuse as a sustainable practice for conserving cultural heritage: The case of the Bukchon Hanok Village in Seoul. *Sustainability, 12*(3), 1103.

Ibrahim, F. S., Esa, M., & Rahman, R. A. (2021). The adoption of IOT in the Malaysian construction industry: Towards construction 4.0. *International Journal of Sustainable Construction Engineering Technology, 12*(1). Advance online publication. DOI: 10.30880/ijscet.2021.12.01.006

Ibrahim, M., Rasli, N., & Ahmad, S. (2020). Material reuse in heritage restoration: Challenges and opportunities. *Built Environment Journal, 11*(3), 56–69.

Ibrahim, M., Salleh, M. F., & Kamaruzzaman, S. N. (2021). Energy efficiency interventions in heritage buildings: Challenges and opportunities. *Built Environment Journal, 12*(1), 35–50.

Ibrahim, R., & Rasli, M. A. (2019). Technical challenges in the adaptive reuse of heritage buildings in Malaysia. *Engineering Journal (New York), 23*(4), 215–228. DOI: 10.4186/ej.2019.23.4.215

Ibrahim, R., Yunos, M. Y. M., & Hamid, A. F. A. (2021). Community Impact of Adaptive Reuse: The Case of the Kuala Lumpur Railway Station. *Journal of Cultural Heritage Management and Sustainable Development, 11*(1), 47–63.

ICOMOS. (2014). The Venice Charter: International Charter for the Conservation and Restoration of Monuments and Sites.

Idris, M. L., Rahman, A. R. A., Ishiyaku, B., & Alias, A. R. (2024). Preserving heritage buildings with building information modelling: Addressing the challenges and opportunities. *Smart and Green Materials, 1*(1), 33–43. DOI: 10.70028/sgm.v1i1.5

IEA. (2017). Towards a Zero-Emission, Efficient, and Resilient Buildings and Construction Sector [Global Status Report.]. *UN Environment and International Energy Agency, 2017*, 2017.

Iffa, E., & Tariku, F. (2016). A review of heat and moisture performance of attic roofs. *Applied Sciences (Basel, Switzerland), 2*(d).

Imenda, S. (2014). Is there a conceptual difference between theoretical and conceptual frameworks. *Journal of Social Sciences, 38*(2), 185–195. DOI: 10.1080/09718923.2014.11893249

Ingebretsen, S., Andenæs, E., & Kvande, T. (2022). Microclimate of air cavities in ventilated roof and façade systems in nordic climates. *Buildings (Basel, Switzerland), 12*(5), 683. DOI: 10.3390/buildings12050683

IPCC. (2022). Climate change 2022: Impacts, adaptation, and vulnerability. Intergovernmental Panel on Climate Change. https://www.ipcc.ch/report/ar6/wg2/

Ismail, A. (2013). *Cooling potential of passive roof treatments on the concrete flat roof in Penang, Malaysia*. Ph.D. Thesis. Universiti Sains Malaysia, Penang, Malaysia.

Ismail, E. D., Said, S. Y., Jalil, M. K. A., & Ismail, N. A. A. (2021). Benefits and Challenges of Heritage Building Information Modelling Application in Malaysia. *Environment-Behaviour Proceedings Journal, 6*(SI4), 179–184. DOI: 10.21834/ebpj.v6iSI4.2917

Ismail, S., Hon, C. K. H., Crowther, P., Skitmore, M., & Lamari, F. (2023). The drivers and challenges of adopting the Malaysia industrialised building system for sustainable infrastructure development. *Construction Innovation, 23*(5), 1054–1074. DOI: 10.1108/CI-05-2021-0088

Iyamu, T. (2013). Underpinning theories: Order-of-use in information systems research. *Journal of Systems and Information Technology, 15*(3), 224–238. DOI: 10.1108/JSIT-11-2012-0064

Jabbour, C., Rey-Valette, H., Maurel, P., & Salles, J. M. (2019). Spatial data infrastructure management: A two-sided market approach for strategic reflections. *International Journal of Information Management, 45*, 69–82. DOI: 10.1016/j.ijinfomgt.2018.10.022

Jadaa, D. A., Raed, A. A., & Taleb, H. (2019). Assessing the thermal effectiveness of implementing green roofs in the urban Neighborhood. *Jordan Journal of Mechanical and Industrial Engineering, 13*(3), 161–174.

Jahanfar, A., Mohamed, O., & Zayed, T. (2021). Multi-criteria decision-making in construction: Analytic Hierarchy Process applications and perspectives. *Journal of Construction Engineering and Management, 147*(9), 04021105. DOI: 10.1061/(ASCE)CO.1943-7862.0002117

Jain, S., Singhal, S., Jain, N. J., & Bhaskar, K. (2021). Construction and demolition waste recycling: Investigating the role of theory of planned behavior, institutional pressures and environmental consciousness. *Journal of Cleaner Production, 263*, 121405. DOI: 10.1016/j.jclepro.2020.121405

Jamal, K. A. A., Harun, S. N., & Said, S. Y. (2024). A Review and Comparative Study of Retrofit for the Energy Efficiency of Malaysian Heritage Buildings. *Insights on Resiliency and Urban Development*, 42-58.

Jamal, K. A. A., Mohammad, M. F., Hashim, N., Mohamed, M. R., & Ramli, M. A. (2019). Challenges of Building Information Modelling (BIM) from the Malaysian Architect's Perspective. In MATEC Web of Conferences (Vol. 266, p. 05003). EDP Sciences.

Jamaludin, M., & Yahya, W. K. (2019). Educational Aspects of Adaptive Reuse Projects in Malaysia. *Procedia: Social and Behavioral Sciences*, *234*, 405–413.

Jambeck, J. R., Geyer, R., Wilcox, C., Siegler, T. R., Perryman, M., Andrady, A., Narayan, R., & Law, K. L. (2015). Plastic waste inputs from land into the ocean. *Science*, *347*(6223), 768–771. DOI: 10.1126/science.1260352 PMID: 25678662

Jamshed, S. (2020). Qualitative research method-interviewing and observation. *Journal of Basic and Clinical Pharmacy*, *9*(2), 87–88. PMID: 25316987

Janssen, M., Luthra, S., Mangla, S., Rana, N. P., & Dwivedi, Y. K. (2019). Challenges for adopting and implementing IoT in smart cities. *Internet Research*, *29*(6), 1589–1616. DOI: 10.1108/INTR-06-2018-0252

Jia, M., Komeily, A., Wang, Y., & Srinivasan, R. S. (2019). Adopting Internet of Things for the development of smart buildings: A review of enabling technologies and applications. *Automation in Construction*, *101*, 111–126. DOI: 10.1016/j.autcon.2019.01.023

Jiancheng, S., Saar, C. C., & Aminudin, E. (2023). Digital Twinning in the Malaysian Construction Industry. In *Digitalization in Construction* (pp. 271–293). Routledge. DOI: 10.1201/9781003408949-15

Jiménez, B., & Maestu, J. (2020). Public-private partnerships in water management: Enhancing sustainability through collaborative infrastructure. *Water Resources Management*, *34*(7), 2291–2302. DOI: 10.1007/s11269-020-02537-4

Johansson-Sköldberg, U., Woodilla, J., & Çetinkaya, M. (2013). Design thinking: Past, present and possible futures. *Creativity and Innovation Management*, *22*(2), 121–146. DOI: 10.1111/caim.12023

Jokilehto, J. (2017). *A History of Architectural Conservation*. Routledge. DOI: 10.4324/9781315636931

Jouan, P., & Hallot, P. (2019). Digital twin: A HBIM-based methodology to support preventive conservation of historic assets through heritage significance awareness. *The International Archives of the Photogrammetry, Remote Sensing and Spatial Information Sciences*, *42*(W15), 609–615. DOI: 10.5194/isprs-archives-XLII-2-W15-609-2019

Judd, S. (2011). *The MBR book: Principles and applications of membrane bioreactors for water and wastewater treatment*. Elsevier.

Jusić, S., Hadžić, E., & Milišić, H. (2019). Stormwater management by green roof. *Acta Scientific Agriculture*, *3*(7), 57–62. DOI: 10.31080/ASAG.2019.03.0516

Kalay, Y. E., & Marx, J. (2005). Architecture and the Internet: Designing places in cyberspace. *First Monday*. Advance online publication. DOI: 10.5210/fm.v0i0.1563

Kallio, H., Pietilä, A. M., Johnson, M., & Kangasniemi, M. (2016). Systematic methodological review: Developing a framework for a qualitative semi-structured interview guide. *Journal of Advanced Nursing*, *72*(12), 2954–2965. DOI: 10.1111/jan.13031 PMID: 27221824

Kallis, G., Varela-Ortega, C., & González-Gómez, T. (2020). Coastal water management under climate change: Integrated approaches and strategies for adaptation. *Global Environmental Change*, *64*, 102147. DOI: 10.1016/j.gloenvcha.2020.102147

Kamaruzzaman, S. N., Salleh, N. A., & Mahayuddin, S. A. (2019). Energy efficiency in the adaptive reuse of heritage buildings in Malaysia. *Energy and Building*, *183*, 284–295. DOI: 10.1016/j.enbuild.2018.11.039

Kamaruzzaman, S. N., Salleh, N. A., & Mahayuddin, S. A. (2019). The Impact of Adaptive Reuse on the Conservation of Heritage Buildings in Malaysia. *Journal of Building Appraisal*, *15*(1), 63–74.

Kang, K., Besklubova, S., Dai, Y., & Zhong, R. Y. (2022). Building demolition waste management through smart BIM: A case study in Hong Kong. *Waste Management (New York, N.Y.)*, *143*, 69–83. DOI: 10.1016/j.wasman.2022.02.027 PMID: 35240449

Kang, K.-D. (2022). A review of efficient real-time decision making in the Internet of Things. *Technologies*, *10*(1), 12. DOI: 10.3390/technologies10010012

Kant, I. (1781). Critique of pure reason (P. Guyer & A. Wood, Trans.). Cambridge University Press, 1998.

Kar, S. (2020). Smart water metering and real-time monitoring systems for water management. *Journal of Water Supply: Research & Technology - Aqua*, *69*(3), 168–179. DOI: 10.2166/aqua.2020.145

Kavinesh, R., Khairul Najmy, A. R., Alawiyah, A. W., Siti Julia, R., Hasliza, A. R., Lee, Y. S., Mohd Hafizi, O., & Khairul Affendi, R. (2023). Design of A 20-bit chipless RFID tag utilizing multiple resonators in UWB frequency range. *Journal of Advanced Research in Applied Sciences and Engineering Technology*, *29*(2), 25–38. DOI: 10.37934/araset.29.2.2537

Kayan, B. (2016). Maintenance management practices for heritage buildings in Malaysia: Addressing the gaps. *Journal of Building Appraisal*, *12*(4), 305–316.

Kaza, S., Yao, L., Bhada-Tata, P., & Van Woerden, F. (2018). What a waste 2.0: A global snapshot of solid waste management to 2050. World Bank Group. https://openknowledge.worldbank.org/handle/10986/30317

Kaza, S., Yao, L., Bhada-Tata, P., & Van Woerden, F. (2018). *What a waste 2.0: A global snapshot of solid waste management to 2050*. World Bank Group. DOI: 10.1596/978-1-4648-1329-0

Kazmin, A. (2016). India's Population Shift Spaks a Rethink of Rural Strategy. *Financial Times*. https://www.ft.com/content/950a3f1c-ef47-11e5-9f20-c3a047354386

Kermanshahi, E. K., Tahir, M. B. M., Lim, N. H. A. S., Balasbaneh, A. T., & Roshanghalb, S. (2020, April). Awareness of Building Information Modeling Implementation in Retrofitting the Existing Residential Building in Malaysia. []. IOP Publishing.]. *IOP Conference Series. Earth and Environmental Science*, *476*(1), 012007. DOI: 10.1088/1755-1315/476/1/012007

Kevin, S. T. W., Shidiki, A. A., & Tchamba, M. N. (2022). Coastal risk management in a context of climate change: A case study of kribi town of the south region of Cameroon. *Journal of Geoscience and Environment Protection*, *10*(5), 111–124. DOI: 10.4236/gep.2022.105009

Khaddaj, M., & Srour, I. (2016). Using BIM to retrofit existing buildings. *Procedia Engineering*, *145*, 1526–1533. DOI: 10.1016/j.proeng.2016.04.192

Khalaf, R. W. (2021). World Heritage on the Move: Abandoning the Assessment of Authenticity to Meet the Challenges of the Twenty-First Century. *Heritage*, *4*(1), 371–386. DOI: 10.3390/heritage4010023

Khalid, A., Arshad, M., Anjum, M., Mahmood, T., & Dawson, L. (2011). The anaerobic digestion of solid organic waste. *Waste Management (New York, N.Y.)*, *31*(8), 1737–1744. DOI: 10.1016/j.wasman.2011.03.021 PMID: 21530224

Khalil, A., Stravoravdis, S., & Backes, D. (2021). Categorisation of building data in the digital documentation of heritage buildings. *Applied Geomatics*, *13*(1), 29–54. DOI: 10.1007/s12518-020-00322-7

Khan, M. S., Khan, M., Bughio, M., Talpur, B. D., Kim, I. S., & Seo, J. (2022). An Integrated HBIM Framework for the Management of Heritage Buildings. *Buildings (Basel, Switzerland)*, *12*(7), 964. DOI: 10.3390/buildings12070964

Khan, M. Z., Hassan, M., & Yu, Z. (2021). Internet of Things (IoT) in smart waste management: A comprehensive review. *Sustainable Cities and Society*, *69*, 102758. DOI: 10.1016/j.scs.2021.102758

Khodeir, L. M., Aly, D., & Tarek, S. (2016). Integrating HBIM (Heritage Building Information Modeling) tools in the application of sustainable retrofitting of heritage buildings in Egypt. *Procedia Environmental Sciences, 34*, 258–270. DOI: 10.1016/j.proenv.2016.04.024

Khoshand, A., Khanlari, K., Abbasianjahromi, H., & Zoghi, M. (2020). Construction and demolition waste management: Fuzzy Analytic Hierarchy Process approach. *Waste Management & Research, 38*(7), 773–782. DOI: 10.1177/0734242X20910468 PMID: 32189588

Kim, L., Lee, J., & Lee, I. (2017). The expansion of virtual landscape in digital games: Classification of virtual landscape through five principles. Graduate School of Culture Technology, KAIST.

Kim, K., Lee, G., & Kim, S. (2020). A study on the application of blockchain technology in the construction industry. *KSCE Journal of Civil Engineering, 24*(9), 2561–2571. DOI: 10.1007/s12205-020-0188-x

Kim, K., & Lee, J. (2022). Convergence of game design and architectural practice: Cross-disciplinary innovation. *Journal of Digital Creativity, 33*(1), 45–59.

Koch, F., Bassetti, L., & Huynh, D. (2020). Performance of rainwater harvesting systems in urban Southeast Asia: Opportunities and challenges. *Water (Basel), 12*(6), 1607. DOI: 10.3390/w12061607

Ko, H., Chang, J., & Li, Y. (2020). San Francisco's zero waste program: Policies, practices, and challenges. *Environmental Management, 66*(1), 85–94. DOI: 10.1007/s00267-020-01314-6

Kouamou Njifen, S. R., Enyegue A Nyam, F. M., Bikoro, M. B. A., Ngouokouo, A. T., & Tabod, C. T. (2023). Modeling groundwater potential zones in the Kribi-Campo region, South Cameroon using geospatial techniques and statistical models. *Modeling Earth Systems and Environment, 9*(1), 913–935. DOI: 10.1007/s40808-022-01544-x

Krstić, M., Agnusdei, G. P., Miglietta, P. P., Tadić, S., & Roso, V. (2022). Applicability of industry 4.0 technologies in the reverse logistics: A circular economy approach based on comprehensive distance based ranking (COBRA) method. *Sustainability (Basel), 14*(9), 5632. DOI: 10.3390/su14095632

Ku, M. (2019, December 27). Challenges and limitations of adopting IoT. IndSights Research. https://www.indsights.sg/industry-perspective/challenges-and-limitations-of-adopting-iot/

Kumar, S., Tiwari, P., & Zymbler, M. (2019). Internet of Things is a revolutionary approach for future technology enhancement: A review. *Journal of Big Data*, *6*(1), 111. DOI: 10.1186/s40537-019-0268-2

Kundu, A. (2003). Urbanization and Urban Governance: Search for a Perspective beyond Neo-Liberalism. *EPW* 38(29), 3079-3087. https://www.epw.in/journal/2003/29/special-articles/urbanisation-and-urban-governance.html

Kunz, J., & Fischer, M. (2020). Virtual design and construction. *Construction Management and Economics*, *38*(4), 355–363. DOI: 10.1080/01446193.2020.1714068

Kurkinen, K. (2017). Case study of a cold attic in a pitched roof with minimal ventilation. *Energy Procedia*, *132*, 466–471. DOI: 10.1016/j.egypro.2017.09.667

Kvale, S., & Brinkmann, S. (2009). *InterViews: Learning the craft of qualitative research interviewing* (2nd ed.). SAGE Publications.

Laeeq, K. (2022). *Metaverse: Why, how and what*. National University of Computer and Emerging Sciences.

Lal, P., & Nair, S. B. (2017). Urbanization in Kerala—What does the Census Data Reveal? *Indian Journal of Human Development*, *11*(3), 356–386. https://www.researchgate.net/publication/326106606_Urbanization_in_Kerala-What_Does_the_Census_Data_Reveal. DOI: 10.1177/0973703018763241

Langston, C., & Shen, L. Y. (2017). Application of the adaptive reuse potential model in Hong Kong: A case study of Lui Seng Chun. *Facilities*, *35*(3/4), 230–245.

Larsen, T. J., Sørebø, A. M., & Sørebø, Ø. (2009). The role of task-technology fit as users' motivation to continue information system use. *Computers in Human Behavior*, *25*(3), 778–784. DOI: 10.1016/j.chb.2009.02.006

Lausten, M. A., Aschenbrenner, A., & Hughes, J. R. (2020). Zero-waste initiatives and urban waste management: Lessons from Kamikatsu, Japan. *Waste Management & Research*, *38*(10), 1167–1177. DOI: 10.1177/0734242X20962921

Lawson, B. (2006). *How designers think: The design process demystified*. Routledge. DOI: 10.4324/9780080454979

LCCF. (2017). Low Carbon Cities Framework Version 2. Ministry of Energy, Green Technology and Water Malaysia (KeTTHA), ISBN 978-967-5893-10-0.

Lee, L., Zhou, P., Braud, T., & Pan, H. (2022). What is the metaverse? An immersive cyberspace and open challenges.

Lee. (n.d.). 3 challenges of IoT adoption and how enterprises can overcome them. Tech Genius. Retrieved January 14, 2023, from https://techgenies.com/3-challenges-of-iot-adoption-and-how-enterprises-can-overcome-them/

Lee, C. C., Cheng, H. K., & Cheng, H. H. (2007). An empirical study of mobile commerce in insurance industry: Task–technology fit and individual differences. *Decision Support Systems*, *43*(1), 95–110. DOI: 10.1016/j.dss.2005.05.008

Lee, H. K., Tan, B. K., & Soh, Y. W. (2020). The Zero Waste Masterplan of Singapore: Strategies for sustainable waste management. *Environmental Science & Technology*, *54*(19), 11872–11881. DOI: 10.1021/acs.est.0c03094 PMID: 32852946

Lee, T., & Chua, A. (2023). Navigating challenges in heritage restoration: The impact of community knowledge on conservation practices. *Asian Journal of Architecture and Building Engineering*, *12*(2), 89–102.

Lee, T., Mohd Yusof, S., & Idris, M. (2023). The shift towards sustainability in adaptive reuse: Lessons learned from recent projects. *Journal of Urban Heritage*, *24*(2), 145–159.

Leon, A. C., Davis, L. L., & Kraemer, H. C. (2011). The role and interpretation of pilot studies in clinical research. *Journal of Psychiatric Research*, *45*(5), 626–629. DOI: 10.1016/j.jpsychires.2010.10.008 PMID: 21035130

LetsBuild. (2023, October, 18). *Why the slow uptake of technology in construction is holding the industry back*. https://www.letsbuild.com/blog/why-the-slow-uptake-of-technology-in-construction-is-holding-the-industry-back

Li, D., Zheng, Y., Liu, C., Qi, H., & Liu, X. (2016). Numerical analysis on thermal performance of naturally ventilated roofs with different influencing parameters. *Sustainable Cities and Society*, *22*, 86–93. DOI: 10.1016/j.scs.2016.02.004

Lifset, R., Atasu, A., & Tojo, N. (2013). Extended producer responsibility. *Journal of Industrial Ecology*, *17*(2), 162–166. DOI: 10.1111/jiec.12022

Li, H., & Wan, G. (2014). Analysis on city design strategy of dry hot climate region——Taking the city design of civic commerce office area in Panzhihua as an example. *Urbanism and Architecture*, *8*, 44–45.

Li, J., Zhao, X., & He, X. (2019). Permeable pavements for stormwater management in urban areas: A review of materials and techniques. *Environmental Science and Pollution Research International*, *26*(5), 4680–4691. DOI: 10.1007/s11356-019-04387-9 PMID: 30565107

Lim, Y. W., Chong, H. Y., Ling, P. C., & Tan, C. S. (2021). Greening existing buildings through Building Information Modelling: A review of the recent development. *Building and Environment, 200*, 107924. DOI: 10.1016/j.buildenv.2021.107924

Lin, B., & Pan, F. (2023). Applications of DGT in coastal sediments: Monitoring and biogeochemical study of trace metals and oxyanions. *Trends in Environmental Analytical Chemistry, 39*, e00207. DOI: 10.1016/j.teac.2023.e00207

Littlefield, D. (2017). Heritage and time: mapping what is not there. In *Heritage Building Information Modelling* (pp. 32–45). Routledge. DOI: 10.4324/9781315628011-4

Liu, Y. (2021). *Research on energy saving reconstruction design of rural residential roof based on the perspective of thermal comfort in winter——A case study of Yuncheng county*. Master dissertation, Qingdao University of Technology,Qingdao, China.

Liu, C., & Liao, Z. (2019). Overcoming barriers to greywater recycling in urban settings: A review of regulations and technological advancements. *Resources, Conservation and Recycling, 142*, 83–92. DOI: 10.1016/j.resconrec.2018.11.018

Liu, J., Chen, Y., & Wang, X. (2022). Factors driving waste sorting in construction projects in China. *Journal of Cleaner Production, 336*, 130397. DOI: 10.1016/j.jclepro.2022.130397

Liu, J., Foreman, G., Sattineni, A., & Li, B. (2023). Integrating Stakeholders' Priorities into Level of Development Supplemental Guidelines for HBIM Implementation. *Buildings (Basel, Switzerland), 13*(2), 530. DOI: 10.3390/buildings13020530

Liu, J., Zhang, Q., & Li, Y. (2018). Enhancing water conservation in urban distribution systems: A study of smart water metering and real-time leak detection. *Journal of Water Resources Planning and Management, 144*(12), 04018087. DOI: 10.1061/(ASCE)WR.1943-5452.0001005

Liu, T., Ma, G., & Wang, D. (2022). Pathways to successful building green retrofit projects: Causality analysis of factors affecting decision making. *Energy and Building, 276*, 112486. DOI: 10.1016/j.enbuild.2022.112486

Liu, Y., Considine, B., & McNabola, A. (2024). Assessment of hygrothermal performance and mould growth risk in roofs of post-retrofitted non-dormer and dormer attic rooms with reduced ventilation. *Building and Environment, 250*, 111172. DOI: 10.1016/j.buildenv.2024.111172

Liu, Z., Osmani, M., Demian, P., & Baldwin, A. (2015). A BIM-aided construction waste minimisation framework. *Automation in Construction, 59*, 1–23. DOI: 10.1016/j.autcon.2015.07.020

Li, Z., Boyle, F., & Reynolds, A. (2016). Rainwater harvesting and greywater treatment systems for domestic application in Ireland. *Desalination and Water Treatment*, *57*(48-49), 22911–22919. DOI: 10.1080/19443994.2015.1127175

Lizana, J., Molina-Huelva, M., & Chacartegui, R. (2016). Multi-criteria assessment for the effective decision management in residential energy retrofitting. *Energy and Building*, *129*, 284–307. DOI: 10.1016/j.enbuild.2016.07.043

Longhurst, R. (2016). Semi-structured interviews and focus groups. In Clifford, N., Cope, M., Gillespie, T., & French, S. (Eds.), *Key methods in geography* (3rd ed., pp. 143–156). SAGE.

Louis, J., & Dunston, P. S. (2018). Integrating IoT into operational workflows for real-time and automated decision-making in repetitive construction operations. *Automation in Construction*, *94*, 317–327. DOI: 10.1016/j.autcon.2018.07.005

Lovell, L. J., Davies, R. J., & Hunt, D. V. (2023). The application of historic building information modelling (HBIM) to cultural heritage: A review. *Heritage*, *6*(10), 6691–6717. DOI: 10.3390/heritage6100350

Lstiburek, J. A. (2024). A crash course in roof venting: Understand when to vent your roof, when not to, and how to execute each approach successfully, 2024 (accessed on 15 Augest 2024). Available from http://beta.buildingscience.com

Lu, W., Webster, C., Chen, K., Zhang, X., & Chen, X. (2017). Computational Building Information Modelling for construction waste management: Moving from rhetoric to reality. *Renewable & Sustainable Energy Reviews*, *68*, 587–595. DOI: 10.1016/j.rser.2016.10.029

Lu, Y., Ge, Y., Zhang, G., Abdulwahab, A., Salameh, A. A., Ali, H. E., & Le, B. N. (2023). Evaluation of waste management and energy saving for sustainable green building through analytic hierarchy process and artificial neural network model. *Chemosphere*, *318*, 137708. DOI: 10.1016/j.chemosphere.2022.137708 PMID: 36621688

Lysagh. (2024). Vent-a-roof. Retrieved October 15, 2024, from https://lysaght.com/vent-a-roof

MacRae, R., McLeod, A., & Green, J. (2021). Community-based water conservation programs: Impacts on urban sustainability. *Environmental Education Research*, *27*(4), 525–539. DOI: 10.1080/13504622.2020.1819331

Maguire, M., & Delahunt, B. (2017). Doing a thematic analysis: A practical, step-by-step guide for learning and teaching scholars. AISHE-J: The All Ireland Journal of Teaching and Learning in Higher Education, 9(3). Retrieved from https://ojs.aishe.org/index.php/aishe-j/article/view/335

Mahadevia, D. (2001). Sustainable Urban Development in India: An Inclusive Perspective. *Development in Practice, 11*(2-3), 242–259. https://www.jstor.org/stable/4029451. DOI: 10.1080/09614520120056388

Maheshwari, D. G., & Gehlot, A. S. (2017). Evaluation of effectiveness of passive cooling (roof cool) tecnique in Indian climate. *n.d*, 3(4), 256-266.

Mahmoud, A., Asif, M., Hassanain, M., Babsail, M., & Sanni-Anibire, M. (2017). Energy and Economic Evaluation of Green Roofs for Residential Buildings in Hot-Humid Climates. *Buildings (Basel, Switzerland), 7*(2), 30. DOI: 10.3390/buildings7020030

Mahmud, S. H., Assan, L., & Islam, R. (2018). Potentials of Internet of Things (IoT) in Malaysian Construction Industry. [AETiC]. *Annals of Emerging Technologies in Computing, 2*(4), 46–52. DOI: 10.33166/AETiC.2018.04.004

Ma, J., Yu, H., Xu, Y., & Deng, K. (2020). CDAM: Conservative data analytical model for dynamic climate information evaluation using intelligent IoT environment—An application perspective. *Computer Communications, 150*, 177–184. DOI: 10.1016/j.comcom.2019.11.014

Malaysia, C. I. D. B. (2021, August, 30). *MCO and Its Impact on the Construction Industry*. https://cidbibs.com.my/mco-and-its-impact-on-the-construction-industry/

Malaysia, C. I. D. B. (2022, March, 31). *The adoption of technology in the construction industry*. https://www.cidb.gov.my/eng/the-adoption-of-technology-in-the-construction-industry/

Malaysia, C. I. D. B. (2023, August, 17). *Unlocking the potential of cloud computing*. https://www.cidb.gov.my/eng/unlocking-the-potential-of-cloud-computing/

Maqbool, R., Saiba, M. R., & Ashfaq, S. (2023). Emerging industry 4.0 and Internet of Things (IoT) technologies in the Ghanaian construction industry: Sustainability, implementation challenges, and benefits. *Environmental Science and Pollution Research International, 30*(13), 37076–37091. DOI: 10.1007/s11356-022-24764-1 PMID: 36565423

Mari, T., Chiat, C. P., Ng, V., Gunasagaran, S., & Kuppusamy, S. (2024). STRATEGIES FOR NET ZERO CARBON IN SUSTAINABLE CONSERVATION PRACTICES FOR HISTORICAL BUILDINGS IN MALAYSIA. *PLANNING MALAYSIA, 22*.

Marshall, C. (1996). *Designing qualitative research* (3rd ed.). SAGE Publications.

Martinelli, L., Calcerano, F., & Gigliarelli, E. (2022). Methodology for an HBIM workflow focused on the representation of construction systems of built heritage. *Journal of Cultural Heritage, 55*, 277–289. DOI: 10.1016/j.culher.2022.03.016

Massoud, M. A., Tarhini, A., & Nasr, J. A. (2009). Decentralized approaches to wastewater treatment and management: Applicability in developing countries. *Journal of Environmental Management, 90*(1), 652–659. DOI: 10.1016/j.jenvman.2008.07.001 PMID: 18701206

Mauzerall, D. L., Wang, J., & Vigdor, N. (2017). A global assessment of urban water efficiency and conservation technologies. *Environmental Science & Technology, 51*(7), 3989–3998. DOI: 10.1021/acs.est.6b06090

Ma, W., Hao, J. L., Zhang, C., Di Sarno, L., & Mannis, A. (2023). Evaluating carbon emissions of China's waste management strategies for building refurbishment projects: Contributing to a circular economy. *Environmental Science and Pollution Research International, 30*(4), 8657–8671. DOI: 10.1007/s11356-021-18188-6 PMID: 35064493

Ma, Z., Cooper, P., Daly, D., & Ledo, L. (2012). Existing building retrofits: Methodology and state-of-the-art. *Energy and Building, 55*, 889–902. DOI: 10.1016/j.enbuild.2012.08.018

Mbaha, J. P., & Etoundi, M. L. B. A. (2021). *Construire une Ville Portuaire Stratégique et Émergente à L'horizon 2035*. Espace Géograph. Soc.

Mbevo Fendoung, P., Tchindjang, M., & Hubert-Ferrari, A. (2022). Weakening of Coastlines and Coastal Erosion in the Gulf of Guinea: The Case of the Kribi Coast in Cameroon. *Land (Basel), 11*(9), 1557. DOI: 10.3390/land11091557

McCaig, I., Pender, R., & Pickles, D. (2018). *Energy Efficiency and Historic Buildings: How to Improve Energy Efficiency*. Historic England.

McGill, R. (2001). Urban Management Checklist. *Cities (London, England), 18*(5), 347–354. https://www.sciencedirect.com/science/article/abs/pii/S0264275101000270. DOI: 10.1016/S0264-2751(01)00027-0

McGinley, O., Moran, P., & Goggins, J. (2022). An Assessment of the Key Performance Indicators (KPIs) of Energy Efficient Retrofits to Existing Residential Buildings. *Energies, 15*(1), 334. DOI: 10.3390/en15010334

MDPI. (2021). The role of community participation in the adaptive reuse of historic buildings. *Sustainability, 13*(7), 3745.

Mehndi, S. M., & Chakraborty, I. (2020, February). Simulation for a cost-effective and energy efficient retrofits of the existing building stock in India using BIM. In 2020 International conference on contemporary computing and applications (IC3A) (pp. 237-243). IEEE. DOI: 10.1109/IC3A48958.2020.233305

Melchor-Ferrer, E., & Buendía-Carrillo, D. (2014). Financial information management for university departments, using open-source software. *International Journal of Information Management*, *34*(2), 191–199. DOI: 10.1016/j.ijinfomgt.2013.12.009

Meleca, D. (2023). *Low Carbon and Future-Proofing Through Adaptive Reuse*. Architect Magazine.

Melville, N., Kraemer, K., & Gurbaxani, V. (2004). Information technology and organizational performance: An integrative model of IT business value. *Management Information Systems Quarterly*, *28*(2), 283–322. DOI: 10.2307/25148636

Mena, M. S., Tchawa, P., Amougou, J. A., & Tchotsoua, M. (2016). Les changements climatiques à travers la modification du régime pluviométrique dans la région de Kribi (1935–2006). *Rev. Ivoire. Sci. Technol.*, *28*, 389–407.

Meng, Q., Zhang, Y., Li, Z., Shi, W., Wang, J., Sun, Y., Xu, L., & Wang, X. (2020). A review of integrated applications of BIM and related technologies in whole building life cycle. *Engineering, Construction, and Architectural Management*, *27*(8), 1647–1677. DOI: 10.1108/ECAM-09-2019-0511

Meoni, A., Vittori, F., Piselli, C., D'Alessandro, A., Pisello, A. L., & Ubertini, F. (2022). Integration of structural performance and human-centric comfort monitoring in historical building information modeling. *Automation in Construction*, *138*, 104220. DOI: 10.1016/j.autcon.2022.104220

Meor Gheda, M. L., Chung, H. W., Abdul Aziz, F. F., & Mohamad Nusran, N. F. (2025). Bridging the digital divide: Overcoming challenges in technology adoption in Malaysia's construction industry. *Malaysian Journal of Social Sciences and Humanities*, *10*(1), 1–15. DOI: 10.47405/mjssh.v10i1.3137

Messner, J., Anumba, C., Dubler, C., Goodman, S., Kasprzak, C., Kreider, R., Leicht, R., Saluja, C., & Zikic, N. (2019). BIM Project Execution Planning Guide, Version 2.2. Computer Integrated Construction Research Program, The Pennsylvania State University, University Park, PA, USA, August, Retrieved from http://bim.psu.edu

Michell, V. (2016). 3D Printing and Additive Manufacturing Capability Modelling. *Sixth International Symposium on Business Modeling and Software Design*. SCITEPRESS. DOI: 10.5220/0006222400730083

Mikkelsen, P. S., Andersen, R., & Mortensen, T. (2019). Copenhagen's waste management and recycling strategies: Waste-to-energy and circular economy. *Waste Management (New York, N.Y.)*, *94*, 218–226. DOI: 10.1016/j.wasman.2019.06.026

Miles, M. B., & Huberman, A. M. (1994). *Qualitative data analysis: An expanded sourcebook* (2nd ed.). SAGE Publications.

Miller, R. E., Thomas, R. L., & Lehman, P. (2017). *Green infrastructure and water management: Exploring the potential of urban ecosystems in coastal cities*. Wiley.

Mills, J. D., & Boersma, D. L. (2021). Big data and AI in waste management: The future of urban resource optimization. *Journal of Environmental Informatics*, *35*(1), 74–89. DOI: 10.1016/j.envinform.2021.05.006

MINEPDED/PNUD. (2009). Révision /Opérationnalisation Du PNGE Vers Un Programme Environnemental (PE), Volume I: Diagnostic De La Situation De L'Environnement Au Cameroun; Ministère de L'environnement et de la Protection de la Nature: Yaoundé, Cameroon, 203p.

MINHDU. (2014). *Plan Directeur D'Urbanisme De La Ville De Kribi. Rapport Justificatif.* Communauté Urbaine de Kribi.

Ministry of Energy. (2015). *Green Technology, and Water*. National Energy Efficiency Action Plan.

Ministry of Energy. Green Technology, and Water. (2015). National Energy Efficiency Action Plan. Ministry of Energy, Green Technology, and Water.

Ministry of Housing and Urban-Rural Development of the People's Republic of China. (2012). *Design standard for energy efficiency of rural residential buildings*. China Architecture & Building Press.

Ministry of Housing and Urban-Rural Development of the People's Republic of China. (2019). *Standard for design of energy efficiency of residential buildings in moderate climate zone*. China Architecture & Building Press.

Ministry of Housing and Urban-Rural Development of the People's Republic of China. (2021). *General code for energy efficiency and renewable energy appliction in buildings*. China Architecture & Building Press.

Mohamad, D., Yusoff, W. Z. W., & Majid, R. A. (2020). Stakeholder engagement in adaptive reuse projects in Malaysia. *Asian Journal of Environment-Behaviour Studies*, *5*(2), 43–54. DOI: 10.21834/ajebs.v5i2.155

Mohamed, R., Chan, E. H. W., & Yung, E. H. K. (2017). Integrated approach for heritage conservation in Malaysia: Towards sustainable maintenance. *Procedia Engineering, 118*, 360–367.

Mohammed, B. H., Sallehuddin, H., Yadegaridehkordi, E., Safie Mohd Satar, N., Hussain, A. H., & Abdelghanymohamed, S. (2022). Nexus between Building Information Modeling and Internet of Things in the construction industries. *Applied Sciences (Basel, Switzerland), 12*(20), 10629. DOI: 10.3390/app122010629

Mohd Nasir, M. A., & Salleh, M. A. (2020). Social Sustainability through Adaptive Reuse: Malaysian Perspectives. *Social Indicators Research, 147*(3), 875–894.

Mohd Yusof, S., & Idris, M. (2023). Craftsmanship in conservation: Navigating the challenges of energy efficiency in heritage buildings. *Asian Journal of Architecture and Building Engineering, 12*(3), 91–105.

Moneta, A. (2020). Architecture, heritage and the metaverse: New approaches and methods for the digital built environment. International Association for the Study of Traditional Environments (IASTE), 32(1), Fall Issue.

Moran, F., Blight, T., Natarajan, S., & Shea, A. (2013). The use of passive house planning package to reduce energy use and CO2 emissions in historic dwellings. *Energy and Building, 75*, 216–227. DOI: 10.1016/j.enbuild.2013.12.043

Mora, R., Sánchez-Aparicio, L. J., Maté-González, M. Á., García-Álvarez, J., Sánchez-Aparicio, M., & González-Aguilera, D. (2021). An historical building information modelling approach for the preventive conservation of historical constructions: Application to the Historical Library of Salamanca. *Automation in Construction, 121*, 103449. DOI: 10.1016/j.autcon.2020.103449

Moreno, A., Austin, M. C., & Mora, D. (2021). A parametric study of implementing green roofs to improve building energy performance in tropical climate. *E3S Web of Conferences*, 312, 02004.

Morse, J. M. (2000). Determining sample size. *Qualitative Health Research, 10*(1), 3–5. DOI: 10.1177/104973200129118183

Mosadeghi, R., Warnken, J., Tomlinson, R., & Mirfenderesk, H. (2015). Comparison of Fuzzy AHP and AHP in a spatial multi-criteria decision making model for urban land-use planning. *Computers, Environment and Urban Systems, 49*, 54–65. DOI: 10.1016/j.compenvurbsys.2014.10.001

Moudingo, J. H., Ajonina, G., Kemajou, J., Wassouni, A., Bitomo, M., Assengze, A., & Tomedi, M. (2020). Sylvio-socioeconomic study of urban mangrove patches and challenges: case of Kribi, Cameroon. In *Biotechnological Utilization of Mangrove Resources* (pp. 79–115). Elsevier., DOI: 10.1016/B978-0-12-819532-1.00004-4

Moyano, J., Carreño, E., Nieto-Julian, J. E., Gil-Arizon, I., & Bruno, S. (2022). Systematic approach to generate Historical Building Information Modelling (HBIM) in architectural restoration project. *Automation in Construction, 143*, 104551. DOI: 10.1016/j.autcon.2022.104551

Murphy, M., McGovern, E., & Pavia, S. (2009). Historic building information modelling (HBIM). *Structural Survey, 27*(4), 311–327. DOI: 10.1108/02630800910985108

Musa, M. F., Mohammad, M. F., Mahbub, R., & Yusof, M. R. (2018). Adopting modular construction in the Malaysian construction industry. *Asian Journal of Environment-Behaviour Studies, 3*(10), 1–9. DOI: 10.21834/aje-bs.v3i10.307

Musarat, M. A., Alaloul, W. S., Hameed, N., Qureshi, A. H., & Wahab, M. M. A. (2022). Efficient construction waste management: A solution through Industrial Revolution (IR) 4.0 evaluated by AHP. *Sustainability (Basel), 15*(1), 274. DOI: 10.3390/su15010274

Mystakidis, S. (2022). Metaverse.

Nair, R., & Goh, K. (2018). Decentralized wastewater treatment systems for small urban communities: Case studies and lessons learned. *Environmental Engineering Science, 35*(4), 240–250. DOI: 10.1089/ees.2017.0324

Nasir, M. A., Rahman, N. A. A., & Salleh, M. A. (2020). Sustainable practices in heritage building adaptive reuse: Malaysian perspectives. *Sustainability, 12*(8), 3221. DOI: 10.3390/su12083221

Nasir, N. M., Rahman, R. A., & Salleh, A. F. (2020). Community perspectives on adaptive reuse of heritage buildings in Malaysia. *Heritage & Society, 11*(3), 120–135.

Nassereddine, H., Hanna, A. S., Veeramani, D., & Lotfallah, W. (2022). Augmented Reality in the Construction Industry: Use-Cases, Benefits, Obstacles, and Future Trends. *Frontiers in Built Environment, 8*, 730094. DOI: 10.3389/fbuil.2022.730094

National Trust for Historic Preservation. (2015). The Impact of Historic Preservation.

National Trust for Historic Preservation. (2023). Adaptive Reuse: A Sustainable Approach.

Nawi, N. H. M., Haron, R. C., & Kamarudin, Z. (2020). Risk cost analysis in Malay heritage conservation project. Planning Malaysia, 18.

NewStudio Architecture. (2023). Adaptive Reuse: A Strategy for Holding History and Carbon.

Ngoran, S. D., & Xue, X. (2017). Public sector governance in Cameroon: A valuable opportunity or fatal aberration from the Kribi Campo integrated coastal management? *Ocean and Coastal Management*, *138*, 83–92. DOI: 10.1016/j.ocecoaman.2017.01.006

Nhamo, L., Rwizi, L., Mpandeli, S., Botai, J., Magidi, J., Tazvinga, H., Sobratee, N., Liphadzi, S., Naidoo, D., Modi, A. T., Slotow, R., & Mabhaudhi, T. (2021). Urban Nexus and Transformative Pathways towards a Resilient Gauteng City Region, South Africa. *Cities (London, England)*, *116*, 103266. https://www.sciencedirect.com/science/article/pii/S0264275121001669. DOI: 10.1016/j.cities.2021.103266 PMID: 37674556

Nieto-Julián, J. E., Farratell, J., Bouzas Cavada, M., & Moyano, J. (2023). Collaborative workflow in an HBIM project for the restoration and conservation of cultural heritage. *International Journal of Architectural Heritage*, *17*(11), 1813–1832. DOI: 10.1080/15583058.2022.2073294

Nieto-Julián, J. E., Lara, L., & Moyano, J. (2021). Implementation of a TeamWork-HBIM for the Management and Sustainability of Architectural Heritage. *Sustainability (Basel)*, *13*(4), 2161. DOI: 10.3390/su13042161

Nishad, A. (2024). Haritha Karma Sena: Transforming Waste Management. *IJEBM* 8(3), 19-25. https://aipublications.com/uploads/issue_files/3IJEBM-SEP20241-Haritha.pdf

Nnaji, C., & Karakhan, A. A. (2020). Technologies for safety and health management in construction: Current use, implementation benefits and limitations, and adoption barriers. *Journal of Building Engineering*, *29*, 101212. DOI: 10.1016/j.jobe.2020.101212

Nor, R. M., Ibrahim, S., & Ng, K. H. (2021). Enhancing energy efficiency in heritage buildings using digital tools. *Energy Reports*, *7*, 33–42.

Noura, M., Atiquzzaman, M., & Gaedke, M. (2019). Interoperability in Internet of Things: Taxonomies and open challenges. *Mobile Networks and Applications*, *24*(3), 796–809. DOI: 10.1007/s11036-018-1089-9

Nowell, L. S., Norris, J. M., White, D. E., & Moules, N. J. (2017). Thematic analysis: Striving to meet the trustworthiness criteria. *International Journal of Qualitative Methods*, *16*(1), 1609406917733847. Advance online publication. DOI: 10.1177/1609406917733847

Noy, C. (2008). Sampling knowledge: The hermeneutics of snowball sampling in qualitative research. *International Journal of Social Research Methodology, 11*(4), 327–344. DOI: 10.1080/13645570701401305

Ntamak-Nida, M. J., Baudin, F., Schnyder, J., Makong, J. C., Komguem, P. B., & Abolo, G. M. (2008). Depositional environments and characterisation of the organic matter of the Lower Mundeck Formation (Barremian? –Aptian) of the Kribi-Campo sub-basin (South Cameroon): Implications for petroleum exploration. *Journal of African Earth Sciences, 51*(4), 207–219. DOI: 10.1016/j.jafrearsci.2008.01.006

Nuissl, H., & Heinrichs, D. (2011). Fresh Wind or Hot Air — Does the Governance Discourse have Something to Offer to Spatial Planning? *Journal of Planning Education and Research, 31*(1), 47–59. https://journals.sagepub.com/doi/10.1177/0739456X10392354. DOI: 10.1177/0739456X10392354

Nyame-Tawiah, D., Attuah, L., & Koranteng, C. (2020). Determination of the Effect of Green Roofs on Indoor Temperature by the Use of Simulation in a Tropical Landscape. *Journal of Scientific Research and Reports*, •••, 12–28. DOI: 10.9734/jsrr/2020/v26i430244

Obeng-Odoom, F. (2012). On the Origin, Meaning, and Evaluation of Urban Governance. *Norsk Geografisk Tidsskrift, 66*(4), 204–212. https://www.tandfonline.com/doi/abs/10.1080/00291951.2012.707989. DOI: 10.1080/00291951.2012.707989

OECD. (2018). *The state of play on extended producer responsibility*. Organisation for Economic Co-operation and Development.

Oke, A. E., & Arowoiya, V. A. (2021). Evaluation of Internet of Things (IoT) application areas for sustainable construction. *Smart and Sustainable Built Environment, 10*(3), 407–422. DOI: 10.1108/SASBE-11-2020-0167

Oke, A. E., Arowoiya, V. A., & Akomolafe, O. T. (2022). Influence of the Internet of Things' application on construction project performance. *International Journal of Construction Management, 22*(13), 2517–2527. DOI: 10.1080/15623599.2020.1807731

Oke, A., Aigbavboa, C., & Raphiri, M. (2020). Cultural considerations in design thinking: Enhancing local relevance in architectural practice. *Sustainable Cities and Society, 55*, 102037.

Olbina, S., & Elliott, J. W. (2019). Contributing project characteristics and realized benefits of successful BIM implementation: A comparison of complex and simple buildings. *Buildings (Basel, Switzerland), 9*(8), 175. DOI: 10.3390/buildings9080175

Omar, A. F., Zainordin, N., Khoo, S. L., & Peng, J. O. H. (2022, November). Barriers in implementing artificial intelligence (AI) and Internet of things (IoTs) among Malaysian construction industry. *AIP Conference Proceedings, 2644*(1), 030002. Advance online publication. DOI: 10.1063/5.0111461

Omar, A. I., Virgone, J., Vergnault, E., David, D., & Idriss, A. I. (2017). Energy Saving Potential with a Double-Skin Roof Ventilated by Natural Convection in Djibouti. *Energy Procedia, 140*, 361–373. DOI: 10.1016/j.egypro.2017.11.149

Onag, G. (2019). Malaysia's Forest City bags IDC Smart City APAC Awards. https://futureiot.tech/malaysias-forest-city-bags-idc-smart-city-apac-awards

Ondoa, G. A., Onguéné, R., Eyango, M. T., Duhaut, T., Mama, C., Angnuureng, D. B., & Almar, R. (2018). Assessment of the Evolution of Cameroon Coastline: An Overview from 1986 to 2015. *Journal of Coastal Research, 81*(10081), 122–129. DOI: 10.2112/SI81-016.1

Onecha, B., Cornado, C., Morros, J., & Pons, O. (2023). New approach to design and assess metaverse environments for improving learning processes in higher education: The case of architectural construction and rehabilitation. Department of Architectural Technology, Universitat Politècnica de Catalunya.

Oreni, D., Brumana, R., Della Torre, S., Banfi, F., Barazzetti, L., & Previtali, M. (2014). Survey turned into HBIM: The restoration and the work involved concerning the Basilica di Collemaggio after the earthquake (L'Aquila). *ISPRS Annals of the Photogrammetry, Remote Sensing and Spatial Information Sciences, 2*(5), 267–273. DOI: 10.5194/isprsannals-II-5-267-2014

Orihuela, P., Noel, M., Pacheco, S., Orihuela, J., Yaya, C., & Aguilar, R. (2019). Application of virtual and augmented reality techniques during design and construction process of building projects. *Annual Conference of the International Group for Lean Construction*. https://doi.org/DOI: 10.24928/2019/0220

Orihuela, P., Pacheco, S., & Orihuela, J. (2017). Proposal of performance indicators for the design of housing projects. *Procedia Engineering, 196*, 498–505. DOI: 10.1016/j.proeng.2017.07.230

Orlikowski, W. J. (2000). Using technology and constituting structures: A practice lens for studying technology in organizations. *Organization Science, 11*(4), 404–428. DOI: 10.1287/orsc.11.4.404.14600

Othman, A., & Mohd Yusof, S. (2020). Non-invasive techniques in the conservation of cultural heritage: A Malaysian perspective. *Asian Journal of Conservation Science, 14*(1), 12–25.

Ouellette, J. W., Woodard, P., & Bekboliev, M. (2022). Information Delivery Manual (IDM) Development for Building Information Modelling (BIM) and Building Energy Modelling (BEM) Workflows. 112.

Pal, D., & Patra, S. (2021). University students' perception of video-based learning in times of COVID-19: A TAM/TTF perspective. *International Journal of Human-Computer Interaction, 37*(10), 903–921. DOI: 10.1080/10447318.2020.1848164

Panzhihua City Renhe District People's government office. (2022), District situation overview. Retrieved October 15, 2024, from http:// http://www.screnhe.gov.cn

Panzhihua Climate. (2023), Weather by month, average temperature (China)—Weather spark. Retrieved December 25, 2023, from https://weatherspark.com

Patel, S. J., Yang, X., & Roy, P. K. (2020). The role of digital technologies in enhancing urban water management: Smart meters and IoT-based solutions. *Water Resources Management, 34*(8), 2557–2571. DOI: 10.1007/s11269-020-02601-1

Patton, M. Q. (2002). *Qualitative research & evaluation methods* (3rd ed.). SAGE Publications.

Pauwels, P. (2015). Analysing how constraints impact architectural decision-making. International Journal of Design Science and Technology.

Peñalvo-López, E., Cárcel-Carrasco, J., Alfonso-Solar, D., Valencia, I., & Hurtado-Pérez, E. (2020). Study of the Improvement on Energy Efficiency for a Building in the Mediterranean Area by the Installation of a Green Roof System. *Energies, 13*(5), 1246. DOI: 10.3390/en13051246

Penjor, T., Banihashemi, S., Hajirasouli, A., & Golzad, H. (2024). Heritage Building Information Modelling (HBIM) for Heritage Conservation: Framework of Challenges, Gaps, and Existing Limitations of HBIM. *Digital Applications in Archaeology and Cultural Heritage*, •••, 00366.

People's Government of Panzhihua Municipal. (2019), An overview of Panzhihua city. Retrieved October 15, 2024, from http://www.panzhihua.gov.cn

Pereira, L. S., Cordery, I., & Iacovides, I. (2012). *Handbook of land and water management*. Springer., DOI: 10.1007/978-1-4614-6435-8

Piercy, J. (2019). Life in Azeroth: Examining place making in virtual environments. (Master's thesis). Appalachian State University.

Pillai, M., Smith, D., & Desai, R. (2021). Waste-to-energy technologies for sustainable waste management. *Renewable & Sustainable Energy Reviews, 135*, 110186. DOI: 10.1016/j.rser.2020.110186

Piselli, C., Guastaveglia, A., Romanelli, J., Cotana, F., & Pisello, A. L. (2020). Facility energy management application of HBIM for historical low-carbon communities: Design, modelling and operation control of geothermal energy retrofit in a real Italian case study. *Energies, 13*(23), 6338. DOI: 10.3390/en13236338

Planning Commission. Govt. of India. (2008). *Eleventh Five Year Plan (2007-2012): Inclusive Growth.* Oxford University Press.

Plevoets, B., & Van Cleempoel, K. (2019). Adaptive reuse as an emerging discipline: an historic survey. In *Adaptive reuse of the built heritage: Concepts and cases of an emerging discipline* (pp. 11–18). Routledge.

Poland, B. D. (2002). Transcription quality as an aspect of rigor in qualitative research. In Gubrium, J. F., & Holstein, J. A. (Eds.), *Handbook of interview research: Context and method* (pp. 629–649). SAGE Publications.

Politi, E., Paterson, S. K., Scarrott, R., Tuohy, E., O'Mahony, C., & Cámaro-García, W. C. A. (2019). Earth observation applications for coastal sustainability: Potential and challenges for implementation. *Anthropocene Coasts, 2*(1), 306–329. DOI: 10.1139/anc-2018-0015

Pradhan, S., Bhatt, M., & Gautam, R. S. (2025). Social Interventions for Liquid Waste Management on Livelihood: A Case of Indore, India. *Journal of Development Policy and Practice.* https://journals.sagepub.com/doi/epub/10.1177/24551333241304538

Prasad, B., & Kumar, V. (2021). Swachh Bharat Mission: Public-private partnership in waste management and sanitation. *Waste Management & Research, 39*(8), 939–948. DOI: 10.1177/0734242X21102194

Praveen, S. R. (2024). Foot Soldiers of Kerala's Silent Revolution against Waste. *The Hindu.* https://www.thehindu.com/news/national/kerala/foot-soldiers-of-keralas-silent-revolution-against-waste/article68106197.ece

ProjectTeam. (2023). *Revolutionizing construction: The impact of technology on industry practices* [Blog post]. ProjectTeam.com. https://blog.projectteam.com/revolutionizing-construction-the-impact-of-technology-on-industry-practices

Punter, J. (1991). Participation in the design of urban space. *Landscape Design, 200*, 24–27.

Purnomo, R. S., Susilawati, C., & Yusof, N. (2019). The role of integrated design approach for sustainable adaptive reuse. *International Journal of Building Pathology and Adaptation, 37*(1), 65–81. DOI: 10.1108/IJBPA-06-2018-0044

Qhawarizmi Architect. (2021). *Sri Perpatih Palace - Metaverse Art Gallery*. YouTube. https://www.youtube.com/watch?v=J2uVm2WsIGY

Qian, J., Siriwardana, C., & Shahzad, W. (2024). Identifying Critical Criteria on Assessment of Sustainable Materials for Construction Projects in New Zealand Through the Analytic Hierarchy Process (AHP) Approach. *Buildings (Basel, Switzerland)*, *14*(12), 3854. DOI: 10.3390/buildings14123854

Qin, Y., Zhang, M., & Hiller, J. E. (2017). Theoretical and experimental studies on the daily accumulative heat gain from cool roofs. *Energy*, *129*, 138–147. DOI: 10.1016/j.energy.2017.04.077

Quattrini, R., Battini, C., & Mammoli, R. (2018). HBIM to VR. Semantic awareness and data enrichment interoperability for parametric libraries of historical architecture. *The International Archives of the Photogrammetry, Remote Sensing and Spatial Information Sciences*, *42*(2), 937–943. DOI: 10.5194/isprs-archives-XLII-2-937-2018

Rahim, A., Hassan, R., & Tan, S. P. (2021). Application of laser scanning technology in the preservation of historical buildings: A Malaysian case study. *Journal of Architectural Conservation*, *27*(2), 147–161.

Rahman, A. S. A., Hamid, Z. A., & Hassan, M. H. (2019). Urban Regeneration through Adaptive Reuse in Malaysian Cities. *Journal of Urban Planning and Development*, *145*(4), 04019020.

Rahman, A. S. A., Hamid, Z. A., & Hassan, M. H. (2020). Structural issues in the adaptive reuse of heritage buildings: Malaysian case studies. *Journal of Construction Engineering and Management*, *146*(11), 04020129. DOI: 10.1061/(ASCE)CO.1943-7862.0001908

Rahman, F., & Wong, M. (2023). Energy-efficient interventions in heritage conservation: Bridging the past and future. *Heritage & Society*, *16*(2), 76–89.

Rai, A., & Tiwari, S. (2020). Robotics and AI in urban waste management: Advancing efficiency and adaptability in waste systems. *Waste Management & Research*, *38*(1), 13–24. DOI: 10.1177/0734242X20923056

Raj, J. S.Jennifer S. Raj. (2020). A novel information processing in IoT based real time health care monitoring system. *Journal of Electronics and Informatics*, *2*(3), 188–196. DOI: 10.36548/jei.2020.3.006

Ramli, M. A., Kamarazaly, M. A., Shi, L. X., Ibrahim, F. I., & Rizal, M. (2021). Building Defects and Restoration Techniques of Heritage Buildings in Malaysia. *Malaysian Construction Research Journal (MCRJ)*, 76.

Ramli, N. N., Ab Wahab, L., & Mohd Isa, H. (2021). A pilot survey on the assessment technique in accessing defects on heritage buildings in Malaysia. In *Virtual Go-Green: Conference and Publication (V-GoGreen 2020)* (pp. 37-42).

Rathore, P., & Sarmah, S. P. (2021). Investigation of factors influencing source separation intention towards municipal solid waste among urban residents of India. *Resources, Conservation and Recycling, 164*, 105164. DOI: 10.1016/j.resconrec.2020.105164

Ravikumar, P. (2024). Public Perception Regarding the Role of Harithakarma Sena in Waste Management at Panchayat Level. *IJRASET* 12(4), 574-578. https://www.ijraset.com/best-journal/perception-regarding-haritha-karma-sena-in-the-sustainable-waste-management

Rawa, D. B., & Alami, H. E. (2023). *Metaverse: Requirements, architecture, standards, status, challenges, and perspectives*. Howard University.

Rawat, M., & Singh, R. N. (2022). A study on the comparative review of cool roof thermal performance in various regions. *Energy and Built Environment, 3*(3), 327–347. DOI: 10.1016/j.enbenv.2021.03.001

Reeves, T., Olbina, S., & Issa, R. R. (2015). Guidelines for using building information modeling for energy analysis of buildings. *Buildings (Basel, Switzerland), 5*(4), 1361–1388. DOI: 10.3390/buildings5041361

Rethinking The Future. (2023). Revitalizing aging infrastructure: Strategies for adaptive reuse of historic buildings. Rethinking The Future. Retrieved from https://www.rethinkingthefuture.com/revitalizing-aging-infrastructure/

Rey-Mahía, C., Álvarez-Rabanal, F. P., Sañudo-Fontaneda, L. A., Hidalgo-Tostado, M., & Menéndez Suárez-Inclán, A. (2022). An experimental and numerical approach to multifunctional urban surfaces through blue roofs. *Sustainability (Basel), 14*(3), 1815. DOI: 10.3390/su14031815

Richter, M., & Dickhaut, W. (2023). Long-term performance of blue-green roof systems—Results of a building-scale monitoring study in Hamburg, Germany. *Water (Basel), 15*(15), 2806. DOI: 10.3390/w15152806

Robinson, O. C. (2020). Sampling in interview-based qualitative research: A theoretical and practical guide. *Qualitative Research in Psychology, 17*(2), 1–22.

Rocha, P. F., Ferreira, N. O., Pimenta, F., & Pereira, N. B. (2023). Impacts of prefabrication in the building construction industry. *Encyclopedia, 3*(1), 28–45. DOI: 10.3390/encyclopedia3010003

Rodrigues de Almeida, P., Zafra Solas, M., Renz, A., Bühler, M. M., Gerbert, P., Castagnino, S., & Rothballer, C. (2016). Shaping the future of construction a breakthrough in mindset and technology. *World Economic Forum.* https://doi.org/DOI: 10.13140/RG.2.2.21381.37605

Rodrigues, C., & Freire, F. (2017). Adaptive reuse of buildings: an eco-efficient life cycle approach. In *Life cycle assessment: Theory and practice* (pp. 1–18). Springer.

Rodrigues, F., Teixeira, J., Matos, R., & Rodrigues, H. (2019). Development of a web application for historical building management through BIM technology. *Advances in Civil Engineering, 2019*(1), 2019. DOI: 10.1155/2019/9872736

Rollings, A., & Morris, D. (2003). *Game architecture and design: A new edition.* New Riders Publishing.

Romeo, C., & Zinzi, M. (2013). Impact of a cool roof application on the energy and comfort performance in an existing non-residential building. A Sicilian case study. *Energy and Building*, *67*, 647–657. DOI: 10.1016/j.enbuild.2011.07.023

Roslan, A. F., Aminudin, E., Lau, S. E. N., & Abidin, N. I. (2022). Construction 4.0 to transform the Malaysian construction industry. *The Ingenieur magazine, 88*, 6-23. https://bem.org.my/documents/20181/268613/5.1.pdf/ea3269cb-ec2e-4d89-884a-33e24d2faa2a

Roslan, Q., Ibrahim, S. H., Affandi, R., Mohd Nawi, M. N., & Baharun, A. (2016). A literature review on the improvement strategies of passive design for the roofing system of the modern house in a hot and humid climate region. *Frontiers of Architectural Research, 5*(1), 126–133. DOI: 10.1016/j.foar.2015.10.002

Rubin, H. J., & Rubin, I. S. (2012). *Qualitative interviewing: The art of hearing data.* SAGE Publications.

Saaty, T. L. (1994). How to make a decision: The analytic hierarchy process. *Interfaces*, *24*(6), 19–43. DOI: 10.1287/inte.24.6.19

Saaty, T. L. (2008). Decision making with the analytic hierarchy process. *International Journal of Services Sciences*, *1*(1), 83–98. DOI: 10.1504/IJSSCI.2008.017590

Saaty, T. L., & Vargas, L. G. (2012). *Models, methods, concepts & applications of the analytic hierarchy process* (Vol. 175). Springer Science & Business Media. DOI: 10.1007/978-1-4614-3597-6

Saaty, T., Vargas, L., & Cahyono, S. (2022). *The Analytic Hierarchy Process*. Springer.

Salcini, S., & Yerlikaya, T. (2022). Metaverse: Technology of the future. Prizren Social Science Journal, 6(3).

Saldaña, J. (2013). *The coding manual for qualitative researchers* (2nd ed.). Sage Publications.

Salleh, M. F., Kamaruzzaman, S. N., & Ibrahim, M. H. (2021). Policy and participatory challenges in heritage adaptive reuse in Malaysia. *Asian Journal of Heritage Studies*, *15*(1), 10–25.

Samarghandi, H., Mousavi, S., Taabayan, P., Mir Hashemi, A., & Willoughby, K. (2016). Studying the Reasons for Delay and Cost Overrun in Construction Projects: The Case of Iran.

Sandanayake, M., Zhang, G., & Setunge, S. (2019). Estimation of environmental emissions and impacts of building construction – A decision making tool for contractors. *Journal of Building Engineering*, *21*, 173–185. DOI: 10.1016/j.jobe.2018.10.023

Sanhudo, L., Ramos, N. M., Martins, J. P., Almeida, R. M., Barreira, E., Simões, M. L., & Cardoso, V. (2018). Building information modeling for energy retrofitting–A review. *Renewable & Sustainable Energy Reviews*, *89*, 249–260. DOI: 10.1016/j.rser.2018.03.064

Santos, A. A., & Simon, A. (2022). The impact of community engagement on the success of heritage conservation projects: A case study of adaptive reuse. *Heritage*, *5*(3), 540–559.

Satterthwaite, D. (1998). Environmental Problems in Cities in the South: Sharing my Confusions. In Fernandez, E. (Ed.), *Grassroots Environmental Action—People's Participation in Sustainable Development* (pp. 62–82). Routledge.

Satyro, J. R., Silva, A. L., & Almeida, S. S. (2020). Managing organic waste through composting and anaerobic digestion: Case studies and sustainable strategies. *Waste Management & Research*, *38*(4), 403–414. DOI: 10.1177/0734242X20914401

Saurabh, S., & Madhura, Y. (2023). A thermal study of roof shapes and materials in hot dry climates: Insights from a simulation from Rajasthan, India. *International Society for the Study of Vernacular Settlements*, 10(12), 597–609.

Schell, J. (2020). *The art of game design: A book of lenses*. CRC Press.

Schoen, R. M., Dzwonczyk, M. K., & Williams, A. D. (2019). Smart water management: Technologies and applications. *Water Resources Management*, *33*(4), 1103–1121. DOI: 10.1007/s11269-019-02248-9

Schumacher, P. (2009). Parametricism: A new global style for architecture and urban design. In Leach, N., & Castle, H. (Eds.), *AD Architectural Design, 79(4)*. July/August. DOI: 10.1002/ad.912

Schumacher, P. (2010). *The autopoiesis of architecture* (Vol. 1). John Wiley & Sons.

Schumacher, P. (2022). *The metaverse as opportunity for architecture and society: Design drivers, core competencies*. Schumacher Architectural Intelligence.

Schweitzer, A., & Brown, J. (2021). The impact of education and public awareness on sustainable waste management practices. *Journal of Environmental Management, 256*, 109885. DOI: 10.1016/j.jenvman.2020.109885

Sedaghat, A., Salem, H., Hussam, W. K., Mahdizadeh, A., Al-Khiami, M. I., & Malayer, M. A. (2023). Experimental Study on Cool Roof Methods for Two Identical Portable Cabins in the State of Kuwait. *ICMAME 2023 Conference Proceedings. International Conference on Mechanical, Automotive and Mechatronics Engineering,* Dubai, UAE., n.d

Segalen, P. (1967). Les Sols et la Géomorphologie du Cameroun. Cahiers ORSTOM. *Série Pédologie., 1967*, 137–187.

Seidel, S., Berente, N., Yepes, G., & Nickeson, J. V. (2022). Designing the metaverse. *Proceedings of the 55th Hawaii International Conference on System Sciences*. DOI: 10.24251/HICSS.2022.811

Selvaraj, S., & Sundaravaradhan, S. (2020). Challenges and opportunities in IoT healthcare systems: A systematic review. *SN Applied Sciences*, *2*(1), 139. DOI: 10.1007/s42452-019-1925-y

Senior Minister of Infrastructure. (2020). *Construction 4.0 Strategic Plan (2021-2025)-Strategic Plan for Construction Industry: Gearing Up for the Fourth Industrial Revolution.* CIDB. https://www.cidb.gov.my/wp-content/uploads/2022/12/CR4_web_compressed.pdf

Shafique, M., Lee, D., & Kim, R. (2016). A Field Study to Evaluate Runoff Quantity from Blue Roof and Green Blue Roof in an Urban Area. *International Journal of Control and Automation*, *9*(8), 59–68. DOI: 10.14257/ijca.2016.9.8.07

Shakeri, S., & Ornek, M. A. (2023). How metaverse evolves the architectural design. Architecture and Planning Journal (APJ), 28(3), Article 35.

Shameerdas, T. (2024). Haritha Karma Sena and Non-Biodegradable Waste Management. *IJIRT* 11(3), 1455-1462. https://ijirt.org/publishedpaper/IJIRT167527_PAPER.pdf

Shan, J., Chai, C., Gui, H., & Xiong, Y. (2021). Enhancing project integration using cloud-based building information modelling: A conceptual model. *International Journal of Advanced Research in Technology and Innovation*, *3*(3), 12–25.

Shan, M., Chen, Y. F., Zhai, Z., & Du, J. (2022). Investigating the critical issues in the conservation of heritage building: The case of China. *Journal of Building Engineering*, *51*, 104319. DOI: 10.1016/j.jobe.2022.104319

Sharifi, A., & Yamagata, Y. (2015). Roof ponds as passive heating and cooling systems: A systematic review. *Applied Energy*, *160*, 336–357. DOI: 10.1016/j.apenergy.2015.09.061

Sharma, M., Joshi, S., Kannan, D., Govindan, K., Singh, R., & Purohit, H. C. (2020). Internet of Things (IoT) adoption barriers of smart cities' waste management: An Indian context. *Journal of Cleaner Production*, *270*, 122047. DOI: 10.1016/j.jclepro.2020.122047

Shehu, Z., Endut, I. R., Akintoye, A., & Holt, G. D. (2014). Cost overrun in the Malaysian construction industry projects: A deeper insight. *International Journal of Project Management*, *32*(8), 1471–1480. DOI: 10.1016/j.ijproman.2014.04.004

Shin, J., Moon, S., Cho, B., Hwang, S., & Choi, B. (2022). Extended technology acceptance model to explain the mechanism of modular construction adoption. *Journal of Cleaner Production*, *342*, 130963. DOI: 10.1016/j.jclepro.2022.130963

Shipley, R., Utz, S., & Parsons, M. (2020). Does adaptive reuse pay? A study of the business of building renovation in Ontario, Canada. *International Journal of Heritage Studies*, *12*(6), 505–520. DOI: 10.1080/13527250600940181

Sholeh, M. N., Fauziyah, S., & Khasani, R. R. (2020). Effect of Building Information Modeling (BIM) on reduced construction time-costs: a case study. In E3S Web of Conferences (Vol. 202, p. 02012). EDP Sciences.

Sighomnou, D. (2004). nalyse Et Redéfinition Des Régimes Climatiques Et Hydrologiques Du Cameroun: Perspectives D'Evolution Des Ressources En Eau. Ph.D. Thesis, University of Yaoundé I, Yaoundé, Cameroon.

Singh, R. (2021). Municipal Solid Waste Management in the City of Indore — A Case Study. *Journal of Civil Engineering and Environmental Sciences*, *7*(1), 8–17. https://www.engineegroup.us/articles/JCEES-7-139.pdf

Sliusar, N., Filkin, T., Huber-Humer, M., & Ritzkowski, M. (2022). Drone technology in municipal solid waste management and landfilling: A comprehensive review. *Waste Management (New York, N.Y.)*, *139*, 1–16. DOI: 10.1016/j.wasman.2021.12.006 PMID: 34923184

Smith, A. (1776). *An Inquiry into the Nature and Causes of the Wealth of the Nations*. Clarendon Press. DOI: 10.1093/oseo/instance.00043218

Smith, J. R., Wong, L., & Tan, K. (2022). Passive design strategies for energy efficiency in heritage architecture. *Energy and Building*, *247*, 111200.

Smith, J., & Chan, K. L. (2019). Energy analysis tools in adaptive reuse of heritage buildings: A comprehensive review. *Architectural Science Journal*, *22*(3), 245–261.

Smith, J., Tan, L. M., & Yusoff, M. (2023). Building Information Modeling (BIM) as a tool for adaptive reuse in heritage sites. *Architectural Review of Southeast Asia*, *15*(1), 45–60.

Smith, R. (2019). *Adaptive reuse: Extending the life of old buildings*. Routledge.

Song, K., Chen, Y., Duan, Y., & Zheng, Y. (2023). Urban Governance: A Review of Intellectual Structure and Topic Evolution. *Urban Governance* 3, 169-185. https://www.sciencedirect.com/science/article/pii/S2664328623000670

Song, Q., Yang, L., & Yin, B. (2014). Building for the climate change:Passive building. *Huazhong Architecture*, *32*(04), 11–15.

Sopery, L. H. M., & Ayob, M. F. (2023). An Investigation on Sustainable Refurbishment Practice of Building Works in Malaysia. Journal of Architecture [JAPCM]. *Planning and Construction Management*, *13*(1), 29–39.

Spalding, M. D., Ruffo, S., Lacambra, C., Meliane, I., Hale, L. Z., Shepard, C. C., & Beck, M. W. (2014). The role of ecosystems in coastal protection: Adapting to climate change and coastal hazards. *Ocean and Coastal Management*, *90*, 50–57. DOI: 10.1016/j.ocecoaman.2013.09.007

Spies, R., Grobbelaar, S., & Botha, A. (2020, April). A scoping review of the application of the task-technology fit theory. In *Conference on e-Business, e-Services and e-Society* (pp. 397-408). Cham: Springer International Publishing. DOI: 10.1007/978-3-030-44999-5_33

Srećković, M., Šibenik, G., Breitfuß, D., Preindl, T., & Kastner, W. (2021). Analysis of design phase processes with BIM for blockchain implementation. In *ECPPM 2021-eWork and eBusiness in Architecture, Engineering and Construction* (pp. 125–131). CRC Press. DOI: 10.1201/9781003191476-17

Stanga, C., Banfi, F., & Roascio, S. (2023). Enhancing building archaeology: Drawing, UAV photogrammetry and scan-to-BIM-to-VR process of ancient Roman ruins. *Drones (Basel)*, *7*(8), 521. DOI: 10.3390/drones7080521

Steenhuis, H. J., & Pretorius, L. (2015). Additive manufacturing or 3D printing and its adoption [Paper presentation]. *International Association for Management of Technology 2015 Conference Proceedings. University of Pretoria, Pretoria, South Africa.*

Strain, L., Rajabifard, A., & Williamson, I. (2006). Marine administration and spatial data infrastructure. *Marine Policy, 30*(4), 431–441. DOI: 10.1016/j.marpol.2005.03.005

Subramaniam, R., & Ramachandiran, C. R. (2020). Emergence of advanced digital technology to increase productivity in civil and infrastructure industry in Malaysia. *PalArch's Journal of Archaeology of Egypt/Egyptology, 17*(7), 5309-5318.

Suchel, J. B. (1987). Rainfall patterns and regimes rainfall in Cameroon. *Doc. Geogr. Trop., 5*, 287.

Sulaiman, A., Mustaffa, M., & Rahman, F. (2023). Innovative engagement methods in heritage conservation: The role of virtual reality. *Heritage & Society, 16*(1), 68–81.

Sutcliffe, J., Hunter, S., & Sim, N. (2018). The role of green infrastructure in sustainable urban drainage systems: Case studies from London. *Urban Water Journal, 15*(1), 50–60. DOI: 10.1080/1573062X.2017.1280226

Swan, W., & Blanchard, M. (2015). *Waste management in coastal cities: Challenges and solutions*. Routledge.

Syed Zainol Abidin, I. (1995). *Pemeliharaan Warisan Rupa Bandar*. Badan Warisan Malaysia.

Szokolay, S. V. (1980). *Environmental science handbook: For architects and builders* (2nd ed.). Architectural press.

Tafesse, S., Girma, Y. E., & Dessalegn, E. (2022). Analysis of the socio-economic and environmental impacts of construction waste and management practices. *Heliyon, 8*(3), e09169. DOI: 10.1016/j.heliyon.2022.e09169 PMID: 35368528

Tahir, M. M., Haron, N. A., Alias, A. H., Harun, A. N., Muhammad, I. B., & Baba, D. L. (2018). Improving Cost and Time Control in Construction Using Building Information Model (BIM): A Review. *Pertanika Journal of Science & Technology, 26*(1).

Tan, K. W. (2023). *Examine digital transformation in the Malaysian construction industry* [Unpublished doctoral dissertation]. UTAR, Perak, Malaysia.

Tan, K. W. (2023, September, 27). *Integrating technology in construction*. https://thesun.my/opinion-news/integrating-technology-in-construction-PB11543936

Tan, A. L., Wong, H. L., & Yap, S. L. (2019). NEWater: Singapore's success story in water reuse and sustainability. *Desalination and Water Treatment, 146*, 207–217. DOI: 10.5004/dwt.2019.24274

Tan, C. P., Tan, T. Y., & Ho, C. S. (2018). *Urban green infrastructure in Singapore: Integrating climate adaptation into water management*. Routledge.

Tan, H. W., & Lee, C. P. (2020). Sustainable approaches in historic building restoration: Techniques and implications. *Built Environment Review*, *15*(4), 180–195.

Tan, S. C., Lee, C. H., & Foo, T. S. (2020). Adaptive Reuse and Urban Regeneration: The Case of Penang, Malaysia. *Planning Practice and Research*, *35*(2), 172–188.

Tan, Y. Y., Ang, H. M., & Kuan, W. S. (2018). Smart water management and sustainable water usage in Singapore. *Water Science and Technology: Water Supply*, *18*(5), 1463–1471. DOI: 10.2166/ws.2018.104

Tariku, F., & Iffa, E., D. (2017). Temperature and air flow patterns in attic roofs. *Journal Of Architectural Engineering*, 23(3), 04017014-1-04017014-11.

Technology Cards. (2019, August, 16). *Autonomous construction vehicles*. https://www.technologycards.net/the-technologies/autonomous-construction-vehicles

Terry, G., Braun, V., & Clarke, V. (2020). Digital preservation practices for heritage buildings: A qualitative perspective. *Heritage Studies Quarterly*, *8*(3), 110–130.

Terry, G., Hayfield, N., Clarke, V., & Braun, V. (2017). Thematic analysis. In Willig, C., & Stainton Rogers, W. (Eds.), *The SAGE handbook of qualitative research in psychology* (pp. 17–37). SAGE., DOI: 10.4135/9781526405555.n2

Terry, G., Hayfield, N., Clarke, V., & Braun, V. (2019). Thematic analysis. In Willig, C., & Rogers, W. (Eds.), *The SAGE Handbook of Qualitative Research in Psychology* (pp. 17–37). SAGE.

Thakkar, J. J. (2021). *Multi-criteria decision making* (Vol. 336). Springer. DOI: 10.1007/978-981-33-4745-8

Tharim, A. H. A.. (2018). Economic Impact of Adaptive Reuse in Malaysian Cities. *International Journal of Built Environment and Sustainability*, *5*(1), 41–49.

The Digital Speaker. (2024). Digital twins are a building block of the metaverse. Retrieved from https://www.thedigitalspeaker.com

Thodukayil, F. S. O., Raseena, R. S., Palaniswamy, U., & Panneer, S. (2024). Challenges of Women Workers in the Solid Waste Management Sector of Kerala: on the Path to Formalization. In Vinodan, A., Mahalakshmi, S., & Rameshkumar, S. (Eds.), *Informal Economy and Sustainable Development Goals: Ideas, Interventions, and Challenges* (pp. 373–397). Emerald Publishing., https://www.emerald.com/insight/content/doi/10.1108/978-1-83753-980-220241020/full/html DOI: 10.1108/978-1-83753-980-220241020

Thomas, E., Nelson, N. S., Saju, N., Menon, A., & Aleena, P. (2023). A Study on the Socio-Economic Profile of Haritha Karma Sena Workers with Special Reference to Cheranalloor Panchayat. http://117.239.78.102:8080/jspui/bitstream/123456789/2856/1/PROJECT%20COMBINED.pdf

Thompson, E., & Qwabe, B. (2020). Sustainability in contemporary architecture: A review of current research and practice. *Journal of Environment & Development, 12*(3), 45–62.

Thompson, J., Jones, M., & Wild, G. (2020). Smart technologies in architectural design: Enhancing adaptive and responsive environments. *International Journal of Architectural Computing, 18*(4), 347–362.

Thravalou, S., Michopoulos, A., Alexandrou, K., & Artopoulos, G. (2023, June). Energy retrofit strategies of built heritage: Using Building Information Modelling tools for streamlined energy and economic analysis. [). IOP Publishing.]. *IOP Conference Series. Earth and Environmental Science, 1196*(1), 012115. DOI: 10.1088/1755-1315/1196/1/012115

Tibaijuka, K. (2009). *Building Prosperity: Housing and Economic Development.* Earthscan.

Tobias, M. (2004, March 18). Elements of Sustainable architecture. Retrieved from https://www.ny-engineers.com/blog/elements-of-sustainable-architecture

Tom, T., & Kamat, R. (2020). A Study on Waste Management of Household with Special Reference to Uzhavoor Panchayat — Kottayam District, Kerala. *Alochana Chakra Journal, 9*(5), 2533–2543.

Tuğba Ercan, M., Tuna Kayili, M., & Sultan Qurraie, B. (2021). The effects of green roof on heat loss and energy consumption in the buildings. *Computational Research Progress in Applied Science & Engineering, 7*(4), 1–8. DOI: 10.52547/crpase.7.4.2422

Tunis, J. (2021, September 7). A detailed guide to real-time monitoring. Netreo. https://www.netreo.com/blog/real-time-monitoring/#:~:text=A%20Quick%20Definition%20of%20Real,applications%2C%20and%20services%20are%20performing

Tze Qing, A. (2019). Exploring the adoption of Internet of Things in Malaysian construction industry [Quantity Surveying, Universiti Tunku Abdul Rahman]. http://eprints.utar.edu.my/3384/1/QS-2019-1425152-1.pdf

Ukwe, C. N., Ibe, C. A., & Sherman, K. (2006). A sixteen-country mobilization for sustainable fisheries in the Guinea Current Large Marine Ecosystem. *Ocean and Coastal Management, 49*(7-8), 385–412. DOI: 10.1016/j.ocecoaman.2006.04.006

UN. (2022). Transforming our World: The 2030 Agenda for Sustainable Development. A/RES/70/1. 2015. Available online: https://sustainabledevelopment.un.org/post2015/transformingourworld (accessed on 3 August 2022).

UNCHS (United Nations Centre for Human Settlements). (1998). *Sustainable Cities Program*. UNCHS.

UNDESA (United Nations Department of Economic and Social Affairs). (2018). *World Urbanization Prospects: The 2018 Revision*. UN.

UNDESA. (2019). *World Urbanization Prospects 2018: Highlights*. UN.

UNEP. (2016). *Global outlook on sustainable consumption and production policies*. United Nations Environment Programme.

UNEP. (2016). Global outlook on sustainable consumption and production policies. United Nations Environment Programme. https://www.unep.org/resources/report/global-outlook-sustainable-consumption-and-production-policies

UN-Habitat. (2011). *Global report on human settlements 2011: Cities and climate change: Global report on human settlements*. United Nations Human Settlements Programme.

UN-Habitat. (2020). *The state of African cities 2020: Inclusive cities for resilient economies*. United Nations Human Settlements Programme.

UN-Habitat. (2020). The state of African cities 2020: Inclusive cities for resilient economies. United Nations Human Settlements Programme. https://unhabitat.org/the-state-of-african-cities-2020-inclusive-cities-for-resilient-economies

UNSDG. (2015). UN General Assembly, Transforming our world: the 2030 Agenda for Sustainable Development, 21 October 2015, A/RES/70/1, available at: https://www.refworld.org/docid/57b6e3e44.html [accessed 1 December 2021]

Vaitis, M., Kopsachilis, V., Tataris, G., Michalakis, V. I., & Pavlogeorgatos, G. (2022). The development of a spatial data infrastructure to support marine spatial planning in Greece. *Ocean and Coastal Management*, *218*, 106025. DOI: 10.1016/j.ocecoaman.2022.106025

Van Eldik, M. A., Vahdatikhaki, F., dos Santos, J. M. O., Visser, M., & Doree, A. (2020). BIM-based environmental impact assessment for infrastructure design projects. *Automation in Construction*, *120*, 103379. DOI: 10.1016/j.autcon.2020.103379

Van Tam, N., Quoc Toan, N., Phong, V. V., & Durdyev, S. (2023). Impact of BIM-related factors affecting construction project performance. *International Journal of Building Pathology and Adaptation*, *41*(2), 454–475. DOI: 10.1108/IJBPA-05-2021-0068

Vegavid. (2023). How does digital twin work in the metaverse? Retrieved from https://www.vegavid.com

Vicat, J.P., Pouclet, A., Nsifa, E. (1998). Les Dolérites du Groupe du Ntem (Sud Cameroun) et des Régions Voisines (Centrafrique, Gabon, Congo, Bas Zaïre): Caractéristiques Géochimiques et Place dans L'évolution du Craton du Congo au Protérozoïque. Géologie Environ. Cameroun Collect. GEOCAM. 305–324.

Villarini, G., Vecchi, G. A., & Smith, J. A. (2019). The role of climate change in the occurrence and intensity of extreme weather events. *Journal of Climate*, *32*(24), 1–15. DOI: 10.1175/JCLI-D-19-0047.1

Wang, X., Zhang, Y., & Chen, L. (2020). Balancing conservation and sustainability in heritage building retrofits. *Journal of Architectural Conservation*, *26*(2), 97–110.

Wang, X., Zhang, Y., & Chen, L. (2021). Leveraging digital technologies for sustainable heritage conservation. *International Journal of Heritage Studies*, *27*(5), 609–627.

Wang, Z., Liu, X., Wang, X., Wang, H., Sun, Y., Zhang, J., & Li, H. (2024). Contrasting benthic bacterial and fungal communities in two temperate coastal areas affected by different levels of anthropogenic pressure. *Marine Environmental Research*, *198*, 106501. DOI: 10.1016/j.marenvres.2024.106501 PMID: 38615486

Waqar, A., Khan, M. B., Shafiq, N., Skrzypkowski, K., Zagórski, K., & Zagórska, A. (2023). Assessment of challenges to the adoption of IOT for the safety management of small construction projects in Malaysia: Structural equation modeling approach. *Applied Sciences (Basel, Switzerland)*, *13*(5), 3340. DOI: 10.3390/app13053340

Water, U. N. (2018). The United Nations world water development report 2018: Nature-based solutions for water. United Nations. https://www.unwater.org/publications/world-water-development-report-2018/

Water, U. N. (2018). *The United Nations world water development report 2018: Nature-based solutions for water*. United Nations.

Watson, J. (2020, October, 30). How cloud can enable collaboration in construction industry. https://www.acecloudhosting.com/blog/cloud-collaboration-in-construction-industry/

Weinberg, A. (2021, October 24). Top 4 challenges in IoT data collection and management. first point. https://www.firstpoint-mg.com/blog/top-4-challenges-in-iot-data-collection-and-management/

Wilkinson, S. (2011). *Sustainable retrofit potential in lower quality office stock in the central business district*. CIB Management and Innovation in the Sustainable Built Environment.

Wilson, D. C., Velis, C. A., & Cheeseman, C. R. (2015). Role of informal sector recycling in waste management in developing countries. *Habitat International*, *31*(1), 6–15.

Wise, F., Moncaster, A., & Jones, D. (2021). Rethinking retrofit of residential heritage buildings. *Buildings & Cities*, *2*(1), 495–517. DOI: 10.5334/bc.94

Wong, J. H., Rashidi, A., & Arashpour, M. (2020). Evaluating the impact of building information modeling on the labor productivity of construction projects in Malaysia. *Buildings (Basel, Switzerland)*, *10*(4), 66. DOI: 10.3390/buildings10040066

Wong, J. K. W., & Zhou, J. (2015). Enhancing environmental sustainability over building life cycles through green BIM: A review. *Automation in Construction*, *57*, 156–165. DOI: 10.1016/j.autcon.2015.06.003

Wong, J., Hoong, P., Teo, E., & Lin, A. (2022, November). Digital twin: A conceptualization of the task-technology fit for individual users in the building maintenance sector. [). IOP Publishing.]. *IOP Conference Series. Earth and Environmental Science*, *1101*(9), 092041. DOI: 10.1088/1755-1315/1101/9/092041

Wong, K. S., Ibrahim, R., & Lim, S. (2021). Advances in rainwater harvesting technologies: Challenges and solutions for urban areas. *Environmental Science & Technology*, *55*(13), 8904–8914. DOI: 10.1021/acs.est.1c01677

Wong, L., & Cheng, S. (2023). Fostering cultural heritage preservation through community involvement: Case studies from Malaysia. *Heritage Management*, *10*(1), 15–30.

Won, J., & Cheng, J. C. (2017). Identifying potential opportunities of building information modeling for construction and demolition waste management and minimization. *Automation in Construction*, *79*, 3–18. DOI: 10.1016/j.autcon.2017.02.002

Won, J., Cheng, J. C., & Lee, G. (2016). Quantification of construction waste prevented by BIM-based design validation: Case studies in South Korea. *Waste Management (New York, N.Y.)*, *49*, 170–180. DOI: 10.1016/j.wasman.2015.12.026 PMID: 26754615

World Bank (2013). *Global Monitoring Report 2013: Rural-Urban Dynamics and the Millennium Development Goals.*

World Bank. (2018). *Solid waste management in developing countries.* World Bank Group.

World Bank. (2018). Solid waste management in developing countries. World Bank Group. https://openknowledge.worldbank.org/handle/10986/30303

World Bank. (2023). *World Bank Urban Development Review: The 2023 Revision.* World Bank.

World Urbanization Prospects, U. N. (2018). Population Dynamics. https://www.un.org/development/desa/pd/sites/www.un.org.development.desa.pd/files/files/documents/2020/Feb/un_2018_wup_highlights.pdf

Wu, B., & Chen, X. (2017). Continuance intention to use MOOCs: Integrating the technology acceptance model (TAM) and task technology fit (TTF) model. *Computers in Human Behavior*, *67*, 221–232. DOI: 10.1016/j.chb.2016.10.028

Wu, Z., Luo, L., Li, H., Wang, Y., Bi, G., & Antwi-Afari, M. F. (2021). An Analysis on Promoting Prefabrication Implementation in Construction Industry towards Sustainability. *International Journal of Environmental Research and Public Health*, *18*(21), 11493. DOI: 10.3390/ijerph182111493 PMID: 34770008

Xu, X., & De Soto, B. G. (2020). On-site autonomous construction robots: A review of research areas, technologies, and suggestions for advancement. *Proceedings of the International Symposium on Automation and Robotics in Construction, 37*, 385-392. https://doi.org/DOI: 10.22260/ISARC2020/0055

Yan, Z., Guo, X., Zhao, Z., & Tang, L. (2024). Achieving fine-grained urban flood perception and spatio-temporal evolution analysis based on social media. *Sustainable Cities and Society*, *101*, 105077. DOI: 10.1016/j.scs.2023.105077

Yap, J. B. H., Chow, I. N., & Shavarebi, K. (2019). Criticality of construction industry problems in developing countries: Analyzing Malaysian projects. *Journal of Management Engineering*, *35*(5), 04019020. Advance online publication. DOI: 10.1061/(ASCE)ME.1943-5479.0000709

Yau, Y. H., & Hasbi, S. J. E. P. (2017). A comprehensive case study of climate change impacts on the cooling load in an air-conditioned office building in Malaysia. *Energy Procedia*, *143*, 295–300. DOI: 10.1016/j.egypro.2017.12.687

Yousif, O. S., Zakaria, R., Aminudin, E., Yahya, K., Mohd Sam, A. R., Singaram, L., Munikanan, V., & Shamsuddin, S. M. (2021). Review of Big Data Integration in Construction Industry Digitalization. *Frontiers in Built Environment*, *7*, 770496. DOI: 10.3389/fbuil.2021.770496

Yuan, J., Patra, I., Majdi, A., Acwin Dwijendra, N. K., Catalan Opulencia, M. J., & Chetthamrongchai, P. (2022). RETRACTED: Fundamental green roof performance of residential building in desert climate: In terms of sustainability and decrease in energy consumption. *Sustainable Energy Technologies and Assessments*, *53*, 102574. DOI: 10.1016/j.seta.2022.102574

Yu, J., Leng, K., Ye, H., Xu, X., Luo, Y., Wang, J., Yang, X., Yang, Q., & Gang, W. (2020). Study on thermal insulation characteristics and optimized design of pipe-embedded ventilation roof with outer-layer shape-stabilized PCM in different climate zones. *Renewable Energy*, *147*, 1609–1622. DOI: 10.1016/j.renene.2019.09.115

Yung, E. H. K., & Chan, E. H. W. (2018). Adaptive reuse: A sustainable solution for historic buildings in Hong Kong. *Procedia Engineering*, *118*, 352–357.

Yunos, M. Y. M., Nasir, N. F. M., & Othman, R. (2018). Preserving Heritage through Adaptive Reuse: A Case Study of George Town and Melaka. *Asian Journal of Environment-Behaviour Studies*, *3*(6), 75–89.

Yunos, M. Y., Nasir, N., & Othman, R. (2019). Evaluating the effectiveness of adaptive reuse projects in George Town, Penang. *Asian Heritage Review*, *6*(1), 78–95.

Yusof, A., Khoso, A. R., Sohu, S., Khahro, S. H., & Chai, C. S. (2021). Improving performance in construction projects: A case study of Malaysian public projects. *Pertanika Journal of Science & Technology*, *29*(4), 2579–2604. DOI: 10.47836/pjst.29.4.19

Yusoff, M. (2022). Digital innovations in architectural heritage: Malaysia's approach. *Heritage Conservation Journal*, *8*(3), 85–98.

Yusoff, M., & Ibrahim, R. (2022). Balancing heritage preservation and sustainability in Malaysian architecture. *Heritage Conservation Journal*, *8*(3), 85–98.

Yusoff, S. N. S., & Brahim, J. (2021). Implementation of building information modeling (Bim) for social heritage buildings in kuala lumpur. *International Journal of Sustainable Construction Engineering and Technology*, *12*(1), 88–99. DOI: 10.30880/ijscet.2021.12.01.009

Yusoff, W. Z. W., Majid, R. A., & Ghazali, A. (2019). Economic Benefits of Heritage Building Conservation in Ipoh, Malaysia. *International Journal of Heritage Studies*, *25*(5), 455–470.

Zaha Hadid Architects. (2023). *Metrotopia: Metaverse virtual communication hub.* Retrieved fromhttps://www.zaha-hadid.com/2023/06/01/metrotopia-metaverse-virtual-communication-hub/

Zaha Hadid Virtual Reality Group. (2024). *Metrotopia.* Retrieved from https://www.zhvrgroup.com/metroto

Zahrizan, Z., Ali, N. M., Haron, A. T., Marshall-Ponting, A., & Hamid, Z. A. (2013). Exploring the adoption of Building Information Modelling (BIM) in the Malaysian construction industry: A qualitative approach. *International Journal of Research in Engineering and Technology, 2*(8), 384–395. https://citeseerx.ist.psu.edu/document?repid=rep1&type=pdf&doi=c2dde67f06be045aadb766e739241ca02c996b70. DOI: 10.15623/ijret.2013.0208060

Zaid, N. S. M., Ghazali, M. I., Azmi, N. F., Kamarulzaman, P. N., & Akashah, F. W. (2021). Environmental sustainability in Malaysian Heritage Building Conservation: Review of Heritage Legistation and Policy. Malaysian Construction Research Journal (MCRJ), 203.

Zainal, A. (2021). Economic benefits of low-carbon restoration in heritage building projects. *Sustainable Architecture Reports, 12*(1), 57–73.

Zainal, R., Ahmad, F., & Hamdan, H. (2018). Enhancing Cultural Awareness through Adaptive Reuse of the Sultan Abdul Samad Building. *Journal of Heritage Tourism, 13*(4), 314–330.

Zhang, L., & Liao, B. (2020). The role of greywater recycling in sustainable urban water management. *Environmental Science & Technology, 54*(15), 9208–9217. DOI: 10.1021/acs.est.0c02238

Zhang, S., Yi, B., Worrell, E., Wagner, F., Crijns-Graus, W., Purohit, P., Wada, Y., & Varis, O. (2019). Integrated Assessment of the Resource-Energy-Environment Nexus in China's Iron and Steel Industry. *Journal of Cleaner Production, 232*, 235–249. https://www.sciencedirect.com/science/article/abs/pii/S0959652619319353. DOI: 10.1016/j.jclepro.2019.05.392

Zhang, Y., Bi, J., & Huang, X. (2020). Circular economy and sustainable waste management in coastal cities: A review of challenges and opportunities. *Environmental Science and Pollution Research International, 27*(9), 9143–9155. DOI: 10.1007/s11356-020-07768-3

Zhang, Y., Wang, X., & He, L. (2020). Circular economy in urban waste management: Concepts, applications, and future perspectives. *Waste Management & Research, 38*(4), 347–357. DOI: 10.1177/0734242X20910101 PMID: 32116140

Zhao, M., Li, W., & Chen, Z. (2020). The role of water-efficient appliances in urban water conservation. *Water Resources Management*, *34*(2), 561–577. DOI: 10.1007/s11269-019-02439-9

Zhao, X., Ma, X., Chen, B., Shang, Y., & Song, M. (2022). Challenges toward carbon neutrality in China: Strategies and countermeasures. *Resources, Conservation and Recycling*, *176*, 105959. DOI: 10.1016/j.resconrec.2021.105959

Zhao, Y., Sun, Y., Zhou, Q., Cui, C. Y., & Liu, Y. (2022). How A/E/C professionals accept BIM technologies in China: A TAM perspective. *Engineering, Construction, and Architectural Management*, *30*(10), 4569–4589. DOI: 10.1108/ECAM-04-2022-0308

Zhou, D., Tian, Y., & Jiang, G. (2018). Spatio-temporal investigation of the interactive relationship between urbanization and ecosystem services: Case study of the Jingjinji urban agglomeration, China. *Ecological Indicators*, *95*, 152–164. DOI: 10.1016/j.ecolind.2018.07.007

Zhou, Y., & Staatz, J. (2016). Projected Demand and Supply for Various Foods in West Africa: Implications for Investments and Food Policy. *Food Policy*, *61*, 198–212. https://www.sciencedirect.com/science/article/abs/pii/S0306919216300185?via%3Dihub. DOI: 10.1016/j.foodpol.2016.04.002

Zhou, Y., Wang, S., & Jiang, C. (2019). Waste-to-energy technologies and their potential in reducing urban waste: A case study of pyrolysis and gasification. *Renewable & Sustainable Energy Reviews*, *109*, 358–368. DOI: 10.1016/j.rser.2019.04.037

Zhou, Z., Zhang, L., & Li, C. (2021). *Desalination: Principles, technologies, and environmental considerations*. Elsevier., DOI: 10.1016/B978-0-12-819138-1.00001-3

Zimmermann, M. (2023, September 26). Industrialized construction (IC), the key to reducing the environmental impact of industry [Blog post]. *Cemex Ventures*. https://www.cemexventures.com/industrialized-construction/

Žurić, J., Zichi, A., & Azenha, M. (2023). Integrating HBIM and sustainability certification: A pilot study using GBC historic building certification. *International Journal of Architectural Heritage*, *17*(9), 1464–1483. DOI: 10.1080/15583058.2022.2042623

About the Contributors

Sujatavani Gunasagaran stands as a distinguished Program Director and Senior Lecturer within the esteemed School of Architecture, Building, and Design at Taylor's University. Her academic journey is marked by excellence, having earned a BSc in Architecture from the University of Malaya, an MSc in Building Technology from the University Science of Malaysia, and a Doctorate of Education from the University of Selangor. Dr. Suja is recognized as a Professional Technologist and holds membership in prestigious organizations such as the Malaysia Board of Technologist (MBOT), Malaysian Institute of Architects (PAM) Education, Malaysia Green Building Council (malaysiaGBC), Society of Building Science Educators (SBSE), and is a Green Building Index (GBI) Facilitator. Her commitment to advancing the realms of architectural education has not gone unnoticed, with several awards honouring her innovative approaches to teaching and learning. At the heart of her teaching philosophy and research lies a fervent dedication to sustainability, creativity and cutting-edge technology use such as Building Information Modeling (BIM) and digital simulation. Dr. Suja extends her expertise to guide and mentor students pursuing advanced degrees, overseeing both master's and PhD programs. Beyond her role in academia, she plays a pivotal role in shaping the academic landscape, contributing significantly as a member of the Committee for Curriculum, Design, Assessment, and Delivery. Her administrative responsibilities involve designing novel programs and ensuring the ongoing quality of existing ones. In addition to her teaching and administrative roles, Dr. Suja is a distinguished Editor for two journal publications, further establishing herself as an active and influential figure in the realm of research and publication in built environement. Her multifaceted contributions underscore her commitment to advancing the field of architecture, construction management and education, making her a respected and integral part of the academic community.

Kenn Jhun Kam stands as a Senior Lecturers within the esteemed School of Architecture, Building, and Design at Taylor's University. He has a Degree in Building Technology, a Master's Degree in Building Technology, and a Ph.D. in Building Quality and Assessment. He is a Chartered Surveyor from the Royal Institution of Chartered Surveyors (RICS), a registered Professional Technologist with the Malaysian Board of Technologists (MBOT) and a registered Engineering Technologist with the Board of Engineers Malaysia (BEM). He has experience in construction for 5 years and has completed 2 high-rise projects from IJM before joining the education industry. He has almost 10 years of experience in academia. He also held a management position such as Associate Dean at his current institution. Most of the subjects that he has been teaching are construction-related subjects such as Construction Technology, Building Services, Research Methodology and Dissertations.

***

Khairool Aizat Ahmad Jamal is a well-rounded professional who is both knowledgeable and passionate about the field of Building Information Modelling (BIM). With a strong background in architecture and construction, he holds a Bachelor of Architecture (Hons.) from Universiti Teknologi Malaysia (UTM), an M.Sc degree in Integrated Construction Project Management from Universiti Teknologi MARA (UITM), and a Post-Graduate Certificate of Teaching and Learning from Taylor's University. He is also registered as a Graduate Architect with the Malaysia Board of Architects (LAM) and a Professional Technologist with the Malaysia Board of Technologists (MBOT). Khairool's prior expertise in the construction industry was honed through his work with both local and international firms, where he gained valuable experience in the process. His role as an academician has allowed him to expand his knowledge and perspectives, working with students from a variety of disciplines within the built environment. His passion for BIM and Integrated Project Delivery (IPD) has inspired him to further pursue a PhD studies in the topic of Historic-BIM Retrofit. He is currently active in research and supervision activities and has made significant contributions to the field through presentations at various workshops and conferences, and by publishing research books, articles, and proceedings in indexed journals and magazines.

Ariventhar Ayahvoo is a Graduate Architect and a postgraduate student at Taylor's University, Malaysia. He holds a Bachelor of Science (Honours) in Architecture, which he completed in April 2021. With a strong interest in adaptive reuse and heritage conservation, his research focuses on innovative strategies for revitalizing historical structures while preserving their cultural significance.

Ariventhar brings practical experience in architectural design and heritage studies, and he is passionate about exploring sustainable approaches to architectural preservation. He is currently contributing to IGI Global as an author.

Hong Cing Cing is a Quantity Surveyor at PMT Enterprise. She holds a Bachelor of Science in Housing, Building and Planning (Quantity Surveying) from Universiti Sains Malaysia. Her professional affiliations include being a Provisional Quantity Surveyor registered with the Board of Quantity Surveyors Malaysia, a Graduate Member of the Royal Institution of Surveyors Malaysia, and a member of the Construction Industry Development Board (CIDB) Malaysia.

Serkan Derici started his academic career as a research assistant at Nevşehir Hacı Bektaş Veli University, Faculty of Economics and Administrative Sciences, Department of Business Administration in 2017. He is currently working as an Assistant Professor Doctor at the same university. Derici received his bachelor's degree from Nevşehir University, Department of Business Administration in 2012. He studied at Erciyes University for one year with the Farabi student exchange programme. He completed his master's degree in Business Administration at the same university and then completed his doctorate. Derici started his academic life as a research assistant at Istanbul Gelisim University School of Applied Sciences, Department of Aviation Management in 2016. His current research mainly focuses on quantitative methods, machine learning, supply chain management, operations management and multi-criteria decision making.

Jipson Joseph has a PhD in Ethics from DVK Bangalore. Currently, he is a Research Scholar in the School of Law at Christ University, Bangalore, India. His research interests are in the fields of ethics, gender studies, constitutional law, human rights, human security, environmental justice, SDGs, developmental policies and global poverty. He has published 2 monographs, co-edited one book, and contributed scholarly articles to many international and national journals and edited volumes including IGI Global Scientific Publishing's edited books.

Myzatul Aishah Hj Kamarazaly is the Programme Director for the Bachelor of Quantity Surveying (Hons.) Programme in the School of Architecture, Building & Design at Taylor's University, Lakeside Campus, Malaysia. She brings on board more than 15 years of academic leadership experience. Myzatul received her Doctoral of Philosophy (in Facilities Management) and Master of Philosophy (in Construction) from Massey University, New Zealand, Bachelor of Quantity Surveying (Hons) and Diploma in Quantity Surveying from UiTM, Malaysia. She is an active Associate Member of the Malaysian Association of Facilities Management (MAFM), a member of the Board of Quantity Surveyor Malaysia (BQSM) and Royal

Institution of Surveyor Malaysia (RISM). Dr Myzatul Aishah has extensive research publications and involvement as International Reviewers and Editorial Board for Journal Publication. Her area of specialization is Facilities Management, BIM, QS education and profession as well as built-environment related-studies.

Noor Azeyah Khiyon, Head of School of the Built Environment at University of Reading Malaysia

Chia Kuang Lee is the Deputy Dean of Research and Postgraduate Studies at the Faculty of Industrial Management, UMPSA. He earned a PhD in Civil Engineering from The University of Auckland, New Zealand, and holds a Bachelor of Quantity Surveying and a Master of Science (MSc. Construction Contract Management) from UTM. Dr. Lee has actively researched the use of Alternative Dispute Resolution (ADR) in construction projects. He has authored more than 100 scholarly works, focusing mainly on dispute resolution, construction and project management, safety, and health management.

Nur Syaimasyaza Mansor is a Senior Lecturer at Universiti Sains Malaysia. She holds a PhD in Built Environment from the International Islamic University Malaysia, a Master's in Construction Contract Management from Universiti Teknologi Malaysia, and a Bachelor's degree in Quantity Surveying from Universiti Malaya. Her areas of expertise and research interest include quantity surveying, construction contracts, procurement, and cybersecurity in the construction industry. She is a registered Professional Quantity Surveyor with the Board of Quantity Surveyors Malaysia and a member of the Royal Institution of Surveyors Malaysia. She can be contacted at syaimasyaza@usm.my.

TamilSalvi Mari is an Associate Professor and the Program Director for the Master of Science in Virtual Design and Construction (MVDC) at Taylor's University, School of Architecture, Building, and Design (SABD). She brings extensive experience in the construction industry, focusing on academic leadership, sustainable design, and digital construction technologies. As a certified Green Building Index (GBI) Facilitator and Council Member of the Malaysia Green Building Council (MGBC), she co-chairs the Sustainable Development Goals (SDG) Subcommittee and leads workshops on sustainability. Her research spans energy efficiency, smart cities, building materials, digital construction, and Virtual Reality applications in construction, with her work widely published and recognized for its contributions to enhancing community well-being and the built environment.

Mohamed El Bakali, senior lecturer at the Higher Institute of Nursing and Health Techniques – Kenitra branch, has been teaching biostatistics, research methodology, and epidemiology since 2024. Previously, starting from 2011, he led the evaluation of social programs targeting vulnerable populations at the central headquarters of Entraide Nationale. Holding a doctorate in epidemiology and biostatistics, a master's degree in neurobiology, pharmacology, and genetics, a bachelor's degree in life sciences, and a diploma in private Arab law, he has authored numerous Scopus-indexed articles in environmental and life sciences. His expertise includes biostatistical analysis (R, Python, SPSS), machine learning and deep learning techniques (1D CNN with self-attention, XGBoost, PySR), and project management. He communicates fluently in Arabic and French and possesses intermediate proficiency in English and Spanish.

Yasine Mouniane is a researcher in sustainable development, with a strong focus on environmental resilience, spatial data infrastructures and geospatial technologies. His academic and field contributions focus on the integration of digital solutions to assess and manage ecological and environmental challenges. Passionate about the intersection between science, technology and territorial planning, he is actively working to create tools to support evidence-based decision-making and promote climate adaptation strategies.

Yufeng Niu is pursuing a Master's in the School of Architecture at Taylor's University, Malaysia.

Ananya Pandey is a Research Scholar in School of Law at Christ University, Bangalore. She has her research interests in the field of Constitutional Law, Human Rights, SDGs, Climate Change, Intellectual Property Law and Jurisprudence.

Chai Chang Saar is an Associate Professor in the School of Architecture, Building & Design at Taylor's University, Malaysia. He holds a Ph.D. in Civil Engineering, an M.Sc. in Construction Management, and a B.Sc. in Construction, all from Universiti Teknologi Malaysia. With over a decade of combined industry and academic experience, his research centres on digital construction, encompassing technology adoption, API development, BIM-driven workflows, construction project management, and sustainability. He has authored numerous refereed journal articles, secured competitive national and international research grants, and serves on several construction-policy drafting committees to advance the field's practice and standards.

Chai Chang Saar is an Associate Professor in the School of Architecture, Building and Design, Taylor's University. He is a passionate academic and researcher with enthusiasm towards internationalisation of his work and service. Chai spent a decade in construction, higher education, information technology and project management, and completed several projects locally and internationally prior to joining academia. His research interest encompasses the digital construction in the areas of technology adoption, API development, project management, and sustainability. His intellectual curiosity and dedication to advancing knowledge have yielded an impressive collection of impactful research findings, prominently featured in distinguished refereed journal articles. Additionally, Chai has proven his mettle by securing competitive research grants from esteemed national and international funding bodies. His commitment to the industry transcends academia. Chai's involvement in shaping national and international construction-related policy drafting committees underscores his influence on the broader narrative of the field. Beyond the academic realm, he remains actively engaged with a growing network of professional bodies and industry collaborations, testament to his enduring commitment to shaping the future of his domain.

Shahrul Yani Said is an Associate Professor in Heritage Conservation. She started her journey in academics with a BSc Architectural Studies (Hons) and a Bachelor Architecture (Hons) from International Islamic University Malaysia, studying heritage design. She continues her pursuit of heritage conservation in the United Kingdom and received an MA in Conservation Studies (Historic Buildings) in 2008 from the University of York. Her investigation into urban heritage and townscape led to her PhD journey from Oxford Brookes University, where she received her Doctor of Philosophy in 2012. She is an inspiring researcher and produces and reviews many journal articles in the field of heritage conservation. Currently an ICOMOS member, her interest in heritage area nomination and digital documentation of heritage is evident in the number of active research grants received.

Li Shun is a PhD student at the School of Architecture and Building Design, Taylor's University. He worked as an architecture teacher in the School of Civil and Architectural Engineering of Panzhihua University in China.

Mervyn Wong began his career at Design Network Architects (DNA) after graduating from Universiti Sains Malaysia (USM) with a Bachelor of Science in Housing, Building, and Planning (HBP) and a Bachelor of Architecture. As his career progressed, Ar. Mervyn became known for his ability to blend innovation with practicality. His designs were characterized by a seamless integration of sustainable practices, cutting-edge technology, and an unwavering commitment to creating spaces that harmonize with their surroundings. Ar. Mervyn has accomplished a large number of residential and commercial architecture projects during the previous decade. He believes that architecture should be a reflection of the people it serves and that every structure should tell a story. His approach is deeply rooted in understanding the needs and aspirations of the clients, allowing him to create bespoke designs that are functional, aesthetically pleasing, and emotionally resonant.

Yaoli Xiong is pursuing a PhD in Architecture at Taylor's University, Malaysia.

Siti Norzaini Zainal Abidin is a Senior Lecturer at School of Architecture and Building Design, Taylor's University. Her contribution on the paper 'Strategic Daylight Visualization for the Green Building Process: Integrating Optimization Into An Airport Terminal Design With Skylights' has obtained her the best paper, PhD student award for ConVR in December 2016 at Regal Kowloon, Hong Kong organised by The Hong Kong University of Science and Technology. Her research interests include passive design in green building, integrated design process, design management, built environment, sustainable living, energy and architecture, daylighting in airport design and green building.

Ahmad Fahmi Zainazlan Graduated from UiTM Malaysia for his Part 1 and 2 Architectural Study, Ahmad Fahmi got recognized as a Professional Technologist specialized in Building Construction and BIM. Experienced in Building Information Modelling in Schematics, Submissions and Site Implementation as well as teaching BIM as academician. Starting discovering BIM in Heritage and Archaeology through a systematic review paper, He continues developing portfolio through Heritage digital BIM reconstruction of a conservation temple building. Through the same spectrum, he was involved in preparing 2 BIM digital reconstruction of federal buildings at Putrajaya. Under UNESCO world heritage project, He is currently works as a Senior Lecturer in Taylors University and involves in International BIM Heritage and Digital Reconstruction Project of a Fort of 4.6 acres in Pakistan under the Directorate of Archaeology and Museum, Government of Khyber Pakhtunkhwa, Pakistan.

Index

U

V

www.ingramcontent.com/pod-product-compliance
Ingram Content Group UK Ltd.
Pitfield, Milton Keynes, MK11 3LW, UK
UKHW050949290926
15470UKWH00005B/1718

9 798337 352787